ENVIRONMENTAL SCIENCE, ENGINEERING AND TECHNOLOGY

BUILT ENVIRONMENT: DESIGN, MANAGEMENT AND APPLICATIONS

ENVIRONMENTAL SCIENCE, ENGINEERING AND TECHNOLOGY

Additional books in this series can be found on Nova's website at:

https://www.novapublishers.com/catalog/index.php?cPath=23_29&seriesp=
Environmental+Science%2C+Engineering+and+Technology

Additional e-books in this series can be found on Nova's website at:

https://www.novapublishers.com/catalog/index.php?cPath=23_29&seriespe=
Environmental+Science%2C+Engineering+and+Technology

BUILT ENVIRONMENT: DESIGN, MANAGEMENT AND APPLICATIONS

PAUL S. GELLER
EDITOR

Nova Science Publishers, Inc.
New York

Copyright © 2010 by Nova Science Publishers, Inc.

NOTICE TO THE READER

The Publisher has taken reasonable care in the preparation of this book, but makes no expressed or implied warranty of any kind and assumes no responsibility for any errors or omissions. No liability is assumed for incidental or consequential damages in connection with or arising out of information contained in this book. The Publisher shall not be liable for any special, consequential, or exemplary damages resulting, in whole or in part, from the readers' use of, or reliance upon, this material.

Independent verification should be sought for any data, advice or recommendations contained in this book. In addition, no responsibility is assumed by the publisher for any injury and/or damage to persons or property arising from any methods, products, instructions, ideas or otherwise contained in this publication.

This publication is designed to provide accurate and authoritative information with regard to the subject matter covered herein. It is sold with the clear understanding that the Publisher is not engaged in rendering legal or any other professional services. If legal or any other expert assistance is required, the services of a competent person should be sought. FROM A DECLARATION OF PARTICIPANTS JOINTLY ADOPTED BY A COMMITTEE OF THE AMERICAN BAR ASSOCIATION AND A COMMITTEE OF PUBLISHERS.

LIBRARY OF CONGRESS CATALOGING-IN-PUBLICATION DATA

Built environment : design, management and applications / editor, Paul S. Geller.
     p. cm.
  Includes index.
  ISBN 978-1-60876-915-5 (hbk.)
  1. City planning. 2. Urbanization. I. Geller, Paul S.
  HT166.B82 2009
  307.76--dc22
                         2009049832

Published by Nova Science Publishers, Inc. New York

CONTENTS

PREFACE

Many cities of the developing countries have, for a long time, been burdened with the sprawling squatter settlements constituted of insanitary spaces in ad-hoc structures, overcrowded and inappropriate to the contemporary standards of living. This book presents a novel approach to transforming a number of old squatter settlement into condominium high rises. The problems and potentials for creating sustainable habitats in place of squatter settlements in cities are highlighted. This book also outlines a quantitative study of the holistic thermal capacity of examples of vernacular from a wide range of climates and locations around the world, comparing the results for a range of climate classificatory scales.

Chapter 1 - It did not take long for the two megacities of Beijing and Shanghai to embrace the fashion of culture-led urban regeneration. Forerunner is the case of Beijing 798, wherein a semi-abandoned factory was gradually turns to the largest cluster of art production and art consumption. The process never runs smoothly as this zone is planned as high-tech industrial zone according to the comprehensive plan. During the past years, artists petitioned for preserving the factory, claiming that the preservation is significant in conserving historical buildings and sustaining the only and largest 'artists' village'. The cultural group in Shanghai is comparatively lucky as the municipal government promptly announced the plan of cultural rehabilitation of industrial buildings in 2004. Within 2 years, around 80 dilapidated industrial sites have been converted to cultural infrastructures. Shanghai Sculpture Space is the first project initiated by the government, who see the project as a model to demonstrate rehabilitation of heritage and to encourage the development of creative industry.

In both two cities, the process started when a cultural group moved to dilapidated industrial plants in the city, reusing them as studios. The spontaneous activities largely inspire many other parties, who soon find it beneficial to actively involve themselves in the process. Nevertheless, authentic heritage conservation by professional ways of conservative interventions is, all of a sudden, advocated and appreciated widely, after dozens of years of fearless deconstruction and reconstruction in China.

Driven by global city making, both two cities see culture as a key to bolster a new economy and to deal with decayed urban sites. Meanwhile, differences are detectable due to the various natures and meanings of local culture and therefore different roles of culture assigned by the two cities respectively. When Beijing has long claimed its orthodoxy in representing Chinese culture, taking art production as one pillar industry; Shanghai hardly hides its absorptive and sometimes eclectic nature, caring more on global standard of art consumption. This paper attempts to analyze the differences of the two culture-led regeneration projects, the spatial outputs of which stem from different cultural circumstances and, respond to power relationships of a variety of actors in the urban regimes.

Chapter 2 - This article considers the significance of the concept of the built environment with respect to rural gentrification, drawing principally on studies of the British countryside. It begins by drawing attention to the importance given to the built environment within early definitional debates over gentrification and in more recent discussions concerning changing forms of gentrification and the emergence of a 'third-wave' of 'new-build gentrification'. The article then considers whether these debates and discussions of gentrification, which have largely been urban focused, can be connected into studies of rural space. Attention is paid to the conceptual distinction between classical and new-build gentrification, it being suggested that this is a far from simple dualism. Eight dimensions of differentiation are identified from the urban literature, which it is then suggested can be applied to the study of rural gentrification as well. Attention is then drawn to the degree to which the concept of built environment could be expanded to encompass much more than buildings and how this may impact on understandings of the dynamics of rural gentrification.

Chapter 3 - Since the late 1980's a range of applications of fractal geometry, for the analysis of the built environment, have been proposed. One of these, the mathematician Benoit Mandelbrot's "box counting" approach, was developed by Carl Bovill who demonstrated a manual method for determining an approximate fractal dimension of architectural plans and elevations. Since that time, a computational variation of the method has been developed and tested providing a mathematical determination of the characteristic visual complexity of a building.

The present chapter records research into the changing formal strategies employed by Frank Lloyd Wright for the design of houses. Wright designed many houses during his almost 70 year professional career. Historians have defined three distinct stylistic periods in his housing oeuvre; the Prairie style (generally the first decade of the 20th century), the "Textile Block" era (the 1920s) and the Usonian era (the 1930s to the 1950s). This chapter undertakes the first mathematical analysis of visual complexity in the domestic architecture of Wright. The research analyzes five houses from each of these three eras using the computational fractal approach to measuring visual properties of architecture. In this way it is possible to provide a mathematical determination of the relationship between the three major stylistic periods in Wright's work.

Chapter 4 - Capturing and modelling 3D information of the built environment is a big challenge. A number of techniques and technologies are now in use. These include EDM (Electronic Distance Measurement), GPS (Global Positioning System), and photogrammetric application, remote sensing and traditional building surveying applications. However, the use of these technologies cannot be practical and efficient in regard to time, cost and accuracy. Furthermore, a multi disciplinary knowledge base, created from the studies and research about the regeneration aspects is fundamental: historical, architectural, archeologically, environmental, social, economic, etc. In order to have an adequate diagnosis of regeneration, it is necessary to describe buildings and surroundings by means of documentation and plans. However, at this point in time the foregoing is considerably far removed from the real situation, since more often than not it is extremely difficult to obtain full documentation and cartography, of an acceptable quality, since the material, constructive pathologies and systems are often insufficient or deficient (flat that simply reflects levels, isolated photographs,..).

Sometimes the information in reality exists, but this fact is not known, or it is not easily accessible, leading to the unnecessary duplication of efforts and resources.

Systems that measure range from the time-of-flight of a laser pulse have been available for about 25 years, so this does not constitute new technology. However, the development of fast measurement (up to 10000 measurements per second) and a scanning mechanism (using rotating mirrors) has only occurred in this decade or so. Packaging these components into a robust and reliable instrument has resulted in the innovation of a 3D laser scanner.

In this chapter, the authors discussed 3D laser scanning technology, which can acquire high density point data in an accurate, fast way. Besides, the scanner can digitize all the 3D information concerned with a real world object such as buildings, trees and terrain down to millimetre detail Therefore, it can provide benefits for refurbishment process in regeneration in the Built Environment and it can be the potential solution to overcome the challenges above. The chapter introduces an approach for scanning buildings, processing the point cloud raw data, and a modelling approach for CAD extraction and building objects classification in IFC (Industry Foundation Classes) format. The approach presented in this section can lead to parametric design and Building Information Modelling (BIM) for existing structures. In this chapter, while use of laser scanners are explained, the integration of it with various technologies and systems are also explored for professionals in the Built Environment.

Chapter 5 - Anthropocentric biocybernetic computing uses machine learning to provide a heightened level of understanding of human perceptions of complex systems. In this chapter, such processes are the catalyst for the development of two new methods for architectural and urban analysis and the revision and expansion of a third, existing method. The two new methods rely on, respectively, the 'Hough transform' algorithm to model line segmentation in images of buildings and a newly authored program to investigate facial pareidolia in façades. The third method is a computational variation of the fractal analysis technique for the classification and review of urban skylines.

Anthropocentric biocybernetic methods are ideal for the analysis of the built environment because architecture evokes a complex emotional and perceptual response in viewers and building users. While architectural and urban spaces may be readily measured and understood in terms of their formal characteristics, the phenomenal and semiotic qualities of the built environment have rarely been investigated using computational means. This chapter responds to this situation by providing examples of new techniques that are ideal for expanding the computational analysis of architecture beyond the simple analysis of form.

Chapter 6 - Many cities of the developing countries have for a long time been burdened with the sprawling squatter settlements constituted of insanitary spaces in ad-hoc structures, overcrowded and inappropriate to the contemporary standards of living. Occupied by the poor who have been trapped in the vicious circle of poverty, lack of access to education, unemployment and underemployment and therefore continued poverty, squatter settlements pose an insurmountable challenge to the city authorities and governments involved in the design and management of the built-environments in cities. While approaches to breaking this cycle has shifted from housing to poverty alleviation often dependant on government subsidies, the need for sustainable development demands that these settlements be transformed in a manner that they contribute positively to social progress, economic growth and environmental improvements by themselves.

This chapter presents a case study of an approach to transforming a number of old squatter settlement in Colombo, by which the squatters have been persuaded and facilitated to voluntarily move from slums and shanties to modern, multi-storey condominium apartments built replacing the squalid housing estates. The program based on an exchange of the real estates occupied by the squatters for apartments with modern conveniences has released the valuable urban land for progressive developments while enabling the poor to get out of the poverty trap and live in decent housing. It discusses the *Sustainable Township Program* (STP), the government-owned development company that spearheaded the program; *Real Estate Exchange Limited* (REEL) and how residents of *Vanathamulla;* one of the most dense and derelict squatter settlements have been moved to the condominium high rise; *Sahaspura.* The chapter highlights the problems and potentials of this approach as a model for creating sustainable habitats in place of squatter settlements in cities.

Chapter 7 - A persistent and prevalent view exists within the architectural and ethnographic literature that the development of vernacular buildings throughout the world has been deterministically driven by a conscious desire amongst humans to create interior spaces that are both thermally 'comfortable' and thermally neutral. This was purportedly achieved through the gradual harmonisation of the buildings' construction and thermal operation with the outside climate. This view is, however, largely erroneous and is the result of an overwhelming focus on only select examples of vernacular. Studies have tended to focus on extreme and easily classifiable climates, such as hot-arid, hot-humid and very cold climates, and have overlooked or ignored other more complex types of climate, such as high altitude and temperate climates. Studies have also tended to ignore traditional buildings that by western standards of thermal comfort are thermally marginal.

A theory that attempts to explain the development of vernacular must, however, be based on a quantitative examination of a wide range of buildings from diverse climates and cultures. It must also examine thermal performance holistically, because the thermal behaviour of structures is the result of the ever-changing interaction between all their built components and thermal features. This paper outlines a quantitative study of the holistic *thermal capacity* of examples of vernacular from a wide range of climates and locations around the world, comparing the results for a range of scales of climate classification.

The paper concludes that the achievement of thermal 'satisfaction' has had a much greater influence than thermal comfort/neutrality on the types of structures that people have traditionally chosen to inhabit. Analysis of the ethnographic record indicates that people who inhabited thermally 'marginal' structures would have been thermally satisfied, if not comfortable, supporting the findings of Adaptive Comfort theory. This theory states that people appear to have a universal preference for possessing the ability (or perceived ability) to make thermal choices and to be able to enact those choices, in preference to being always thermally neutral, regardless of social and climatic differences. This has important implications for contemporary buildings and implies that a building's adjustability and *thermal capacity* is of much greater importance than thermal comfort/neutrality in achieving occupant satisfaction and, ergo, their capacity to support social and functional change.

Chapter 8 - There is considerable support for the development of the secondary mortgage market in Ghana. A secondary mortgage market, proponents argue, would be a panacea to the housing problems in Ghana. However, the studies on which a secondary mortgage market has been recommended have focused only on the process and structural pre-requisites needed for establishing such a market. As such, the authors do not know what the outcome of a secondary mortgage market would be in the Ghanaian context. This study takes up the challenge by analyzing the possible outcome of a secondary mortgage market in Ghana. It finds that a secondary mortgage market in Ghana would not resolve the housing problems of the poor and would condemn a majority of people to homelessness. It presents alternative approaches likely to offer a more lasting solution for the housing crisis in Ghana.

Chapter 9 - Access to activity opportunities depends, among other factors, on land-use mix and density at an individual's place of residence. Travel mode choice varies considerably depending on distances to facilities as well as on realised trip distances. This chapter presents findings from an empirical study conducted in the region of Cologne. Standardised data collected from 2,000 residents are used for an in-depth analysis of the interrelation between trip distances and mode choice. Based on structural equation modelling, life situation (sociodemographics), car availability, location preferences and the built environment are taken into account. In addition, lifestyles receive consideration in the study of leisure trips. The findings show the impact life situation, car availability and the built environment have both on trip distances and mode choice, as well as the interrelation between the latter two. What is more, significant effects of lifestyles and residential self-selection on travel behaviour are found. The chapter closes with discussion of consequences for urban planning, making a case for more balanced, mixed-use patterns in urban development.

Chapter 10 - Unauthorized building works (UBWs), particularly those attached to external walls of buildings, have posed serious threats to the safety of the community in Hong Kong. As estimated by the government, there exist around 0.75 million UBWs in some 39,000 private buildings throughout the territory. In spite of the plenteous literature on this topic, associated empirical studies have been relatively rare, and this is not constructive for the exploration of the causes of UBW proliferation. To straddle the existing research gap, this study empirically explores what types of unauthorized appendages dominate and what factors affect the degree of proliferation of unauthorized appendages in multi-storey residential buildings in Hong Kong. Given that the majority of Hong Kong's populace live in this type of housing, residents and the general public are prone to UBW hazards in their environments. Therefore, an empirical study like this one is necessary for providing valuable insights to public administrators for making more informed decisions. Through appraising 429 multi-storey residential buildings in Yau Tsim Mong, the Eastern District and Kowloon City, a profile of unauthorized appendages in these buildings were obtained. The analysis results indicated that over 98 percent of these appendages were not constructed purportedly for increasing usable floor space, but for enhancing the amenities enjoyed by the residents. Besides, building age, development scale and management characteristics have a strong bearing on the number of unauthorized appendages present in a building. These findings have far-reaching implications on the formulation of government policies regarding building safety in Hong Kong.

Chapter 11 - As the number of aged persons in the population grows, it is important to identify and understand those lifestyle components that contribute to an improved quality of life and maintenance of functional independence. Benefits of regular physical activity include reduced risk for diabetes, obesity, cardiovascular disease, cognitive dysfunction, and other physical and psychological health outcomes [Rejeski et al., 2002; Taylor et al., 2004; Weinstein et al., 2004; Kramer et al., 2003; McAuley, Kramer, & Colcombe, 2004]. Despite these benefits, a large proportion of adults fail to engage in any physical activity [United States Department of Health and Human Services (USDHHS), 2000] with women, the elderly, and minority populations reporting the greatest levels of inactivity.

Increasingly, links between elements of the built/physical environment and physical activity are being examined. One potentially useful approach to understanding physical activity behavior is through the use of ecologic models [Berrigan & Troiano, 2002]. An ecologic model examines both environmental and individual determinants of behavior, conceding both direct and indirect influences of each on behavior, and the tendency of their respective influences to shift over time. Individual characteristics interact with behavior, and both are influenced by the social and physical environments in a reciprocally determining manner; contributing to overall health. This triadic form of reciprocal determinism is central to Bandura's social cognitive theory [Bandura, 1986, 1997]. To date, very few studies have utilized transdisciplinary theories and research designs to determine the utility of individual and environmental interactions in predicting physical activity [Li, Fisher, Brownson, & Bosworth, 2005; Satariano & McAuley, 2003; King, Stokols, Talen, Brasington, & Killingsworth, 2002]. Consequently, there exist numerous conceptual, theoretical, and definitional ambiguities associated with this literature.

This chapter provides a critical review of the literature on physical activity and the environment in older adults. Specifically, attention is given to the large degree of variability that exists in the operationalization of common environmental characteristics (e.g., population density, access to recreational facilities, land-use mix) as well as physical activity behavior (e.g., step counts, self-report activity behavior, walking for transport) and how these different viewpoints influence the types of outcomes assessed and the measurement tools utilized. Moreover, the role of theory in the design and implementation of research into the environment-behavior relationship and directions for future research are discussed.

Chapter 12 - The industrial heritage of the UK has given rise to around 100,000 sites, being classified as contaminated. In particular the former Avenue Coking Works, Chesterfield, UK in the 1990s was classified as one of the most contaminated sites in Western Europe. The site was left derelict after its working life and was never decommissioned. Chemical tanks, pipe work, sumps, gantries and contaminated lagoons were never drained down; the condition of these deteriorated. This coupled with site operation has caused severe ground pollution. Major contaminants identified included PAHs, phenols, mineral oil, BTEX, cyanides, tyres, metal, asbestos and spent oxide. In 2001 a series of remediation techniques were trialled to determine the superlative form of remediation. These trials included bioremediation, thermal desorption, soil washing, cement stabilisation and co-incineration. The enhanced thermal conduction (ETC) was the thermal desorption technology used which was found to be the most effective technology for treating this type of soil. Contamination levels such as PAHs (total 16 USEPA) were significant reduced from 76,000 mg/kg to 500 mg/kg at a cost of £60 per cubic metre.

In addition to the reduction in contamination levels the ETC process also significantly improved the geotechnical properties/parameters of the soil. The CBR value increased from 11% to 107%; this was related to the high uniformity coefficient of the soil where excellent particle interlock would have been obtained. Humus would have also been destroyed by the thermal process, which in turn would relate to a greater interlock between soil particles. The thermal process also enhanced the shearing resistance by a factor of two from 17° to 34°. Thus, the residual soil has the potential to be reused as a fill material (Grade 1A) under the Specification for Highway Works, UK (2007). The application for its reuse has been demonstrated in the chapter for an engineered embankment.

Chapter 13 - Today, when the prevailing current of individualism is coming in for well-warranted criticism, one expanding phenomenon deserves to be singled out for attention. This is intergenerational housing, which, as its name intimates, brings a couple of senior citizens and a younger household under the same roof. It differs from formulae such as "Béguine convents" or residential communities (for example, the Abbeyfield "Entre Voisins" house in Etterbeek) that bring together senior citizens *only* (Dossogne & Simon). Intergenerational housing appears to be a particularly stimulating initiative in the galaxy of collective housing, in any event, as it illustrates the fact that, beyond providing physical shelter in the strict sense of the word, housing also helps to *(re)forge social ties*. Although the occupants of such houses do not necessarily share a common space, they nevertheless take part in a lifestyle based on mutual assistance. The philosophy behind the founding of intergenerational housing, in which informal solidarity is the driving force, consists in offering a minimum degree of availability to the other household with a view to providing one or the other service. Yet for all that, this living arrangement does not require the occupants to modify their lifestyles substantially. For example, the younger members are not at all expected to serve as home nurses or family aides. Nor are the occupants required to live as a community. Intergenerational housing (also called "kangaroo houses" in reference to the kangaroo's sheltering and nurturing pouch for its young), which was devised in a spirit of pooling assistance and support, is in any event coming in for more and more attention from the population and authorities alike (Bonvalet). That is why in this article the spotlight is trained on this emerging alternative form of accommodation, which appears to be heavy with promise, despite the many legal obstacles to its official recognition.

Chapter 14 - Building Information Modelling (BIM) enables intelligent use and management of building information embedded in object-oriented CAD (Computer-aided Design) models. The widespread use of object-oriented CAD tools such as *ArchiCAD*, *Revit* and *Bentley* in practice has generated greater interest in BIM. A number of BIM-compliant applications ranging from analysis tools, product libraries, model checkers to facilities management applications have been developed to enhance the BIM capabilities across disciplinary boundaries, generating research interests in BIM servers as tools to integrate, share and manage the model developed in distributed collaborative teams. This chapter presents a computational model for applying Augmented Reality (AR) as a novel approach to interfacing BIM servers for design collaboration in the architecture, engineering and construction (AEC) industries. The chapter firstly elicits technical requirements for a BIM server as a collaboration platform based on: (1) a case study conducted with a state-of-the-art BIM server to identify its technical capabilities and limitations, and; (2) an analysis of features of current collaboration platforms used in the AEC industries. The findings are classified into three main categories including: (1) BIM model management related requirements; (2) design review related requirements; and (3) data security related requirements. An integration framework is then proposed for adopting AR as the primary interface for a BIM server, supporting the above technical requirements for design collaboration in the AEC industries. The adoption of AR opens up new opportunities for exploring alternative approaches to data representation, organisation and interaction for supporting seamless collaboration in BIM servers.

Chapter 15 - The mayors in the newly elected Mills government have now been appointed to help the government to deliver on its promise of making Ghana 'better'. What is the strategy of these mayors? How does this strategy impact different groups in the city? Does this distribution matter? This brief paper attempts to answer these questions.

Chapter 16 - It is argued that, formation of foreign enclave in Shanghai, at its core, is gentrification, which is generally described as the process by which higher status residents displace those of lower status. Superficially, many features associated with gentrification are also observed in foreign enclaves in Chinese cities: beside displacement of the poor defined by traditional gentrification, multiple rounds of displacements have been observed here as many foreign households have moved from one cluster to another, replacing the previous replacers – a change called super-gentrification in the West.

In: Built Environment: Design, Management and Applications   ISBN: 978-1-60876-915-5
Editor: Paul S. Geller, pp. 1-31   © 2010 Nova Science Publishers, Inc.

Chapter 1

THE RHETORIC AND REALITY OF CULTURE-LED URBAN REGENERATION- A COMPARISON OF BEIJING AND SHANGHAI, CHINA

Jun Wang and *Shaojun Li*

National University of Singapore, Singapore
The University of Hong Kong, Hong Kong

ABSTRACT

It did not take long for the two megacities of Beijing and Shanghai to embrace the fashion of culture-led urban regeneration. Forerunner is the case of Beijing 798, wherein a semi-abandoned factory was gradually turns to the largest cluster of art production and art consumption. The process never runs smoothly as this zone is planned as high-tech industrial zone according to the comprehensive plan. During the past years, artists petitioned for preserving the factory, claiming that the preservation is significant in conserving historical buildings and sustaining the only and largest 'artists' village'. The cultural group in Shanghai is comparatively lucky as the municipal government promptly announced the plan of cultural rehabilitation of industrial buildings in 2004. Within 2 years, around 80 dilapidated industrial sites have been converted to cultural infrastructures. Shanghai Sculpture Space is the first project initiated by the government, who see the project as a model to demonstrate rehabilitation of heritage and to encourage the development of creative industry.

In both two cities, the process started when a cultural group moved to dilapidated industrial plants in the city, reusing them as studios. The spontaneous activities largely inspire many other parties, who soon find it beneficial to actively involve themselves in the process. Nevertheless, authentic heritage conservation by professional ways of conservative interventions is, all of a sudden, advocated and appreciated widely, after dozens of years of fearless deconstruction and reconstruction in China.

Driven by global city making, both two cities see culture as a key to bolster a new economy and to deal with decayed urban sites. Meanwhile, differences are detectable due to the various natures and meanings of local culture and therefore different roles of

---

* Corresponding author: E-mail: wang.junesalie@gmail.com

culture assigned by the two cities respectively. When Beijing has long claimed its orthodoxy in representing Chinese culture, taking art production as one pillar industry; Shanghai hardly hides its absorptive and sometimes eclectic nature, caring more on global standard of art consumption. This paper attempts to analyze the differences of the two culture-led regeneration projects, the spatial outputs of which stem from different cultural circumstances and, respond to power relationships of a variety of actors in the urban regimes.

Keywords: industrial heritage, culture-led regeneration, authentic conservation, cultural industry, Chinese cities

1. INTRODUCTION

The era of globalization reinforced the widespread promotion of culture-led urban regeneration (Evans, 2003; Kanai & Ortega-Alcazar, 2009; Miles & Paddison, 2005; Yeoh, 2005). Culture is suggested to be the solution to all, serving not only as "a source of prosperity and cosmopolitanism", "a means of spreading the benefits of prosperity to all citizens", and also "a means of defining a rich, shared identity and thus engenders pride of place (COMEDIA, 2003). Mainly, the mission of cultural policy normally falls into hosting international hallmark events and constructing flagship architectures, development of cultural industrial sectors from the perspective of production, and strategic means of city brandings to increase profile and name recognition (Yeoh, 2005).

Argument in favor of cultural-driven economic growth is a response to the intensification of inter-city competitions, or so-called "place-wars" (Silk, 2002, p. 777), for highly mobile investors, professional talents and élites and visitors by many policy-makers. Advocators suggest the urgent issue is to attract the *Creative Class* as "capitalist development today has moved to a distinctive phase, in which the driving force of the economy is no more technological or organizational, but human" (Florida, 2000; Vanolo, 2008). Florida specifies this social group to be imperative group for cities and regions that expect to succeed in this economy increasingly driven by creativity. Cities possess higher overall structural competiveness when the territory develops to a hub of knowledge-intensive business sectors such as creative industries and commercial leisure and tourism. To achieve this goal, it is thus imperative to forging an inclusive and rich multi-cultural air that is claimed to be appreciated by the creative class. The idea has gained prominence among many entrepreneurial mayors who attempt to accelerate economic growth and finally project their cities to higher tier in the global city hierarchy.

City identity needs to be established to distinguish a given territory from others, with a promise to offer a unique living or visiting experience. Recalling the word "to imagineer" coined by the Walt Disney Studio to describe its way of "combining imagination with engineering to create the reality of dreams," the thesis of urban imagineering is introduced at its core as a political act turning to the question of what and how to build at the local level in a slimier manner as Disney does (Paul, 2004, p. 574, 2005). For many, decision makers resort to local culture, tradition and history to produce a distinctive hybrid identity.

Critics mainly approaches two perspectives, the conscious and deliberate manipulation of culture and tradition in the simplistic and stereotyped hard-branding (Evans, 2003), and the social consequence of creative policy (Atkinson & Easthope, 2009). The Frankfort School coined the term of "cultural industry' in the early industrialization age to uncover the nature of so-called mass culture, which might be akin to a factory producing standardized cultural good, aided by media, to manipulate the masses into passivity. The worry is triggered again in this era of cultural regeneration, when many scholars doubt the rush to *commodification and exploitation of culture and history*. Many culture-led urban regeneration projects, might merely "begin with poetry and ends with real estate" (Klunzman, November 19 2004, cited by Evans, 2005, p959). Disney was not the "first to pioneer the idea of replicating places of the 'other' for people to experience." However, it "was the first to recognize the permanent, continuing commercial potential of such installation" (Alsayyad, 2001, p. 9).

The promotion of a particular set of values through themed built environment and spectacles reflects *social divide and unequal relationship* (Atkinson & Easthope, 2009). The aestheticization of archaic buildings in the picturesque style of heritage conservation is often claimed to be a new type of space tailored for a creative group. The conscious manipulation of image for a given place may respond to the large-scale social transformation from a Fordist to a Post-Fordist society, namely, the birth of the new middle class which seeks out the stylization and aestheticization of life. Meanwhile, neglecting uncreative class is sanitized and social inequity is legitimized. As Bourdieu points out, "art and culture consumption are predisposed, consciously and deliberately or not, to fulfill a social function of legitimating social differences" (Bourdieu & Johnson, 1993, p. 2).

Regional practices differ (Atkinson & Easthope, 2009; Vanolo, 2008). This transformational movement has been the subject of various research works, more so in the aspect of the politico-economic realm. Beauregards and Haila commend that, actors are not "simply puppets dancing to the tune of socioeconomic and political logics but rather relatively autonomous agents" (Beauregard & Haila, 1997, p. 328). Cities are governed by regimes, as put by Stone (1989), an internal coalition of socioeconomic forces pulls the strings in the urban regime. These influential actors with direct access to institutional resources hold a significant impact on urban policymaking and management, and this often results in the urban landscape's contingent spatial transformation. The spatial outcomes of development and policy spawn continuing social and material consequences infused with the coalition's vested interests. The power relationships among different agents within a governing regime vary and are usually dynamic. In this light, the transformation of urban landscapes needs to be explored from the internal structure of socioeconomic actors and their negotiations in the process. Nowadays, culture-led urban regeneration is observed beyond the original circle of advanced cities in North America and Europe (Kanai & Ortega-Alcazar, 2009; Yeoh, 2005). It did not take long for the two Chinese megacities of Beijing and Shanghai to embrace the fashion. Forerunner is the case of Beijing 798, where former semi-abandoned factory compound started its transformation when lecturers form the Central Academy of Fine Arts (CAFA) reused one plant as contemporary classrooms. The process never runs smoothly as this zone is planned as high-tech industrial zone according to the comprehensive plan. During the past years, artists petitioned for preserving the factory, claiming that the preservation is significant in conserving historical buildings and sustaining the only and largest 'artists' village'. The cultural group in Shanghai is comparatively lucky as the municipal government promptly announced the plan of cultural rehabilitation of industrial buildings in 2004. Within 4 years, around 80 dilapidated industrial sites have been converted to cultural infrastructures. Shanghai Sculpture Space is the first project initiated by the government, who see the project as a model to demonstrate rehabilitation of heritage and to encourage the development of creative industry.

After hundreds of years' deviations in terms of local cultures and correspondingly expressed by local urbanization, it is interesting to see both two cities deploy culture as a key to bolster economic growth and to deal with decayed urban sites, during which each selectively learns from the other. Outputs in these two cities look familiar physically. Meanwhile, differences are detectable due to the various natures and meanings of local urban culture and therefore different roles of culture assigned by the two cities respectively. In a very concise sense, the key difference may lie in the different attitudes towards culture and marketization. When Beijing has long claimed its orthodoxy in representing Chinese culture, always attempting to keep its status as a source of cultural products, with a vision to project Chinese culture to the world market; Shanghai hardly hides its absorptive and sometimes eclectic nature, caring more on economic restructuring through inviting influx of investments, firms and international élites with a global standard. It is thus interesting to explore the processes of culture-led regeneration along the threshold of urban culture. By comparing two representative cases in the two cities, this paper attempts to analyze the spatial outputs by putting them back into their cultural circumstances and dynamic relationships of a variety of actors in the urban regimes, which, again, function in respective urban culture whilst also influence the urban culture in return.

2. BEIJING'S CULTURAL DEVELOPMENT: UNDERGROUND CULTURE AND URBAN SPACE

2.1 The Rise of Underground Culture

In 1271, the most prominent and arguably the most influential emperor Kublai Khan selected Beijing to be the capital of his empire. Since then, this city has retained its status as the political center of China through all the dynasties that followed. It is not surprising then why the symbolic role of culture at the ideological level has been envisioned and stressed in Beijing for the past 800 years. Crossing the lengthy historical spectrum, the culture of Beijing has always been stamped with orthodoxy, classicism, high culture, solemnity, aristocratism, authoritativeness, and bureaucracy (C. Li, 1996; Luo, 1994; Weston, 2004).

As capital, Beijing's position in the hierarchical administrative system makes it the center of cultural institutions at the top tier (Weston, 2004). This is also the place where intellectuals who wanted to become officials must come to attend the final stage of the imperial examination system (*keju*). After the examination, which was directly supervised by the emperor, the top three candidates would be awarded official titles by the emperor and would then start their careers in the bureaucratic system. Beijing, the capital city where top officials gathered, held the congregation of intellectuals whose talents or specialties had been endorsed by the imperial evaluation system. The old and rigid education system was challenged by the young emperor of Guangxu who planned the Late Qing Reform movement with the purpose of introducing Western civilization to China. Although it failed after about 100 days, the movement brought about the first university in China, the Peking University (originally called *Jingshi Daxuetang,* which literally means Jingshi University), which was also then the highest administrative institution in education (Hua, 2004). Its impact on the Imperial Government was discernible, and it led to the establishment of the Tsinghua Imperial College (later changed to Tsinghua University) about 10 years later with the purpose of breeding excellent students before sending them off to the U.S. for overseas education. After 1949, more state-level universities were set up to uphold education in a wide variety of disciplines, such as the Central Academy of Drama, the Central Academy of Fine Arts, China University of Geosciences, Central University of Finance and Economics, and so on. The top authorities that managed cultural issues, namely, the Ministry of Culture and the Ministry of Education were also located in Beijing. With so many universities, authorities, and institutions at the highest tier clustered in Beijing, many renowned scholars and a huge cultural community whose members struggled to make their dreams come true naturally gathered there. Although the *keju* exam was abolished a long time ago, Beijing has kept its charm as a magnet for talents, intellectuals, and artists of all kinds. These individuals constitute a social group known as *Beipiao* (literally, people floating in Beijing) who come to the city from all places hoping to have their talents discovered and recognized by authorized institutions or individuals, and consequently to have a promising future. According to the Beijing Statistics Yearbook (BSB 2000), in the mid-1990s, there were a total of 76 units of theatrical performance organizations, 22 units of city-level and district-level cultural centers, and 342 units of community cultural stations. Cultural workers, a sizable number of them being in art performance, film, and TV industry, come from every corner of the country to the city, seeking a chance to be discovered by famous people and to start a promising future. By 2005, the number of the cultural industrial workforce in Beijing reached 552,000, accounting for 6.3 percent of the total workforce. In 2006, 895,000 people were employed in the cultural and creative industries in Beijing, the total assets of which reached 616.1 billion Yuan and the added value amounting to 81.2 billion, accounting for 10.3 percent of the city's GDP. (BSB 2007)

Beijing's prolonged history as the capital of China distinguishes it from the other cities, giving it the rightful social status of the symbol of orthodoxy of the Chinese culture. The Beijing School, the genre specified for the culture of Beijing, naturally inherits and also functions to sustain the value of hierarchy, orthodoxy, and classicism. No other place is more authentic than Beijing in elaborating the ethos of Confucianism and Chinese imperial value, all of which have penetrated into the urban landscape and everyday lives of the people inhabiting it. The layout of the city is planned in a hierarchical way, where concentric city-walls demarcate the inner zone as the location of the Forbidden City. In the inner zone, other layers of walls and corridors further demarcate the very core as the site of the emperor's palace. Houses, not only those of the royalty but also of those of the populace are arranged in exact rectangle grids in the north–south orientation. Overall, the urban landscape explicitly sends out an image of order, authenticity, and obedience through the clear demarcation of the structure of the spaces and the hierarchical power attached to them. This was especially true during the socialist period when culture was used as a tool to educate the populace about the *right* values, world view, and thus the way of life. The focus on ideology pulls culture away from the modernization of a society characterized by industrialization and commodification. Most cultural institutions in Beijing are classified as social organizations (*Shiye Danwei*), which are financially supported by the public funds and are under the direct supervision of the State Ministry of Culture. The long history of the city and its cultural institutions, the focus on ideology, and the dominance of the established elites helped Beijing establish its authority on the one hand, but made it reluctant to listen to different voices on the other hand. About 40 years later, after the booming of contemporary art in North America in the 1950s, the central government finally officially acknowledged the existence of a "cultural market" in the Notice to Enhance the Management of Cultural Market in 1988 (Tuo, 10 Oct. 2008).

Nevertheless, different genres emerged from the overwhelming power and influence of classic high culture. There came the "red scarf" in the 1950s, "Young Pioneers" in the 1960s, "Red Guards" in the 1970s, and the cosmopolitan intellectuals in the 1980s (D.-P. Yang, 1994). It is difficult to explain why Beijing, under the nose of the Central Government, has become a magnet of various off-mainstream culture and art. Wu (2004) suggests that this may be due to multiple reasons such as the congregation of "universities and art schools that attract the rougher edges of Chinese culture — from punk rockers to deconstructionist thinkers" (Fackler, 07 July 2002), coupled with the looser political atmosphere because the "ministries overseeing cultural activities are perhaps more preoccupied with running the country" (W. Wu, 2004, p. 174). The rise of the art market is also another factor that frees artists from their financial dependence on bureaucratic institutions, which keep a tight rein on the nature of arts. Tension between art and politics became slack to a certain extent, because since the 1980s, a growing openness to Western contemporary art, especially avant-garde art, has become apparent. A series of cultural exchanges were carried out, from inviting foreign artists to exhibit their works to people actively exploring the Western world to study. Furthermore, an intimate relationship between artists and embassy members established a platform to promote the Chinese contemporary art world (Gao, 2006). China's contemporary art scene soon found its popularity amid international arts collectors who sensed the exotic *aura* from socialist legacies. Contemporary art mushroomed but mainly in the form of underground culture.

2.2 Actors: From Yuanmingyuan Village to Beijing 798

The "Artists' Village," a kind of artist enclave that houses artists, has gradually emerged in remote areas in Beijing since the early 1990s. The nickname indicates the isolated situation of the art community, both tangibly and intangibly. Its story can be traced back to the mid-1980s when a group of university and art school graduates gave up their officially assigned jobs and decided to become freelance artists. It was a ground-breaking action, which meant that they were excluded from the then dominating Danwei system and were not eligible for any social welfare for urban citizens. They went to Fuyuanmen village, which was adjacent to the ruins of Yuanmingyuan, and started their "bohemian" life relying merely on selling their paintings (Huan, 07 Aug. 2005; W. Yang, 2005). Although the pioneering people eventually left the village later on to go abroad or other places, their brave pursuit of creative freedom and utopia for art attracted many followers, some of whom became famous painters (Anonymous, August 01 2006). Choosing Fuyuanmen Village might be partly linked to the previous artistic functions held around Yuanmingyuan, and this somewhat reflected the need for low-cost living essential for struggling painters. However, to a larger extent, it might have been its remote location that made these painters feel as if they have escaped from the city. This bohemian enclave was not welcomed by the government, who saw these artists as a group of rogues that had no stable jobs, led alternative lives, and could start riots. In the following years, the paintings made by these artists were mostly in the name of the

Yuanmingyuan Artists' Village, signifying an established community. Their works were approved to be displayed in many official art museums and in Peking University; but at the same time, they were rejected by China People's University and banned by some authorities. The gradually established fame of the Yuanmingyuan Village challenged the nerve of the local government, who eventually "persuaded" all the artists to leave the village with the aid of the police in 1995 (Anonymous, August 01 2006; Huan, 07 Aug. 2005). Some of them relocated to Songzhuang, another village that is even farther away from the city.

In 1995, the Central Academy of Fine Arts was given 200 acres of land in Huajiadi, then a patch of less developed suburban land northeast of Beijing, to relocate the whole campus from its original site in Wangfujin – the famous commercial core of Beijing. Before the new campus opened, an old factory in Huajiadi was used as the temporary base. However, the Department of Sculpture found the temporary base not proper for teaching and learning sculpture, which needed a large space with a high ceiling. The Department Head, Sui Jianguo, then had to search nearby for a studio. Eventually, he found an abandoned workshop in Factory No. 798.

Like many state-owned factories in Chinese cities, Factory No. 798 was also close to bankruptcy at that time, with many nonfunctional workshops. This dilapidated factory, however, had a unique history. In the late 1950s, a big project called "Joint Factory No. 718," also known as the "Northern China Radio and Accessories Plant," was proposed as a joint project with direct involvement and assistance from then East Germany in its design, construction, and production line. In 1964, the giant plant was divided into several factories that ran independently, namely, Factories No. 718, 798, 706, 707, 797, 751, and the Research Institute No.11. Naming the factories by numbers in the socialist China indicated an *ad hoc* military nature of the factory or hidden links to national defense. This unique history did not stop the decaying of this plant, whose employees dropped from over 10,000 at the peak time to around 1,000 (Huang, 2008). In 2000, except for Factory No. 751, the rest of the factories and research institutes were merged to form the Seven Star Group, with the goal of reorganizing the industrial resources for a better performance. While the group worked to keep some workshops running by introducing products like the latest technical equipment and military industrial electronic products, it had to lease out idle workshops owing to the great burden that arose from its responsibility to support social welfare input for the huge community of employers.

The workshop found by Sui Jianguo used to be the Workshop I in Factory No.798. When he walked into this dilapidated building, he saw nothing more than dust and broken glass (Huang, 2008). However, the building had everything a sculpture studio needed: vast cathedral-like space, high ceiling, and abundant and stable daylight from high clerestory windows along with dormers running across the whole workshop. The department thus rented two workshops with an area of 1,000 square meters for the mega-project "Lugouqiao Anti-Japanese Portraits" (Anonymous, 2008). After years of working in this plant, Sui was so used to working in the large workshops that he came back again in 2000 after the new campus was opened because the "spaces (in the new building) were not as large as those in Factory No. 798" (J. Li, 2005). This time, he rented a place to open his personal studio. Other artists followed his move. That same year, designer Lin Jing and publisher Hong Huang also moved in. One year later, the number of artists occupying the workshop increased. There were established artists like Professors Yu Fan and Jia Difei from art schools, famous musician Liu Suola, and struggling contemporary artists as well. Followers who joined the bandwagon also included artistic organizations in different forms, ranging from art galleries like Season Gallery, to bookstores like "Timezone 8," and to complexes like "Time Space," which offers an exhibition space and a café. In 2003, around 30 artists and organizations set up studios or offices in the area, while 200 more were reported to be on the waiting list (Huang, 2008).

Figure 1. artists standing in front of Yuanmingyuan Artists' Village.

Figure 2. layout of Joint Factory 718.

Just like their predecessors in the Yuanmingyuan Village, artists, mostly in the genre of contemporary art or avant-garde art, were frowned upon by the government. The community thus followed the tradition to seek places in the fringes of the city, expecting to establish an enclosed colony of their own. This was partly an imperative factor that could explain the attractiveness of 798 together with the proximity to the new campus of the CAFA, which represented the most sacred land for art in many artists' mind. However, the artists in this era developed more mature skill in networking, not only within their own art realm but also far beyond. They were quite active in organizing exhibitions of all kinds. At the same time, they also showed their talent in managing and expanding their social networks. In 2003 alone, five exhibitions were held: the first Performance Art Festival of "Trans-border Language" in March, the "Reconstruction [of] 798" was kicked off in April, "Blue Skies Exposure - Anti-SAS Exhibition" was conducted in the following month, and the "First 798 Biennale" in September together with the "Left Hand Right Hand." Aside from the efforts targeting the artists and their peers, almost equal endeavors were put on social events targeting foreign politicians, celebrities, and royal members. From 2003 to 2004 alone, banquets, tours, and exchanges were offered to French Musician Jean-Michel Jarre and then French minister of Culture Jean-Jacques Aillagon, European Union Commissioner of Culture and Education Ms. Vivienne Reding, French Minister of Culture Renaud Donnedieu de Vabres, and former German Chancellor Gerhard Fritz Kurt Schröder Factory No. 798 became the focus of media spotlight, and stories about it were featured in many influential newspapers like China Youth Daily and even the so-called party mouthpieces China Daily and China's Central TV (CCTV). Eventually, news about 798 traveled abroad and appeared on overseas newspapers like the New York Times and South China Post in 2003 (Ye, 2008). Its frequent appearances in the media was also a sign of the transformation of the artists themselves, from "escaping from the city" to being proactive, while having selective exposure to the public, particularly those with international influence.

The instant creation of a popular art zone worried the property owner of the Seven Star Group who started to take action to drive out these artists or at least get them under control. Beginning June 2003, the Group froze the rental of new spaces and prohibited all renewals. The reason they gave was that the compound had been planned to become a high-tech industrial base as early as 1993. Threatened by the fact that their safe haven would be demolished, artists petitioned repeatedly to preserve the factory, exploiting every element to make a compelling proposal. Half a year later, the artistic community "consigned" tenant sculptor Li Xiangqun, professor at the School of Fine Arts at Tsinghua University and deputy of Beijing Municipal People's Congress, to submit a bill to the annual congress on 14 February 2004 (Anonymous, 2008, p. 170). The bill, which called for "an immediate suspension of the planned large-scale destruction" and "a re-evaluation of the area's potential worth as a cultural centre instead of a copied Zhongguancun electronic zone on the basis of extensive investigations" (Xiao, July 05 2004), listed down the value of the compound from five perspectives, namely, scientific, historical, cultural, economic, and the potential impact on the forthcoming Olympic Games. The bill was approved one month later. Further, a special committee was set up to carry out the investigation (Anonymous, 2008). The following month happened to be the time for the artist community to host the first month-long Beijing Dashanzi International Art Festival. Just three days before the opening ceremony, the Seven Star Group posted a public notice to announce that the festival would be banned if no "Large Scale Social Activity Permit' would be obtained in advance. The artists fought back quickly. They sought help from the Chaoyang District Government and sent out a public letter three days later. In the letter, the artists reminded the Seven Star Group that the forthcoming festival had been "widely known and was paid close attention to by global mainstream media, government, NGOs, artists, and the populace". In this light, they further questioned whether the Seven Star Group was capable of "tak[ing] the responsibility for [any negative] *international impact* of their sudden measures"(Ye, 2008, p. 31). It was a political trick to deploy the term of "international impact," which is an umbrella term commonly used by China's bureaucratic system that works almost every time. As such, the festival was held smoothly.

After several rounds of similar confrontations between the two sides, the debate on "conserving the art district or redeveloping it to be a high-tech industrial park" finally came to an agreement. In 2005, the Beijing Municipal Tourism Bureau published their research report on the 798 Art District, suggesting a solution called "Cultural Tourism and Consumption Zone." In September of the same year, the Seven Star Group signed a contract with Ullens Art Foundation, making the largest deal of renting out a 5,000-square meter space to a single tenant. This deal was interpreted as a compromise made by the property owner, who realized that how to treat the compound was already beyond their control. They had to accept the reality of an established art district and more important, the greater power possessed by the "determined campaign" (Xiao, July 05 2004) that made good use of the local and international media, and of the accessibility to higher level of government organizations.

The rent market in this compound boomed fast not only in terms of the number of tenants but also in terms of the growing diversity of the nationalities of the tenants. The number of artistic individuals, organization, and agencies from France, the United States, Belgium, the Netherlands, Australia, South Korea, Singapore, and other countries and regions increased continually. Cultural entertainment services also emerged; film showings, restaurants, bars, cafés, and small courtyard gardens provided a variety of social space for artists, if not for more visitors. In a very short span of time, the compound developed into an integrated cultural complex that took care of the needs of the various population segments, from artists and art aficionados who sought artistic work, journalists who explored eye-catching news or Bohemian-style fashion, to general visitors who simply came for curiosity, to people who were attracted by the fancy atmosphere, and to sensitive businessmen who detected great potential for money making from tourists. Factory No. 798 became more and more popular; thus, people began to refer to it as encompassing all the six factories inside the compound.

2.3 From Practicality to Heritage Conservation with a BoBo Style

Although it is easy and fairly natural to connect artists' actions with appreciation for industrial heritage, the transformation of these spaces may have been motivated by practicality. These decrepit buildings command a small price for such a large space, which is imperative for struggling artists needing flexible space in an expensive city. The exclusive nature of these spaces, in other words, used as living and working spaces for those artists themselves, were demanded more for their function than their aesthetics. This is especially true in the preliminary period when the artists at that time worried more about tomorrow's food and the 12-inch black-and-white TV set that was already taken as a luxury decoration in the Yuanmingyuan Village (Huan, 07 Aug. 2005; W. Yang, 2005). When Sui Jianguo worked with his peers and students in the village in the 1990s, what they appreciated most were the vast space and stable daylight. Constrained by their economic capability, most artists did more reduction than addition. In other words, their frequent renovation was nothing more than garbage clearance and division of interior space to demarcate living and working zones (Huang, 2008). Many artists collected and reused junk material and subsequently transformed "junk into antiques, rubbish into something rich, strange, expensive, and amusing" (Raban, 1974, p95, cited by Ley, 2004). For some comparatively wealthier tenants, renovations were made on the physical structure; in many cases, they displayed their own works to express their personal style. In 2003, artist Huang Rui designed the 798 Space Gallery for Hutong photographer Xu Yong. On the same day of the opening ceremony, the two artists initiated the "Reconstruction [of] 798" involving all the artistic studios in the compound. The previous concern on the functionality of space only gradually shifted to rhetoric and aesthetic aspiration of the built environment.

In his book, Huang Rui was introduced as the first person who paid serious attention to the visual beauty of the physical structures and also as the first person who classified them into the architectural genre of "Bauhaus" (Anonymous, 2008). The whole compound was planned and designed by German architects in the late 1950s; Germany was where the Bauhaus architecture originated, and it was the Germans who popularized the Bauhaus style worldwide. It was thus natural to label those buildings as Bauhaus architecture. In line with the key concepts of Bauhaus architecture, the design of the workshops was characterized by radically simplified forms, rational functionality, and the promotion of standardized forms by mass-production (Gropius, 1956). These key concepts had already been partly developed before the founding of the Bauhaus and were widely adopted afterwards in modern architecture (Curtis, 1996). At the same time, it should be noted that the factory was an industrial complex built during the socialist era, which was depicted by the slogan "Production First, Life Second." It is no wonder then that the manufacturing compound bore a certain spirit of the "forms follow[ing] functions" (Sullivan, March 1896), but it was far from enough to make itself a representative of Bauhaus. The conclusion that Factory 798 is one of the few remaining Bauhaus building clusters may be a little bit hasty, and the motivation behind the hard branding remains questionable.

Nevertheless, its branding as "Bauhaus Architecture" was immediately cited as truth without cautious challenges.[1] This declaration spread out via newspaper articles, personal blogs, and even the mouths of professionals and officials (Shu, 2004). The measures and interventions conducted by these artists to renovate the old structures were cheered upon, frequently advocated as proof of heritage conservation. Further, the campaign of historic conservation was initiated by the participation of architects and architectural organizations. Professors from architecture schools such as Beijing's Central Academy of Fine Arts and the Southern Californian Institute of Architecture (SCI-Arc) proposed various development plans for the area involving the preservation of the buildings, even if it was not especially profitable financially. In 2006, the 3rd Dashanzi Art Festival set aside a session for Architectural Design involving the participation of many architects with global reputation.

---

[1] Architect Chang Yung Ho might be the only one who publically challenged the classification of "Bauhaus" in his interview by Shu Kenwen (Shu, 2004).

Figure 3. Interior space of Time Space.

However, the majority of the artists were greenhorns when it came to heritage conservation. With regard to the interventions made to the built environment, the designers differed in their methods to a certain extent. However, the design plans were mainly derived from personal appreciation of history, and they aimed to personalize the space for concise distinction. As early as 2001, architect Chang Yung Ho was commissioned to design the place for Xinchao Magazine (which literally means new fashion). Stemming from his respect for the building, Chang limited his additions to a minimum level, that is, what were only necessary for functionality. This style was evident in his second exhibition in 798. Huang Rui and Xu Yong's 798 Space Gallery was located in the workshop whose multiple-arched ceilings and skylights sustained its status as the symbolic center of the compound. In the vast space of the 1200 m² floor, prominent Maoist slogans on the arches were preserved at the artists' request. Several machines that were part of the production line during the socialist era were kept, and a glass-fronted café was set up in the former office section at the end of the galley, adding an air of hybrid and ironic "Mao kitsch" (Pocha, Februrary 26, 2006).

Time moved on, and the advocacy of historical conservation gradually shifted to another fancy action of forming a *BoBo* (Bourgeois-Bohemian) community (Brooks, 2000). Artists were no longer underclass or proletariats. Artist Zhao Bandi, who used to be called the Starving Zhao, purchased the first Alfa Romeo convertible in Beijing (Huang, 2008). Some resident artists became comparatively more affluent than the others were. The former somewhat criticizes many among the latter struggling group, despising their commercial manner driven by the desire to copy the success of the former. The charm of these industrial facilities became more grounded in their deviation and distance from the present. Through the eyes of avant-garde artists, the swooping arcs and soaring chimneys had an uplifting effect, a kind of post-industrial chic. Trickling in was the establishment of galleries, lofts, publishing firms, design companies, high-end tailor shops, cafés, and fancy restaurants. Nonetheless, the resulting space married aesthetics with functionality, housing such works of art as avant-garde sculptures and paintings. Lending more charm to the place were randomly placed pieces that were both finished and unfinished, and the walls, floors, and doorways served as canvases for graffiti. Its renovation work was commissioned to professional architects, who regarded the endeavor involving such alternative concepts as *experimental*, playing with different shapes and materials that extremely contrasted with one another visually. Expensive materials were used, from structural glass curtain walls to aluminum surfaces.

3. SHANGHAI'S CULTURAL DEVELOPMENT: COMMERCE, COSMOPOLITAN AND URBAN SPACE

3.1 A Review of Culture and Derivatives in Shanghai

In comparison, Shanghai is a metropolis whose urbanization process was accompanied by commercialization and a fusion of a wide variety of forces and population segments. Before the signing of the Nanking Treaty in 1842, Shanghai was merely a remote county. After the county was opened as a free trading port, the colonial powers sent their representatives to grab tangible roots of this land through joint authorities and establishment of foreign concessions. The Western market mechanism was applied to this new colonial city, and the logic of capital accumulation was burgeoning. During the early period, the comparatively loose governance and the emphasis on free trading and marketing developed this place into a "wonderland for adventurers", as well as a home for all races. The city soon earned its reputation as a paradise for immigrants: the pioneer group was mainly composed of merchants who were later joined by former farmers coming to seek jobs and new lives in the instantly modernizing city (Zou, 1980); colonialists made up another major force that intruded the land, including foreign diplomatic officials, entrepreneurs, and merchants. For instance, around 18,000 Jews from Germany, Austria, and Poland immigrated to the safe haven of Shanghai in the late 1930s and early 1940s, as the city at the time was one of few cities open to Jewish refugees during the Holocaust.

The semi-feudal and semi-colonial characteristic of that time played an important role in forming the so-called "*Haipai*" culture (literally sea-culture or Shanghai School), which roughly outlined the fusion of Western and Eastern cultures, as well as a joint culture of varying domestic traditions (W. Wu, 2004). Continuous absorption of, and prompt adaption to, the market demands were the second character of the sea culture in this market-driven city. For instance, novelists often produced romantic love stories to meet the demands of the readers, and magazines released photos of charming movie stars on the cover page to attract the potential buyers (Lee, 1999). These phenomena were rarely seen in Beijing where literature of all kinds tended to appear more critical, and the cover pages of journals were usually designed as simple as possible (F. Wu, 2006). These commerce-oriented characteristics were evident in the urban landscape of Shanghai as well. Showing the face of a modern Shanghai, art deco buildings in monumental dimensions were constructed along the Bund, where foreign commercial ships would enter and have their first sight of the city along the Huangpu River; industrial facilities were clustered along the Suzhou Creek, the back lane functioning as the main transport corridor for raw materials, products, and immigrant labor from the hinterland. Commercial streets with modern department stores stretched from the Bund towards the west and constituted the main fabric of the foreign concessions. Filling the blocks in-between were the Lilong House (*shikumen*) that evolved from a Chinese traditional courtyard house to an efficient form of residential real estate development. Pragmatism penetrated the spatial layout of the city as well as the building design.

3.2 The Story of Shanghai: From Suzhou Creek to Red Town

The mid-1990s witness a series of spontaneous rehabilitations of dilapidated production facilities and warehouses in Shanghai. One of the pioneers is Taiwanese Architect Teng Kun-Yen who rented a warehouse at the south bank of Suzhou Creek and refurbished it as his studio in 1998 (Teng, 2006). Teng' attempt to give new life to 'junks' (Ley, 2003) seems to be resonated by UNESCO, which gave him an award for demonstrating what an individual can do by recycling a building for new uses. Following Teng's case, similar cases sprouted, albeit inadvertently, such as Tianzifang and M50. In around 1999, famous artists like Chen Yifei and Er Dongqiang moved into Tianzifang, a Shikumen block that used to house manufacturing workshops during the socialist period but decayed afterwards. The subsequent opening of their studio attracted many followers and the place gradually developed to a cluster of artistic workshops (CBRE, 2007). Meanwhile, in 2000, the first artist who moved into a factory located on Moganshan Road was Xue Song, after he failed to secure lodging at the "red house" where his peers set up their studios. When the red house was demolished two years later, his peers joined him (Zhong, forthcoming). The place later became known as M50, which remains to be the largest cluster of artists and art galleries until today. The fame of such developments spread by word of mouth, as a number of artists who either were members of or possessed connections to the academe (interview with artists in M50, 2009). Its growing renown, however, was expedited upon capturing commercial interest. Regardless of the motivation behind the renovated spaces, the output maintained its public appeal, luring not only tourists but the media as well. The developments were then given sufficient coverage in newspapers, magazines, and TV programs, particularly in segments offering advice on the latest in art and fashion (Shu, October 29 2002). The artists' personal taste as reflected in their works, albeit off-stream, was soon emulated by the general public, especially the younger set.

It is also the same period when conservationists who have been long struggled for legislation on conservation finally see fruition. The city of Shanghai acted even more promptly than its counterparts in the country. Following a year-long poll and corresponding on-site study and evaluation by field experts and administrative officials, the Shanghai municipal government in 1998 announced the list of Excellent Heritage Buildings, which included 15 industrial sites constructed during the post-Opium War era. In 2003, Wu Jiang, the professor major in architectural history in Tongji University, was appointed as the Deputy Head of Shanghai Municipal Planning Bureau, which is tasked to take care of historical buildings and sites. In the same year, industrial heritage was acknowledged in official documents such as the "Notice on Enhancing the Conservation of Historic Areas and Excellent Heritage Buildings in Shanghai". The document explicitly stated that "workshops, shopping premises, factories, and warehouses which were built over thirty years ago and representative of the historical episodes of Chinese industrial development shall be listed as heritage buildings and effective protection measure shall be applied to them" (SHMG, September 11 2004). By the end of 2005, a total of 43 industrial sites earned the title of Industrial Heritage Sites (Zhang, 2007).

Despite this achievement, which in their eyes remained merely paperwork, conservationists held admiration for the artists' concrete works and resulting influence. The popularity of these spaces provided hope to conservationists who had long lobbied for legislation on conservation. After all, successful industrial rehabilitation is the most effective way of promoting heritage conservation. Figures of conservationists began to show up in many projects such as M50 and Tianzifang. After assistant in the survey of industrial heritages in 1997, Ruan Yishan, another leading scholar in conservation of historical cities in Tongji University, set up a center entitled "National Research Center of Historic Cities". His team is an active player in the battle against the demolition of M50, benefited from his connection with officials at both the local and central levels, intentional individuals and authorities like UNESCO; and more importantly, his smart deployments of these connections to provoke public attention through a wide variety of channels, from academic conferences, meetings with senior officials and party members, workshops with artists and communities, to publishing newspaper articles and books (interview with professors in his team, 2009). Knowledge is power, conservationists' high profile actions demonstrates it by transferring the piecemeal reuses of archaic industrial buildings to a heritage crusade.

The mushrooming of cultural rehabilitation of industrial buildings also caught the attention of the municipal government, who detects the economic potential. After 'closing, terminating production in, merging or transforming' polluting enterprises in inner city in the 1990s (SSOAASC, 2009), the municipal government was burdened by the slowly decaying mammoth structures. They became a stumbling block to a much-needed economic restructuring. Following years of experimental exploration, the importance of integrating spatial planning into economic restructuring was recognized and eventually included in the 10th Five-Year Plan (SHMG 2001; X. Wang & Fang, 2004). The reuse of abandoned plants and warehouses sparked a host of possibilities. Not satisfied with the discrete effect, the government pondered ways to generate employment and revenue through the revitalized areas, aiming for a full regeneration that sustains (Lou, July 06 2006). It sought a model that could be applied at a major scale to facilitate the inner city's industrial restructuring, which was envisioned to result in higher gross domestic product (GDP) and employment opportunities in the service industry. The ultimate goal was to bring Shanghai closer to the vision of a world city. To achieve this goal, the Shanghai Municipal Economic and Information Technology Committee (SHEITC) proposed the creation of the Creative Industrial Agglomeration Area, a zone which restores and reuses industrial legacies to accommodate creativity-based firms (SCIC, 2008). Following a survey on non-functional industrial sites in the city proper, SHEITC pursued the creation of 70 to 80 Creative Industrial Agglomeration Areas by end of 2007 (SCIC, 2008).

To facilitate the project's promotion at the city level, the Shanghai Creative Industry Center, a semi-governmental organization, was established in 2004. Further, the Three Unchanging Principles were proposed (SCIC, 2008; SSOASAC, 2009) to iron out legislative hurdles. Under the principle, the rehabilitation of nonfunctional production and warehouse facility may be conducted if "ownership of Land Use Right," "the major structure of the building," and "nature of the land use" remained unchanged (SCIC, 2008; SSOASAC, 2009). It is worth mentioning that issues related to Land Use Right fall under the watch of the

Shanghai Municipal Housing, Land and Resources Administration (SHMHLRA), while issues pertaining to land use are under the jurisdiction of the Shanghai Municipal Planning Bureau (SHMPB). In other words, SHEITC has no legal right to evaluate whether a rehabilitation project of a former factory site could change the nature of land use, and it does not possess the right to judge whether the transaction of Land Use Right is a prerequisite for a given case. However, if the change of land use is interpreted to be a change from 'manufactory industry' to 'creative industry', and the relationship between land owner (the factory) and developer is understood as that between owner and tenant, the two governmental departments, SHMHLRA and SHMPB, acquiesce the "rationality" of the Three Unchanging Principles and offer their full support after the "care" expressed by the municipal government (interview with a former managerial staff at SHMHLRA, 2009).

By its very nature, the principle generated an informal category of rentable industrial spaces on the state-owned land, which otherwise commands land premium at market price for the transfer of Land Use Right (Yeh & Wu, 1996; Zhu, 2004). In return, a portion of the profit will be returned to the land owner, which is usually a struggling state-owned enterprise facing financial crisis, in the form of a monthly rental fee. The requirement specifying "no major change of the building structure" is appended as evidence to demonstrate the rehabilitation project's nature as a refurbishment of rented space rather than a development project on a piece of urban land. Moreover, the definition of "no major change" is a bit loose and open to interpretation. it is difficult to conclude that the principle is based on heritage conservation, but rather, it more serves as an incentive offered by the government to address the problem of fragmented ownership (SCIC, 2008).

Nevertheless, the term of 'industrial heritage' spread like wildfire, oft-quoted by both politicians and the media. It likewise shone the spotlight on the creative industry. The year of 2004 is destined to be a turning point that witnessed intensive actions around these two issues. In April 2004, the first project to be undertaken by a professional developer commenced (Lai, October 01 2007), involving renovation work for the Shanghai Automotive Brake Systems Co. Ltd.'s former compound. Developer Tony Wong, who used to work on the Xintiandi project, led the undertaking. In August of the same year, then mayor Han Zheng introduced the slogan, "New construction is (a kind of) development, (whilst) conservation and renovation is also (a kind of) development," following a meeting with the National People's Committee (NPC) deputies and China People's Political Consultative Conference (CPPCC) members on heritage conservation (Shi, August 01 2004). A month later, Chen Liangyu, then secretary general of Shanghai China's Communist Party (CCP), gave a speech on Shanghai's cultural development, emphasizing that equal attention should be accorded cultural facility and cultural industry, and formally acknowledging the creative industry (Anonymous, September 14 2004). In the same month, the Master Plan for Cultural Spaces was published. In November, the Shanghai Creative Industry Center was formally established, and the first forum on creative industry was conducted a month later (SCIC, 2008).

Following extensive discussions on culture and conservation, the plan for the Shanghai Sculpture Space was put on the table in 2004. It was an offshoot of the Shanghai Municipal Planning Bureau's urban sculptural development promotion for Expo 2010, which was based on the Master Plan for Urban Sculpture approved by the municipal government in July of the same year (SHMPB, 2004; 2005). The idea of building a Shanghai Sculpture Space by renovating a non-functional industrial building thus emerged, as the bureau is tasked to manage the heritage buildings and sites. The renovated structures were envisioned as "platform of city sculpture," a place "embracing functions such as exhibition and communication, sculpture production and transaction, sculpture reserve and artistic education" (SSS, 2006; L. Wang, 2006, p. 101). Further, the project was designed to explore ways to restore industrial heritage and showcase innovation by reusing Expo facilities after the event (Guo, January 19 2006 ; S. Wang, January 19 2006). Through creative salvaging of old industrial structures, the government aimed for the site and neighboring areas' transformation into "a public art center with the most dynamics" in the urban area. It was likewise expected to serve as one necessary element that would symbolize Shanghai's world city status (SSS, 2006; S. Wang, January 19 2006).

For the project, the former cold rolling workshop in the No. 10 Steel Factory was chosen for its accessibility. The compound, established in 1956 as a branch of the Shanghai Steel Company, was situated in the eastern part of Changning District, occupying a sprawling portion of the downtown area on West Huaihai Road. The site was in close proximity to such establishments as the Hongqiao Economic and Technical Development Zone and the Xujiahui business and commercial center. It was likewise adjacent to the Xinhua Road Historical and Cultural Heritage Area. The plant was abandoned following the company's restructuring in 1989.

Tender was called shortly in September 2004. After nine months' evaluations and modifications, Dingjie Investment Ltd bested nine aspiring developers. Interviews with a Municipal Planning Bureau official (interview, 2008) and company chief executive Mr. Zheng (interview, 2008, 2009) revealed varying reasons for the winning bid. While the official attributed it to the developer's eagerness to embark on heritage conservation, Zheng credited the success to financial resources. For example, other bidders such as Shanghai Grand Theater and Shanghai Art Museum were ruled out because these state-owned enterprises relied on government funding. Among the private developers, Zheng's firm boasted of readily available capital resources. Further, the developer was backed by experience in heritage renovation projects and the CEO's personal background in stage design. Zheng taught at two universities before moving to Hong Kong, where he shifted to the Department of Investment at The Xinhua News' Hong Kong branch. Prior to the Red Town project, he was involved in a series of conservation projects such as Sinan Garden on Mid Huaihai Road, Meiquan Villa inside the Xinhua Road Historical and Cultural Zone, and the municipal historic building No. 201 on Anfu Road.

The contract stipulated the developer's responsibility for the renovation work, after which the company get the right of management for the following 20 years. This role included the staging of two high-level exhibitions per year. The Shanghai Municipal Planning Bureau would subsidize the company by paying rent to the land owner (No. 10 Steel Plant) for the first two years (interview with Zhang, 2009). In November 2005, the Shanghai Sculpture Space (Zones A and B on the master plan) was completed and opened to the public. Later on, the company won the contract to renovate further all the buildings left in the compound to develop a relatively private project, which is known as the Shanghai Red Town International Cultural and Business Community (Zones C to H). The team, constituted by Dingjie Investment Co. Ltd., Shanghai Realize Consulting Co. Ltd., W & R Group and some other assisting institutes, formally changed the company name to Red Town Property Management Co. Ltd in 2006.

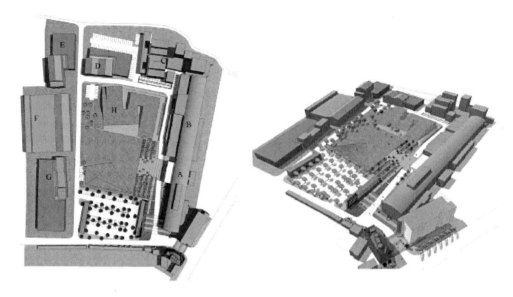

Figure 4. The master plan of the project (source: Liu, 2008; L. Wang, 2006, pp., reproduced by the author).

3.3 Playing the Cards of Industrial Past in the Era of Deindustrialization

Prior to rehabilitation, the building structure in question must first be established as an industrial heritage. The fact that the building is not considered a historical site (SHMUPB & SHHLRA, October 18 2004) renders the approach of historical value or architectural value difficult. The project then resorted to social value, a spotlight in media nowadays (Anonymous, 2007c). Advocates argue that industrial legacies form an honest reflection of the way a community lives and grows. This comment certainly makes sense, but, it is not welcomed by the developer. The community, if referred to the physical labors who used to working in the factory, has been laid off and left this compound in 1989 when the factory was closed. Instead of adding merit, such a history might taint the rehabilitation project's image and rouse social concern. The final solution is then a link made to the steel industry sector, which is emphasized as the main revenue generator in the socialist period. In effect, the Red Town project's Web site[2] waxed poetic in depicting the "red" era, marked by red-hot flames on the stove glowing in the faces of stout-hearted workers. The buildings thus were depicted as a vessel of a glorious era.

Working within the frame of value-centered conservation lends freedom in restoring an industrial building, with no strict rules for physical design. Nevertheless, the renovation of the cold rolling workshop demonstrates professional ways to treat a historical building but further, may goes to unnecessary excess. For most parts, preservation and restoration are deployed for architectural fabric, particularly the attrition and decay. On the other hand, eclectic attitude is evident when one looks at the renovation of the space. The giant interior

---

[2] Please see www.redtown570.com/, but the section on "history of Red Town" has been deleted when the author accessed it one month ago.

space of the plant, which may bear the ethos of muscular industrialization and the efficiency of the utilitarianism of mass production, was altered boldly and roughly.

In the case of cold rolling workshop, treatment of the building's fabric was professionally accomplished, owing to the experience of both designer and builder. Typical elements that distinguish industrial buildings from others are identified and conserved: from the big steel truss and the light pitched roof above, to pillars with concrete brackets to support the moving vehicle for manufacturing process. Original components and materials are preserved without effort to make them tide and neat. However, the attention on the authentic texture seems exceeding its value. Almost all materials are left in a way that they show their surfaces and textures as they were. The rough surfaces of concrete pillars are left nude; many dirty patches on side walls are not cleaned; the rots on steel trusses are not polished; the brick walls, which used to be covered by paint, were scrubbed to reveal the texture of red brick inside. The employment of preservation of these fabrics, as argued by the author, is to forge a distinctive visual impact of 'the past'. It is not the scientific or architectural beauty of these structural components that is respected; rather, it is the characteristic of dilapidation, agedness, and decay that are appreciated. The estimation given to traces left by time even spreads to cover many small installations: for instance, rotten steel nails on the concrete pillar are left untouched, non-functional electrical appliances installed on the truss remain, and other similar situations.

Figure 5.(left) A view from the corridor .

Figure 6.(right) Most weight was accorded to those that collectively forge an atmosphere of the past like an aged industrial structure that has experienced a series of ongoing movements, including attrition and decay.

Meanwhile, the cautious attention on authenticity fades when it comes to the interior spatial organization. A large internal space with a long-span steel truss supporting the light roof is one dominant feature of manufacturing plants, which may be the primary base to support the ethos of the industrial architecture and its "muscular modernism," "glorified in the efficiency of utilitarianism of mass production" (Ley, 2003, p. 2529). This is also the case for the cold rolling workshop in No.10 Steel Plant, which used to be a giant space 180 meters in length, 35 meters in width, and 15 meters in height before drastic alterations were made (L. Li, 2008; L. Wang, 2006). This huge plant was divided into two parts, Zone A was designed to keep the original interior spatial structure, and zone B is being divided further to sub-levels. A series of rectangular shaped concrete boxes are constructed along the long axis of the workshop right in the middle, dividing the whole space into three subzones horizontally. The monster-like addition that dominates the interior space was reserved for office use whilst the corridor and pocket spaces are left for sculptures. As it is, only small sculptures can be accommodated by these small, discrete, and mostly dim and negative spaces, where artificial lighting is necessary. The re-organization of the space is more like inserting a box which offers ordinary office-scale spaces into the so-called 'inspiring' large industrial space, leaving the latter serving as a decorative canopy or man-made landscape for the former.

Upon the complement of the construction work, spaces for creative industry were about to put into the market. The Red Town Company launched a branding campaign to attract the intended market. For this purpose, the company leveraged print media by publishing its own magazine called "Red Town" and arranging interviews with newspapers[3]. A website was also established to introduce the place from all perspectives, highlighting its history and its function as a center of art. Obviously, the difference between Red Town and Shanghai Sculpture Space is blurred in both the magazine and the website, wherein Red Town is described to be the overall development, within which Shanghai Sculpture Space is merely one phase.

The articles focused on how Shanghai had joined the global trend in industrial building rehabilitation, joining the ranks of such icons as the Tate Modern in London, Musée d'Orsay in Paris, and the Massachusetts Museum of Contemporary Art. Such projects, which involved the conversion of old industrial buildings and railway stations into functional spaces, are globally renowned for demonstrating a sustainable way of urban development while preserving history. Association with the said structures infused the Red Town project with global flavor, strongly suggesting its adherence to international standards.

In addition, the quarterly magazine disperses comments on culture-led rehabilitation of industrial sites to a much larger filed, such as review of art development, offering its readers general pictures of China's contemporary art, biographies of artists, and key individuals and authorities in art auctions. The magazine also involves shots of avant-garde garments with models wearing weird and heavy makeup standing in the cool setting of aged and archaic production plant. The audience who read hard copies or online versions is delivered a variety of pragmatic knowledge of what is happening in the field of contemporary art and related. As expected, these media idealize the transformation of the former industrial site to a hub that welcomes creative class, naming it "progress and advancement of civilization" (Cheng, 2007, p. 6). Fancy slogans like 'moving back to industrial sites in the ear of de-industrialization', or

---

[3] Representative newspaper articles include (Guo, January 19 2006 ; Sheng, June 10 2006 ; S. Wang, January 19 2006)

terms like 'post-industrial space' and 'post-modern' are introduced to the general public, who is educated that this is the trend of urban development and inevitably the direction of Shanghai (Zhou, 2007, p. 1).

4. THE LATER DEVELOPMENT

4.1 Exploring the Strategy Towards a New Economy

Four months after the opening of the Shanghai Sculpture Space, both cities presented their blueprints for building creative cities at the annual National People's Congress in 2006.

It was the first time for the city of Beijing to officially adopt *Cultural and Creative Industry* as the city development strategy, depicting a vision of the city to become a creative city by 2010. The presentation by Wang Qishan, then mayor, reported that the city would give priority to six sectors: *art performances, publications and copyright trade, film and TV production and transactions, animation and network gaming, cultural conventions and exhibitions, and antique and art articles transactions*. In September 2007, the Municipal Government endorsed the "Plan of Beijing's Cultural and Creative Industry Development for the 11th Five-Year Period." According to the plan, the developmental foci were extended from the original six sectors to nine sectors, adding three more sectors on *creative design, cultural tourism, and cultural and sport industry* (BMG, 2006). Further, it was urged by many scholars and then stressed by the official documents that the very goal was to project "Chinese Cultural Product" to the world market. To facilitate the work, a giant corporative organization called "the Cultural & Creative Industry Leading Group" was set up, with its team members coming from 23 government departments covering almost every perspective of urban development. The involved authorities included authorities related to economic restructuring such as the Municipal Development and Reform Commission; authorities related to the nine specified sectors such as the Municipal Science & Technology Commission and the Department of Cultural Affairs; and more importantly, authorities supervising a wide variety of incentives such as the Municipal Finance Bureau, Municipal Personnel Department, Municipal Bureau of Land and Resources, Beijing Local Taxation Bureau; and the Legislative Affairs Office under the Beijing Municipal Government. In December 2006, the Beijing Cultural and Creative Industry Promotion Center (hereinafter referred to as "the Beijing Center") was established in to function as the permanent institution of the Leading Group and its General Office.

The municipal government's response disappointed avant-garde artists who overestimated the influence of the case of 798. There was an obvious endeavor by the Beijing government to become orthodox, impartial, and comprehensive when it included the nine sectors in the cultural and creative industry. The compound of 798 is now officially authorized as one of the 10 Agglomeration Areas of Cultural Industry in the Five-Year Plan but is only one of the two areas that focus on art and design, with six others related to the high-tech industry (BMG 2006). Although Li Xiangqun's bill successfully saved 798 from becoming another Zhongguancun high-tech park, it seems that the success was not influential enough to be taken to the policy level. In the plan, as many as 23 departments were mobilized to formulate preferential policies mainly in support of cultural branches whose products were produced in a form for selling at a massive scale and whose products had a typical label of the Chinese culture. First, Beijing's definition of *cultural and creative industry* is closer to the mechanism of cultural industry in the industrialization society, which boomed in the 1940s in North America but has now been surpassed. Today, the transition to the post-industrial society introduces a more specified cultural industry based on originality and inspiration, as capital circulation in the late phase of consumerism has found a niche in disjunction and alternativeness. Beijing's plan, however, failed to add some chic to creativity. Except for *cultural conventions and exhibitions*, all the other eight sectors were directly linked to commodity and commerce which, however, hardly registered in high personal taste Second, it is unfair to criticize the attempt to make full use of the city's orthodoxy in Chinese culture; however, overemphasis on tradition and history could eliminate the efforts of the off-mainstream art community. As mentioned earlier, Beijing's authorized character of Chinese-ness is a double-edged sword; a balanced way of handling the two still needs further exploration.

The municipal government of Shanghai was even more aggressive at the annual congress. According to their report, the city will attempt to become one of the most influential Centers of Creative Industry in Asia in 10 years and then one of the most influential centers of Creative Industry in the world in 20 years. Upon the past experiences, the main strategies still put a lot of attention on development of the hardware, and through which, to attract incoming of creative firms from the global. The resorting to attracting incoming firms and investment is natural when one takes into account of the city's long tradition of openness to influx of population, capital and ideology, and the high adaptation expressed its citizens. The seemingly win-win situation of Shanghai Sculpture Space also generated significant mileage. The planning bureau, which used this project to demonstrate its support to the scheme of Creative Industry Agglomeration Area, is satisfied with the attention on authentic heritage conservation. After the tirades of critics against many pseudo-conservation projects in Shanghai, the case gives conservationists a hope against demolishment. Renowned scholars become highly tolerable about flaws in many forms, as at least some old buildings are saved from the bulldozer. The project's commercial success was emulated by other developers who recognized the potential for profit by investing in low-cost properties. In the year of 2006, Red Town was put into the market. In the same year, the real estate developers tried to adapt to the 11th Five-year Plan, where the industrial sector of Real Estate was replaced by Creative Industry on the list of Six Pillar Industrial Sectors (SHMG, 2006). One after another, developers copy the way through which a low-cost property can be developed to be spectacular through the marriage of creative industry and industrial heritage in the property market at a "warrior" period (Lai, October 01 2007). Indeed, enterprises attracted by the restored spaces powerful image proved the projects viability, and what used to be spontaneous artistic activities became a city-wide endeavor. The economic authority is eager to move on after this experiment, claiming that, not only is the transformed space functional and aesthetically pleasing, it is economically viable as well. Both the government and developer were hailed for the innovative model, where the "government sets up the platform whilst enterprises perform the opera" (e.g., J. Wang & Jiang, 2006; S. Wang, January 19 2006).

The Shanghai Creative Industry Center started its own practices and demonstration of constructing such industrial parks. In 2006, one private enterprise entitled "Shanghai Creative Industry Investment Co. Ltd." was collectively established by the three shareholders of

Shanghai Creative Industry Center, Shanghai Automotive Asset Management Co. Ltd., and John Hawkins - consultant of the center (interview with managerial staffs at the Shanghai Creative Industry Center, 2009). This one-project developer started to renovate one former slaughter house constructed in 1933, which is now the asset of Shanghai Automotive Industry Corporation. Here, the model of "government sets up the platform whilst enterprises perform the opera" was further modified with more intensive involvement of the government agency. The structure after renovation, now dubbed 1933 Old Millfun, is open for lease with intention to build up a high-end cultural and entertaining hub (Anonymous, 2007a). From 2006 to 2007, 75 sites were endorsed by the scheme of Creative Industry Agglomeration Area. In June 2008, the "Notice: Advisory Comments on Accelerating the Development of Creative Industry" was promulgated collectively by the Shanghai Economic and Information Technology Committee and Shanghai Department of Propaganda. The regulation explicitly stated that, "(for these rehabilitation projects), Land Use and Ownership of Land Use Right can remain unchanged at this moment" in Article 1.3. Since then, the Three Unchanging Principles have been formally legalized (SHEITC & DPSHMCCPC, June 30 2008).

On the other hand, the municipal authority that manages cultural issues, that is, Municipal Administration of Culture, Radio, Film & TV, remained quiet until recently. This is also noticed by Wu Weiping (2004, p. 174), who finds that Shanghai cultural authorities are not creative enough, instead they normally "keep a tight rein on the arts", characterized by conservative attitude. This might explain the absence of cultural authorities in this scheme and probably the biased concern on cultural infrastructure in nature. Before the year 2008, cultural authorities in Shanghai mostly remain a working manner of local authorities during the central-planned era. In other words, the cultural bureau functioned as a spokesman to deliver policies from the central state and an agent to implement the policies, rather than developing any municipal policies based on the circumstances of Shanghai's urban context. In February 2008, the cultural bureau made some process, but again, more constrained in building the hardware. The scheme by the cultural bureau, entitled Shanghai Cultural Industrial Park, firstly endorsed 15 cases. Many cases endorsed in the first phase of scheme showed similar nature with those in Beijing - that is, mainly themed around classic domain of culture, such as Zhangjiang National Digital Publishing Base, Jing'an Contemporary Opera Valley, Jinshan Peasant Painting Village, and Cangcheng Film and Television Studio. Managerial staffs emphasized the big differences between the two schemes of Creative Industry Agglomeration Area and the Cultural Industrial Parks, spotting out the latter's nature of incubator for culture. But in the following phases, many projects that wan the plague of "Creative Industry Agglomerating Area" are also awarded the other plague of "Cultural Industrial Park".

4.2 Social Divide and Gentrification

Regeneration is beyond bricks and mortar. It pertains to an area's physical, social, and economic well-being (ODPM, 2003). Celebrating authentic conservation may be truly acceptable only when carried out as an artistic expression and considered a way of life, one that can only be appreciated by those who possess a cultural competence to decode, or in other words, taste. As Bourdieu pointed out, "Taste (is) a marker of social class" (Bourdieu, 1984, p. 1). If the spontaneous rehabilitation of old industrial buildings in 798 is a process wherein the artists' distinctive taste "cappuccinos" the physically dilapidated structures, making the latter fancy and valuable in the eye of the populace, the action taken in Red Town is a inverted process, wherein decay is cautiously preserved and then packaged as a tailored product to meet the distinctive cultural demands of artists. Once the link between artists and archaic industrial buildings is underscored and legitimized, the resulting space becomes commercialized and, to an extent, discriminatory. The transformation of a place may generate economic returns, but in the process it results in gentrification and social exclusion.

The development of hardware in Shanghai was visually appealing. The daring intervention of the construction of the Creative Industrial Agglomeration Area of the government impressed its counterpart in Beijing. On the website of the Chaoyang Research Institute of Cultural and Creative Industry, Shanghai's practices are envisioned as creative because they "hardly bother to defend the ideological critiques;" they are "not constrained by existing rigid regulations;" they realize that the "role of government is more like an assistant that offers services in finance, marketing, and so on," and the "direct involvement of the government through its financial support is later proven to be profitable after the rise of rent" (Anonymous, March 2006). In March 2006, the Chaoyang District government established the Construction and Management Office of 798 Arts District, eagerly applying what they had learned to the compound of 798.

That same year witnessed the formal establishment of the Beijing 798 Art Zone Administration & Development Office, which started a large-scale and overall renovation of the whole compound that immediately changed the outlook of the place. Different from the previous small, fragmented, and particular interventions of individual buildings by artists, this overall renovation was planned and designed with a clear target on the entrance and public spaces. Parking lots were expanded; roads were repaved; infrastructure and facilities like street lights, greenery, telephone systems, fire exit systems, and security equipment were upgraded; and signage and boards were erected for tourists. The office was also in charge of organizing cultural functions and events, whose nature was becoming increasingly diversified. In September 2007, the festival called "In Memory of the 50th anniversary of Joint Factory 718" was kicked off. Later on, the former annual "Danshanzi International Art Festival" now called the "Beijing Contemporary International Art Festival" was held. At the same time, international big names were also welcomed despite their commercial nature. The opening ceremony for the "Nike 706 Space" that featured the "25th Anniversary of Nike Air Force" was held on 5 January 2007. Two days later, BMW Concessionaires Ltd. displayed four models in their Art Car Collection in 798 at the same site where five artworks on the "Automatic Transmission (Auto) in Aesthetic" were displayed. In July 2008, Hong Kong's biggest IT fashion group initiated a show for "Comme des Garçons", the off-beat but established fashion label by Rei Kawakubo.

Thousands of tourists are lured in every day, and the total number reached 1.5 million in 2007 alone. On 798's Opening Day for the Olympic Games, Chen Gang, general secretary of the Chaoyang District Government, introduced his slogan "Beijing - The Great Wall, Peking Duck, and 798!" (Shen, August 25, 2009) The number of visits leaped to more than 10,000 per day during the Olympic Games, including many important personalities. Security guards working in 798 proudly claimed, "I met three presidents in one single day" (D. Yang, Feburary 05, 2009). The tide of tourists hardly shows signs of ebbing. In March 2009, the author stayed near the entrance to calculate the number of tourists. The number was impressive, with around 30-40 visitors per minutes during the peak hours from 13:00 to

17:00. Along with the tourists, businessmen who opened the so-called tourism consumption places, from souvenir shops, restaurants to café houses would also come in. The busy streets became even busier during weekends and were crowded with vendors. The popularity of 798 as an art village even went to the extent that "sale prices of new properties' nearby are far higher than the old buildings in the neighborhood," after the advertisement of "proximity was sieved by speculative developers" (Currier, 2008).

Contrary to the increasing popularity of 798 among art-related commercial companies, tourists, and businessmen, artists started a different wave of retreat. As early as 2004, the very pioneer Sui Jianguo removed his studio from 798. Three years later, the other leading personality, Huang Rui, left as well. By March 2009, the number of artists among the pioneer group was lower than 30. Aside from the complaint that the district became "more about a show rather than serious art," another fundamental factor that drove the artists out was the rise in rent, which ranged from 2 to 7 RMB per sq.meters per day, more than doubled that in 2002. Without any financial assistance, many independent artists had to search for a more affordable workplace. Another annoying factor was the increased number of tourists. The attention of the general public, which used to be actively sought and appreciated by artists during the petition against the demolition of 798, became a burden. After all, there was a part for private life and a part for private working place in the semi-opened SOHO model. Tourists usually took for granted that every room there was open for visits. Thus, the artists opted to shut their doors and turn their backs on the public again.

In the case of Red Town, image is built by incorporating cultural pieces such as artworks by renowned talents and the presence of multinational design firms. These efforts, however, are not geared towards building space for artists' consumption. Instead, they are targeted mature and wealthier multinational firms seeking a creative identity. The renovated spaces were a purposive exploitation of artists' taste and the creative use of space.

Figure 7. commercialization in Beijing 798 (left: A café house at 798, upper right: Nike Space at 798, lower right: BMW show at 798).

The renovated space does not offer an appropriate spatial setting for artists, and it is not designed to do so. In fact, the project targets at building an up-market of office stocks, together with the so-called associated spaces of wine-and-dine and retail-and-entertainment. Before the implementation of Phase 1, Redtown Company has already foreseen a promising future, with office space rent at Zones B and C pegged at 6.8 RMB per m^2 per day, even higher than the average rent of Shanghai's class-A office towers, which was pegged at 6.73 RMB per m^2 per day (Jones Lang LaSalle, 2007). Zone A, a large exhibit hall, is offered for a fee commensurate to its "distinctiveness." Zone A may be rented for 10,000 RMB per day, with a minimum period of 15 days. Functions may be staged for 120,000 RMB on the day of the function and an additional 20 RMB per m^2 per day for ingress and egress.[4]

Collectively, these firms belong to the very loosely defined creative industry, including architectural design, information technology, education, mass broadcasting, and advertisement, among others, who are mostly established in the industry but desperately need an identity to join the so-called reactive industry(Table 1). Electrolux, a company producing electric appliances, is a tenant as well (Cheng, 2007). Another character of intended targets is that most of them are local branches of Multi-national Companies (MNCs). Tenants include Singaporean firm Amoeba Digital Technology Co. Ltd., Italian brand Leissie, and Spanish-operated Art in Capital, as well as those belonging to the category of cultural consumption such as art galleries, bookstores, and tutorial schools. The plant serves as venue for the "annual release of international brands" such as automotive giants BMW and Porsche, and hosts such events as the "Anniversary of Omega," "Celebration of Swarovski," and "Chivas Milan Fashion Party" (Anonymous, 2007b). With its motley attractions and high-profile tenants and visitors, Red Town is naturally a place for entertainment where people can wine and dine. In fact, it is said to be reminiscent of Moulin Rouge, which in the 1820s served as a gathering place for Europe's impressionists. Following the logic, Red Town is likewise hoped to exude a casual atmosphere in terms of food and entertainment to encourage artists from all walks of life. Red Town thus features a culinary ensemble ranging from Hong Kong-based nightclubs such as CICI Club and Children's Club to Swedish Fan Town, to French restaurant BECA.

In the case of Red Town, the influx of creative professionals guaranteed a steady revenue stream, and "[their] taking over the space after the rehabilitation work might provide a vivid testimony of the continuing evolution of urbanization, from industrialization to deindustrialization" (L. Wang, 2006, p. 4). However, such "advancement in civilization," as phrased in "Red Town" magazine, distracts the public from the occurrence of social displacement (L. Wang, 2006, p. 4). History and memory are romanticized and sanitized, and the concept of art and artists is redefined. The irony is that while the project was initially based on artists' space consumption, these artists themselves – particularly the struggling ones – were not truly welcome in the new creative community. The project, which might be an application of Florida's thesis on how to forge an air that is attractive to the Creative Class, reveals itself the consequence of social divide and social exclusion. Not only the whole society is divided into creative group and non-creative group, the category of creative class is further divided by the market, that is, those who industrializes creativity and make it profitable through massive consumption, and those who are not yet.

---

[4] The figure here was obtained from a telephone interview with the Director of the Marketing Department of Red Town Company in 2008.

REFERENCES

Alsayyad, N. (2001). Global Norms and urban Forms the Age of Toursim: Manufacturing Heritage, Consuming Tradition. In N. Alsayyad (Eds.) *Consuming Tradition, Manufacturing heritage: Global Norms and Urban Forms in the Age of Tourism* (1-33). London, New York: Routledge.

Anonymous. (2007a, June 18 2008). *The Concept.* 1933, from http://www.1933-shanghai.com/index_cn.html.

Anonymous. (2007b). Exhibition Planning & Chronicle of the Red Town. *Red Town, 2,* 54-55.

Anonymous. (2007c). Features. *Red Town, 1(1),* 18-19.

Anonymous. (2008). 798 Art District: Narrative. In R. Huang (Eds.) *Beijing 798: Reflections on "Factory" of Art,* (168-174). Beijing: Sichuan Press, Sichuan Fine Art Press.

Anonymous. (August 01 2006). Chronics of Yuanmingyuan Village. *News of Art in China,* Retrieved March 05 2009, from http://www.cnarts.cn/yszx/12380.html.

Anonymous. (March 2006). dWorking report on Cultural and Creative Industry in Chaoyang District, Beijing

Anonymous. (September 14 2004). Planning Cultural development in Shanghai. *Shanghai,* from http://www.shszx.gov.cn/epublish/big5/paper120/2/class012000001/hwz667762.htm

Atkinson, R. & Easthope, H. (2009). The consequences of the creative class: The pursuit of creativity strategies in Australia's cities. *International Journal of Urban and Regional Research, 33(1),* 64-79.

Beauregard, R. A. & Haila, A. (1997). The Unavoidable Incompleteness of the City. *American Behavioral Scientist, 41,* 327-341.

BMG, (Beijing Municipal Government). (2006). Part Three Development Focus and Policy Orientation. In B. M. Government (Eds.) *The 11th Five-Year Plan Programme of Beijing.* Beijing.

Bourdieu, P. (1984). *Distinction: A Social Critique of the Judgment of Taste.* Cambridge, Mass.: Harvard University Press.

Bourdieu, P. & Johnson, R. (1993). *The field of cultural production : essays on art and literature.* Cambridge: Polity Press.

Brooks, D. (2000). *Bobos in Paradise: The New Upper Class and How They Get There.* New York, London, Toronto, Sydney, Singapore: Simon & Schuster.

BSB, (Beijing Statistic Bureau). (2000). *Beijing Statistics Yearbook 1999.* Beijing: Beijing Statistic Publishing House.

BSB, (Beijing Statistic Bureau). (2007). *Beijing Statistics Yearbook 2006.* Beijing Beijing Statistic Publishing House.

CBRE, (CB Richard Ellis). (2007). On the waterfront: Recasting Shanghai's industrial heritage. *CBRE Research - Asian(4).*

Cheng, S. (2007). Industry structure, art blood: Interview with Mr. Zheng Peiguang, CEO of Red Town (in Chinese). *Red Town, 1,* 20-27.

COMEDIA, (2003). *Releasing the cultural potential of our core cities: culture and the core cities.*

Currier, J. (2008). Art and power in the new China: an exploration of Beijing's 798 district and its implications for contemporary urbanism. *Town Planning Review*, *79(2-3)*, 237-265.

Curtis, W. J. R. (1996). Walter Gropius, German Expressionism, and the Bauhaus. In W. J. R. Curtis (Eds.) *Modern Architecture since 1900*, (3rd Edition, (Eds.) 183-200). London: Phaidon Press.

Evans, G. (2003). Hard-branding the cultural city - from Prado to Prada. *International Journal of Urban and Regional Research*, *27(2)*, 417-440.

Fackler, M. (07 July 2002). High-rises will erase Shanghai's SoHo. *The Seattle Times*.

Florida, R. (2000). *The Rise of The Creative Class: and How It's Transforming Work, Leisure, Community and Everyday Life*. New York: Basic Books.

Gao, M. L. (2006). *The Wall: Reshaping Contemporary Chinese Art (in Chinese)*. Beijing: China Renmin University Press.

Gropius, W. (1956). *The New Architecture and the Bauhaus* (P. Morton, Trans.). London Faber and Faber.

Guo, Y. (January 19 2006). *No.10*, Steel Plant converted to the Red Town Cultural and Business Community (in Chinese). *Business News*,

Hua, S. (2004). The Meiji Restoration and the Late Qing Reform revisited: Strategies and philosophies. *East Asia*, *21(3)*, 3-22.

Huan, D. (07 Aug. 2005). A Visit to Yuanmingyuan Artists' Village. *Xin'an Evening Post*.

Huang, R. (Eds.) (2008). *Beijing 798: Reflections on a "Factory" of Art*. Beijing: Sichuan Publishing Group, Sichuan Fine Art Publishing House.

Jones Lang LaSalle. (2007). *Corporate Occupiers Guide (Asian Pasific) - China*.

Kanai, M. & Ortega-Alcazar, I. (2009). The prospects for progressive culture-led urban regeneration in Latin America: Cases from Mexico City and Buenos Aires. *International Journal of Urban and Regional Research*.

Klunzman, K. (November 19 2004). *Keynote speech to Intereg III Mid-term Conference*. Paper presented at the Regeneration and Renewal.

Lai, C. (October 01 2007). Bridging the Gap. *South China Morning Post*.

Lee, L. O. F. (1999). *Shanghai Modern: The Flowering of a New Urban Culture in China*, 1930-1945. London: Harvard University Press.

Ley, D. (2003). Artists, aestheticisation and the field of gentrification. *Urban Studies*, *40(12)*, 2527-2544.

Li, C. (1996). Rediscovering urban subcultures: The contrast between Shanghai and Beijing. *The China Journal*, *36*, 139-153.

Li, J. (2005). *The Site: 798 Art Zone (in Chinese)*. Beijing: Culture and Art Publishing House

Li, L. (2008). Design creates value: the growth of Shanghai Sculpture Space (in Chinese). *Time + Architecture*, *25(2)*, 102-107.

Liu, C. (2008). The Shanghai Sculpture Space (in Chinese). *Chinese and Overseas Architecture(1)*, 82-89.

Lou, J. (July 06 2006). When industrial heritage dances with the creative industry (in Chinese). *China Daily*,

Luo, S. (Eds.) (1994). *A Critique of the Beijingnese?! (in Chinese)*. Beijing: China Society Press.

Miles, S. & Paddison, R. (2005). Introduction: The rise and rise of culture-led urban regeneration. *Urban Studies*, *42(5.6)*, 833-839.

ODPM. (2003). *Assessing the impacts of spatial interventions. Regeneration, Renewal and Regional Development.* London: Office of the Deputy Prime Minister.

Paul, D. E. (2004). World cities as hegemonic projects: the politics of global imagineering in Montreal. *Political Geography, 23(5)*, 571 - 596.

Paul, D. E. (2005). The local politics of 'going global': making and unmaking Minneapolis-St Paul as a world city. *Urban Studies, 42(12)*, 2103 - 2122.

Pocha, J. S. (Februrary 26, 2006). How now, Mr. Mao? T-shirts and kitsch Chairman's images are new China cool. San Francisco Chronicle.

SCIC, (Shanghai Creative Industry Center). (2008). Profile. *Shanghai Creative Industry Center,* Retrieved March 12, 2008, from http://www.shec.gov.cn/

SHEITC, (Shanghai Economic and Information Technology Committee). & DPSHMCCPC, (Department of Propaganda of Shanghai Municipal Committee of Communist Party of China). (June 30 2008). *Notice: Advisory Comments on Accelerating the Development of Creative Industry in Shanghai (in Chinese).* from http://www.shec.gov.cn/shec/jsp/zfxxgk/zfxxgkml-1.jsp?id=27917.

Shen, Z. (August 25, 2009). "The Great wall, Peking Duck, 798" - Debates on the tourism promotion of 798 Art Zone. *China Tousirm News,* from http://review.artintern.net/html.php?id=7183

Sheng, F. (June 10 2006). Industrial heritage: wiping the dust off your face (in Chinese). *Sunday Morning Post.*

Shi, H. (August 01 2004). Han Zheng: Conservation and renovation is also development (in Chinese). *Labor,* from http://sh.eastday.com/eastday/shnews/fenleixinwen/qita/userobject1ai412028.html

SHMG, (Shanghai Municipal Government). (2001). *Outline of the Tenth Five-Year Plan for National Economic and Social Development in Shanghai (in Chinese).* Shanghai.

SHMG, (Shanghai Municipal Government). (2006). *Outline of the Eleventh Five-Year Plan for National Economic and Social Development in Shanghai.* Shanghai.

SHMG, (Shanghai Municipal Government). (September 11 2004). *Notice on Enhancing The Conservation of Historic Areas and Excellent Heritage Buildings in Shanghai (in Chinese).* from http://218.242.144.41:82/gate/big5/www.sh.gov.cn/shanghai/node2314/node2319/node10800/node11407/node12940/userobject26ai2166.html.

SHMPB, (Shanghai Mucipal Planning Bureau). (2004). *Main focuses of urban planning in Shanghai in 2004.* from http://218.242.144.41:82/gate/big5/www.shanghai.gov.cn/shanghai/node2314/node11019/node11030/userobject31ai1114.html.

SHMPB, (Shanghai Mucipal Planning Bureau). (2005). Background information of the Planning of Urban Sculpture in Shanghai (in Chinese). Retrieved July 10 2009: http://shgtj.gov.cn/xwdt/ztxx/csds/shcsds/200812/t20081223_146699.htm.

SHMUPB, (Shanghai Municipal Urban Planning Bureau). & SHMHLRA, (Shanghai Municipal Housing, L. a. R. A. (October 18 2004). *List of Historical Buildings (Phase IV) in Shanghai.* from http://www.shanghai.gov.cn/shanghai/node2314/node2319/node12344/userobject26ai2359.html.

Shu, K. (2004). "Factory 798" and Urban Landscapes. *Sanlian Life Week, 20.*

Shu, K. (October 29 2002). Stepping into the Suzhou Creek, twice (in Chinese). *Sanlian Life Week.*

Silk, M. (2002). 'Bangsa Malaysia': a global sport, the city and the mediated refurbishment of local identities. *Media, Culture and Society, 24,* 775-794.

SSOASAC, (State-owned Assets Supervision and Administration Commission of Shanghai Municiple Government). (2009). The development of creative industry should follow the logic of market - Interview of Le Jingpeng, the director of Economic Committee of Shanghai People's Political Consultative Conference (CPPCC) *Capital Shanghai (Shanghai Guozi, in Chinese)*, 11(9).

SSS, (Shanghai Sculpture Space). (2006). Introduction of Shanghai Sculpture Space Retrieved March 12, 2008, from http://www.sss570.com/

Stone, C. N. (1989). *Regime politics: Governing Altanta, 1946-1988.* Lawrence, KS: University Press of Kansas.

Sullivan, L. h. (March 1896). The Tall Office Building Artistically Considered. *Lippincott's Magazine.*

Teng, K. Y. (2006). *Spatial Revolution: Flame from Suzhou Creek to Huangpu River (in Chinese).* Shanghai: China Eastern Normal University.

Tuo, Z. (10 Oct. 2008). Transformation of mission: Cultural market management in the past 30 years Retrieved 10 March 2009, 2009, from http://www.ccnt.gov.cn/sjzz/whscs/zhxw/200810/t20081010_58412.html

Vanolo, A. (2008). The image of the creative city: Some reflections on urban branding in Turin. *Cities, 25(6)*, 370-382.

Wang, J. & Jiang, N. (2006). Adaptive reuse of industrial heritage in the era of deindustrialization (in Chinese). *Architectural Journal (Jianzhu Xuebao)*, 8, 8-11.

Wang, L. (2006). Memory and revival of a city: practice in Shanghai Sculpture Space (in Chinese). *Red Town*, 1, 4-8.

Wang, S. (January 19 2006). Conversion of Shanghai *No.10* Steel Plant: Exploring sustainable urban development (in Chinese) *Daily Economic News.*

Wang, X. & Fang, L. (2004). Development of urban industry and the construction of urban industrial parks in Shanghai, China (in Chinese). *Urban Research(1)*, 42-45.

Weston, T. B. (2004). *The Power of Position: Beijing University, Intellectuals and Chinese Political Culture.* Berkley and Los Angeles: University of California Press.

Wu, F. (2006). *In-between the Two Cities (Youzou Shuangcheng, in Chinese).* Beijing: The People's Literature Publishing House.

Wu, W. (2004). Cultural strategies in Shanghai: regenerating cosmopolitanism in an era of globalization. *Progress in planning*, 61, 159-180.

Xiao, C. (July 05 2004). End of an era? *China Daily*, from http://english.sohu.com/2004/07/05/42/article220864288.shtml

Yang, D. P. (1994). *Monsoon passing Cities: Comparison between Cultures of Beijing and Shanghai (Chengshi Jifeng: Beijing he Shanghai de Wenhua Jingshen).* Beijing: Dong Fang Press.

Yang, D. (Feburary 05, 2009). Escaping 798: Artsits' instict or mechants' logic? *News of The Chinese People's Political Consultative Conference*, from http://big51.chinataiwan.org/wh/tp/200902/t20090205_824886.htm

Yang, W. (2005). Talk about the "Yuanmingyuan Artists' Village" on the ruin of Yuanmingyuan Retrieved 15 March 2009, 2009, from http://yule.sohu.com/20050817/n226698747.shtml

Ye, Y. (2008). Weightlessness - 798 in the present continuous tense. In R. Huang (Eds.) *Beijing 798: Reflections on "Factory" of Art* (26-37). Beijing: Sichuan Publishing Group, Sichuan Fine Art Publishing House.

Yeh, A. G. O. & Wu, F. (1996). The new land development process and urban development in Chinese cities. *International Journal of Urban and Regional Research*, *20*, P330-353.

Yeoh, B. S. A. (2005). The global cultural city? Spatial imagineering and politics in the (multi)cultural marketplace of South-east Asia. *Urban Studies*, *42(5/6)*, 945-958.

Zhang, S. (2007). Conservation and adapitive reuse of industrial heritage in Shanghai. *Front. Archi. Civ. Eng. China*, *1(4)*, 481-490.

Zhong, S. (forthcoming). From fabrics to Fine Arts: Urban restructuring and the formation of an art district in Shanghai. *Crtical Planning*.

Zhou, Y. (2007). Post-industrial space. *Red Town*, *1(2)*, 1.

Zhu, J. (2004). From land use right to land development right: Institutional change in China's urban development. *Urban Studies*, *41(7)*, 1249-1267.

Zou, Y. (1980). *A Demographic Study on the Flux of Population in Old Shanghai (Jiu Shanghai Renkou Bianqian de Yanjiu)*. Shanghai: Shanghai People's Publishing House (Shanghai Renmin Chubanshe).

In: Built Environment: Design, Management and Applications   ISBN: 978-1-60876-915-5
Editor: Paul S. Geller, pp. 33-61   © 2010 Nova Science Publishers, Inc.

Chapter 2

RURAL GENTRIFICATION AND THE BUILT ENVIRONMENT: EXPLORING THE CONNECTIONS

Martin Phillips

Reader in Social and Cultural Geography, University of Leicester, Leicester, UK

ABSTRACT

This article considers the significance of the concept of the built environment with respect to rural gentrification, drawing principally on studies of the British countryside. It begins by drawing attention to the importance given to the built environment within early definitional debates over gentrification and in more recent discussions concerning changing forms of gentrification and the emergence of a 'third-wave' of 'new-build gentrification'. The article then considers whether these debates and discussions of gentrification, which have largely been urban focused, can be connected into studies of rural space. Attention is paid to the conceptual distinction between classical and new-build gentrification, it being suggested that this is a far from simple dualism. Eight dimensions of differentiation are identified from the urban literature, which it is then suggested can be applied to the study of rural gentrification as well. Attention is then drawn to the degree to which the concept of built environment could be expanded to encompass much more than buildings and how this may impact on understandings of the dynamics of rural gentrification.

INTRODUCTION

This article looks at the significance of the built environment to rural gentrification, drawing principally on studies of the British countryside. The term gentrification is a "congested and contested" concept (Phillips, 2005a, p. 477), with a wide variety of definitions having been advanced and debated, without there being clear consensus as to its meaning, and indeed the whole value of the concept itself has been questioned. The concept of the built environment is, it will be argued, quite central to many interpretations of gentrification and to contemporary debates surrounding the term. Much of the work on gentrification has had an urban focus, as is clearly demonstrated by the titles of many of the leading texts on the subject, such as Smith and William's (1986) *Gentrification and the city*, Caulfield's (1994) *City form and everyday life*, Ley's (1996) *The new middle class and the remaking of the central city* and Smith's (1996) *The new urban frontier: gentrification and the revanchist city*. This focus itself can be seen to connect to concept of the built environment, with people such as Harvey (1985, p. 14) arguing that "whatever else it may entail, the urban process implies the creation of a material physical infrastructure ... [or] built environment". The focus of this article will, however, be on the built-environment in rural areas, outlining some of the transformations to it that might be associated with processes of gentrification, as well as considering how hitherto urban focused debates and discussions connect into studies of rural space. The article will also reflect on the degree to which studies of rural gentrification may raise some significance questions about the concept of the built environment, both with respect to studies of gentrification and more widely.

THE BUILT ENVIRONMENT AND THE CONCEPT OF GENTRIFICATION: LESSONS FROM URBAN STUDIES

As noted above, gentrification has been the subject of considerable and long running debate. The term itself is often traced back to Glass (1964) who made passing reference to it in her study *London: aspects of change*, suggesting that:

> "One by one, many of the working-class quarters of London have been invaded by the middle classes ... Shabby modest mews and cottages ... have been taken over ... and become elegant, expensive residences. Larger Victorian houses, down graded in an earlier or recent period ... [to] lodging houses or ... multiple occupation - have been upgraded again ... Once this process of 'gentrification' starts in a district it goes on rapidly until all or most of the original working-class occupiers are displaced and the whole character of a district is changed" (Glass, 1964, p. 33).

Here one finds a clear sense that gentrification is about the refurbishment or doing up the buildings, or built environment, of an area, in conjunction with changes in the social composition of the inhabitants stemming from both middle-class in-migration and working class out-migration or displacement. Subsequent years have seen considerable debate over the definition of gentrification, with Slater et al (2004, p. 1143), for example, arguing that "the extent and meaning of gentrification have changed remarkably since 1964", although Glass' original emphasis on alteration of the built environment and social change through in-migration and displacement is retained in many more recent definitions of gentrification. Davidson and Lees (2005), for example, claim that the defining features of contemporary gentrification are: reinvestment of capital, social upgrading of locale by incoming high-income groups, landscape change, and direct or indirect displacement of low-income groups.

Whilst there are clear strands of continuity between early and recent definitions of gentrification, there has also been considerable variability in understandings of the term, with a range of different conceptual perspectives having been advanced and debated. Much of this variability can be linked to differences in philosophical position (see Phillips, 2005b), although it has also been argued that the concept of gentrification has changed to reflect changes in the nature and processes of gentrification itself (see Lees, et al., 2008; Slater, et al.,

2004; N. Smith, 2002). Amongst the changes seen as potential drivers of redefinition have been spatial and substantive expansion. Slater et al (2004, p. 1143, original emphasis), for instance, argue that there has been a "widening in the *extent* of gentrification", which is now to be observed in cities "down the urban hierarchy". They also note that gentrification has been identified in "rural locations", although they suggest that this, along with "the new construction of new-build luxury housing developments in city centres" (ibid), represents an expansion in the meaning of the term gentrification. Neil Smith makes very similar claims, suggesting that gentrification has become 'generalised', both through lateral and vertical expansion through the settlement hierarchy - such that it has a reported presence in spaces as geographically distant as Tokyo and Tenerife, Sao Paulo and Puebla Mexico, Cape Town and the Caribbean (see also Phillips, 2005b) and is evident not only in global cities like London, New York, Paris and Sydney but also "in more unlikely centres such as the previously industrial cities of Cleveland or Glasgow, smaller cities like Malmö or Grenada, and even small market towns such as Lancaster, Pennsylvania or Ceské Krumlov in the Czech Republic" (N. Smith, 2002, p. 439) – and through sectoral combination such that gentrification "involves much more than simply … housing" but has "evolved" into the formation of "new landscape complexes" that "integrate housing with shopping, restaurants, cultural facilities …, open space, employment opportunities" (ibid, p. 443). For Smith gentrification has changed in form, such that earlier definitions focused on the rehabilitation of residential properties in inner city neighbourhoods seems overly narrow, and indeed somewhat "quaint" (ibid, p. 439), while Lees et al (2008) talk of the 'mutation of gentrification', suggesting that the meaning of the term has changed to reflect "temporal and spatial changes in the process".

The concept of the built environment was, as noted before, quite central to the initial delimitation of gentrification by Glass, and has also been quite significant in subsequent evolutions/mutations of the term. In its formulation by Glass there was, as previously noted, a clear focus on the refurbishment of existing buildings, an issue which was given particular prominence by advocates of a so-called 'production-side' theorisation of gentrification which emerged from the late 1970s, principally through the Marxist inspired work of Neil Smith (e.g. Smith, 1979b; 1982; 1987, 1996; Smith & DeFilippis, 1999). Smith repeatedly characterised gentrification as a process of 'investment in the built environment', variously suggesting that gentrifiers should be seen not so much as occupiers of residential properties but as developers of such properties who invested productive capital – that is money, resources and labour power – as a consequence of the formation of rent-gaps, whereby actual rents – viewed in the sense of general financial returns accruing to property owners – are significantly lower than the potential rents available from converting the property for occupation by middle-class gentrifiers.

Neil Smith's concept of rent-gap implied that gentrification was a process of reinvestment in an already constructed built environment, albeit one which had experienced some devalorisation such that the realised value of its current form differed significantly from its potential value if redeveloped for gentrified land-use. Processes of devalorisation reflected both physical and social depreciation related to the physical degradation of buildings and stylistic, technological and functional changes which mean that new buildings tend to be of higher value than older ones, unless there was some reinvestment of capital to refurbish, restyle and restructure the buildings.

Neil Smith's conception of rent-gaps has been much debated and criticised (e.g. see Badcock, 1989; Bourassa, 1990; 1993; Clark, 1988; 1991; 1995; 1999a; Hammel, 1999b; Ley, 1986; 1987), with concerns being expressed variously about: the relationship between the concept and other, largely more neo-classical economic, concepts; the empirical validity of the concept; the motivational assumptions and implications of the rent-gap thesis, which might suggest that gentrification activity would be stimulated by profit motives; and the range of agents of gentrification that the concept seemed to encompass. With respect to the last issue, for instance, Neil Smith argued that gentrification did not simply, or necessarily predominately, involve members of residential households buying a property but could also involve a range of professional/institutional development agencies, including property development companies, builders, architects, financial institutions, estate agents and planning organisations.

The inclusion of large institutional agencies within the ambit of gentrification was a subject of concern to some researchers. Warde, for example, suggests,

> "There is a world of difference between the activities of large-scale property developers who buy a large tract of land to build a condominium and those of individual households that buy an old house in an 'improving neighbourhood' and set about restoring it" (Warde, 1991 p. 224)

He adds that failing to recognise this difference risks making gentrification a chaotic concept conflating unrelated processes together. Similar arguments have also been made by Lambert and Boddy (2002, p. 23) who questioned whether the term gentrification should be applied to "corporate new build or conversion of former commercial or industrial buildings" which they suggest has dominated the reurbanisation of 'second-tier' cities such as Bristol in the 1990s, claiming that whilst there are there are parallels, such as "new geographies of neighbourhood change, new middle-class fractions colonising new areas of central urban space, and attachment to a distinctive lifestyle and urban aesthetic", the agents of gentrification are quite different from "the pioneer gentrifiers of earlier decades" who bought up "individual housing units … renovating and restoring them for their own use". They further add that, in contrast to the earlier phase of gentrification, "there is no direct displacement of other social groups and lower income households", although there may be "secondary processes of displacement" due to rising rental levels in areas having undergone gentrification (ibid, p. 21). They conclude that applying the term gentrification to the reurbanising housing developments is "stretching the term and what it set out to describe too far" (ibid, p. 23), a claim that is further elaborated in Boddy (2007, p. 98) who questions whether 'developer-led new-build' developments should not be described as gentrification as doing so involves strengthening the concept "beyond the point at which it remains useful and credible as a means of understanding the processes at work".

Boddy (2007) develops his arguments in part through a critique of the work of Davidson and Lees (2005) which explicitly advances the notion of 'new-build gentrification', suggesting that it represents "one of the mutations of the gentrification" which has occurred in a 'third-wave' of gentrification' which they suggest, drawing on the work of Hackworth and Smith (2001), has emerged from the second half of the 1990s. They add that this form of gentrification is not the only form of gentrification occurring in the third wave, there also being gentrification in its "traditional or classic form" in which individual householder renovate "disinvested or derelict old housing through sweat equity or by hiring builders and interior designers", "state-led gentrification", often in combination with large-scale developers, "commercial gentrification", and "regentrification" through "supergentrification", whereby workers with very high incomes gentrify properties in already gentrified areas (ibid, pp. 1167-1168), an argument which can also be seen to have influenced Shaw's (2008) contention that there are a range of forms of gentrification which can be placed on a continuum of social and economic geographic change.

More generally it is possible to suggest that there have been three broad responses to the apparent substantive and spatial widening in the application of the term gentrification. The first response, represented by the likes of Neil Smith (2002), Lees et al (2008), Davidson and Lees (2005) and, albeit perhaps to a lesser extent, Shaw (2008), might be characterised as 'ontology led modification', in that it basically argues that the object of gentrification studies – that is gentrification – has been transformed and that the concept therefore needs modification to establish a new conceptualisation of this object, or a new ontology of gentrification. A second response, exemplified here by Lambert and Boddy (2002) and Boddy (2007), recognises many of the same spatial and substantive expansions identified in the first approach, but suggests that the concept of gentrification cannot be modified to incorporate these changes without it loosing all meaning. In a rather similar vein, Bondi (1999, p. 255) has suggested that gentrification research may have lost creative momentum and be failing to "open up new insights", in part because of the increasing and conflicting burdens the concept has been asked to carry, to such an extent that she suggest that it might be "time to allow it to disintegrate under the weight". This response might be term labelled 'ontology led abandonment'.

A third response might be characterised as a more 'epistemological response', in that it suggests that many of these seeming new features reflect long established uncertainties surrounding the delimitation of gentrification which require further consideration. This point is recognised by Davidson and Lees (2005, p. 1168), who whilst generally championing ontological modification lines of argument, also comment that an examination of gentrification literature from the early 1990s onwards reveals that "authors have in fact been thinking about these issues for some time now". Clark (2005, pp. 256-257), however, makes the argument more explicitly, arguing that expansion in the substantive and spatial extent of gentrification actually requires the adoption of a "broader definition of gentrification than is commonly found in the literature" and "more deep probing" into its "basic conditions of existence". As with Smith, Slater et al, Lambert and Boddy, and Shaw, Clark highlights both spatial and substantive extension, noting that gentrification not only occurs in inner city locations but also "occurs in other places as well" and commenting:

> "For years I have waited for the convincing argument why renovated buildings can be sites of gentrification, but not new buildings replacing demolished buildings. With as much anticipation, I have awaited the succinct delineation between rehabilitation and clearance/new construction, wondering in which category the cleared lot with braced and girded façade will fall" (ibid, p. 258).

Clark adds that in his view, many of the attempts to delimit gentrification have been based on 'sloppy abstractions' centred upon contingent rather than necessary relations, suggesting, for instance, that:

"Central location may be one important cause of the process [of gentrification] in some cases, but abstracting this relation to define the process leads to a chaotic conception of the process, arbitrarily lumping together centrality with gentrification. What becomes of gentrification in rural areas? Calling it something else would involve just another form of chaotic conception based on another form of bad abstraction that arbitrarily divides gentrification" (ibid, p. 259).

In making these arguments Clark draws directly on the realism of Sayer (2000) and its distinctions between abstract/concrete theorisations and necessary/contingent relations, although more broadly Clark's arguments suggest that concerns over the widening spatial and substantive application of the term gentrification may not simply be a reflection that gentrification may have changed its form since the initial coinage of the term by Glass, but rather that some of the central distinctions and delimitations made within gentrification research may themselves have long been quite problematic, including that between re-building existing build and creating new-build on sites from which any existing buildings have been demolished.

Davidson and Lees (2005, p. 1169) do go some way in seeking to establish potential grounds for such a distinction, suggesting that new-build gentrification "contrasts with previous rounds of gentrification because different groups of people are involved, different types of landscapes are being produced, and different sociospatial dynamics are operating". It is, however, possible to identify at least eight lines of difference that are drawn into debates about the distinctiveness, or otherwise, of new-build gentrification (see Figure 1).

The commonly used criterion of distinction is that the scale of change, with studies frequently drawing contrasts between the individual household/building focus of 'classical' gentrification and large-scale 'third-wave' new-build developments. Reference has, for example, been made to Neil Smith's (2002) characterization of classical gentrification as 'quaint' and his explicit contrasting of this with the formation of gentrified 'landscape complexes' involving housing, shopping, restaurants, cultural facilities, business employment and open space. He also makes much of a second widely used line of differentiation, namely the agent of change, suggesting, for instance, that the formation of these new landscape complexes involves a weaving of together of global finance with "large- and medium-sized real-estate developers, local merchants, and property agents with brand-name retailers, all lubricated by city and local governments" (ibid, p. 443). In other words, third-wave new-build developments involve corporate agents, as opposed to the pioneer, first-wave, gentrification centred around individual householders buying up-properties to refurbish, often through the expenditure of their own labour, or what Smith (1979a) identifies as 'sweat-equity'. Whilst Neil Smith's early work was keen to highlight that these actions were themselves bound to the actions of a range of other professional agents of gentrification providing and servicing the provision of productive capital, by Smith (2002) he is highlighting the difference in scale and form of the agents involved in third-wave new-build gentrification, a point also emphasised by critics such as Warde (1991), Lambert and Body (2002) and Boddy (2007) who all emphasise the differences between corporate and householder gentrification.

Whilst differences in scale of development product and the agents of gentrification have been widely emphasized in debates over new-build gentrification, a series of other differences are often also employed. For example, a third line differentiation, closely related to the issue of agents of gentrification, is the mechanism of change, with third-wave new-build gentrification often being identified as a more commodified form of gentrification. Davidson and Lees (2005, p. 1182), for instance, claim that the recent new-build development along the River Thames is "a commodified, mass-produced, and niche-marketed product" in which "[t]he traditional gentrifier's lifestyle has been appropriated and sold commercially to more gentrifiers" who "buy or rent, as opposed to create, a lifestyle", making use of "interior-design services, through which the resident simply orders the interior of the apartment when he or she buys the property". Similar arguments are advanced in works by Butler and Robson which have contrasted areas of gentrification according to the degree to which that have utilized commodification, as opposed to social networking and moral regulation: or as they describe it, the degree to which use is made of economic, social and cultural capital (e.g. see Butler & Robson, 2001, 2003a, 2003b). Not all of the areas examined by Butler and Robson were subject to new-build gentrification, although there was a clear sense that areas such as Docklands where new-build predominated were amongst the areas where commodified relations/economic capital held strongest sway: Butler and Robson (2003a, p. 182), for instance, remark that in this area "what is valued is what is marketed" and that "flight from social obligation in most cases appears to be utter".

Butler and Robson's work not only emphases mechanisms of social interaction between gentrifiers and various constituents of the areas they inhabit, including other residents, but they also suggest that gentrification is undertaken by different elements or fractions of the middle class(es), which constitutes a fourth line of argument for differentiating new-build gentrification. In particular they make use of an 'asset based' theorisation of class (Savage & Butler, 1995) to suggest that gentrification is linked to the particular composition of economic, social and cultural capital assets available to particular social groups. They, along with several other gentrification researchers (e.g. Bridge, 2001a; 2006; 2007; Jager, 1986), suggest that classical or pioneer gentrification was associated with householders who held relatively limited amounts of economic capital but considerable amounts of cultural capital and who therefore sought out dilapidated housing which they could buy relatively cheaply and refurbish in a way which would not only enhance their liveability but would also demonstrate cultural taste and social distinctiveness. However, as gentrification has proceeded in these and other areas, it is suggested that it has increasingly involved middle class residents who are richer in economic capital and poorer, or at least less concerned to demonstrate, cultural capital. Lambert and Body (2002, p. 22), for instance, comment that people "buying in to the new central area developments and conversions in Bristol and other second tier cities ... have much higher levels of economic capital and ... are buying in to a highly commodified form of urban lifestyle".

While several description of third-wave new-build-development make reference to the reproduction of the aesthetics of classic/pioneer gentrification, through, for instance, the production of "mock-Georgian townhouses and New-York-style apartments" (Davidson & Lees, 2005, p. 1181), studies have also highlighted aesthetic difference between classic/pioneer gentrification and third-wave new-builds, even amongst those which include elements of reproduction. Boddy (2007, p. 90), for example, suggests that new-build developments exhibit a range of styles, from a "neomodernist emphasis on concrete, glass, and stainless steel" through to historically referenced constructions, although the latter are "typically adapted and are often incorporated into uncompromisingly modernist styles of development" creating "postmodern pastiche" which is "a long way from the small-scale, traditional 'authentic' refurbishment of Victorian neighbourhoods".

A sixth line of argument bound up with differentiating new-build gentrification from classic forms of gentrification is the type or sector of capital involved. Within Judith Glass' original description, the focus was quite clearly on investment in residential housing. Even when studies began to identify gentrification as involving the conversion of non-residential properties, as in the establishment of gentrified apartments or lofts in former industrial and warehousing spaces (e.g. Hamnett & Whitelegg, 2007; Short, 1989; Zukin, 1982), the focus was principally on the formation of residential properties. However, as already discussed, work on new-build third-wave gentrification has often highlighted how this has involved not only the formation of new residential properties, but also the establishment of a new built environment for retailing, leisure and commercial activities, as in Smith's (2002) discussion of the formation of landscape complexes. A series of studies have indeed highlighted non-residential aspects of gentrification, including transformations in food-consumption (Beauregard, 1986; May, 1996; Zukin, 1990), leisure activities (Mills, 1988), consumer good purchasing (Bridge & Dowling, 2001; Zukin & Kosta, 2004; Zukin, et al., 2009) and cultural provisions (Deutsche & Ryan, 1984; Zukin, 1982), although much of this work has focused on small-scale 'boutique' development rather than the formation of large-scale landscape complexes. However, studies of inner-city shopping-centre development by the like of Raco (2003) and Lowe (2005) have highlighted impacts which closely parallel those associated with gentrification (see Phillips, 2007).

Many of the studies of non-residential aspects of gentrification highlight issues of displacement, and this issue has formed a distinct seventh line of argument relating to new-build gentrification. As previously mentioned, critics of the concept such as Lambert and Boddy (2002) have suggested that there is little or no direct displacement associated with new-build schemes, and as a result have questioned whether the term gentrification can reasonably be applied. Other researchers, such as Davidson and Lees (2005), have argued that displacement is very much associated with new-build developments, and hence the term gentrification is quite appropriate, and indeed in a political sense quite critical, to apply. However, even they remark that much inner city new-build development is being built on "brownfield sites or on vacant and/or abandoned land" and, as such, may "not displace a pre-existing residential population in the same way as classical gentrification has done" (ibid, p. 1169).

New-build gentrification	Classical gentrification
Large (landscape) scale	Small (household) scale
Corporate development	Householder development
Commodified/mass-produced	Sweat-equity/crafted
Economic capital-rich middle class	Cultural capital-rich middle class
Postmodern pastiche	Pre-modern
Cross-sectoral	Residential
No direct displacement	Direct displacement
New-build	Re-build

Figure 1. Binary categories in dualistic constructions of new-build and classical gentrification.

In these debates, the issue of displacement and new-build gentrification has been approached largely from an empirical angle, but it also connects to theoretical issues which connect to an eighth line of argument concerning new-build and gentrification, which concerns the relationship between new-build and preceding environments. Some of the early production-side studies of gentrification, for instance, made much of gentrification being about the refurbishment of a built environment rather than the construction of new buildings. Smith (1982, p. 139), for example, argued that there was an important theoretical distinction to be drawn between "gentrification and redevelopment" as the latter "involves not the rehabilitation of old structures but the construction of new buildings on previously developed land". Harvey (1987; 1989) develops similar arguments, albeit couched in a discussion of the emergence of postmodernism, suggesting that the 1970s witnesses a movement from 'urban redevelopment' focused on the replacement of non-modernised urban buildings with modern buildings, to 'urban regeneration' involving the modernization of existing buildings and new-build developments that integrated into existing built environments. As noted earlier, notions of third-wave landscape-scale new-build construction has rather over-turned these arguments, suggesting that gentrification may indeed involve the construction of a built environment which supplants rather than incorporates previous built-forms.

Studies espousing the notion of new-build gentrification, and indeed many of those critiquing it, can be seen to draw on many of the eight strands of argument, although most of these arguments can bc scc to have somewhat wider applicability. Hence, not only have "authors been thinking about" the issue of new-build gentrification "for some time" (Davidson & Lees, 2005, p. 1168) but many of these thoughts have actually been developed in relation to subjects only tangentially connected to the developments which are now being described as third-wave gentrification. There is a sense that rather dualistic constructions are being built through a coalescence of a range of quite distinct lines of argument and difference, such that third-wave new-build gentrification, or indeed the alternative constructions of new-build residentialisation or reurbanisation being promoted by those objecting to this mutation of the gentrification concept, is seem to encompass an amalgam of binary characteristics which serve to separate it from more classical forms of gentrification (see Figure 1).

Dualistic constructions which amalgamate a series of binary distinctions have been widely critiqued for ignoring and conflating all manner of differences (e.g. see Berg, 1994; Giddens, 1976; 1984; Sayer, 1989; 1991; Soja, 1996), and the dualistic construction of new-build and classical gentrification outlined in Figure 1 might well be open to extensive deconstruction. However, for the purposes of this paper I want to focus attention of the lines of differentiation listed in Figure 1 and consider their significance in respect to expressions of concern about the spatial and substantive expansions in applications of the term gentrification. More specifically I want to consider these new-build/re-build distinctions in relation to use of the term gentrification to encompass change in rural as opposed to urban areas, and then to use this to ask some critical questions about the concept of new-build gentrification through consideration of the concept of the built environment.

NEW-BUILD AND RURALITY

The debate over new-build gentrification has been almost exclusively urban in focus, although Phillips (1993) noted that it was also an issue of significance to rural studies:

"in the rural context it is often hard to differentiate between rebuilds and new builds. In rural areas existing building stock is often used merely as a shell or facade behind which a totally new dwelling is effectively constructed. This is particularly true when barns, sheds and other agricultural outbuildings are used as the basis for construction" (Phillips, 1993, p. 129).

Source: top photograph reproduced courtesy of A McKenna, lower photograph by the author

Figure 2. Rustification of 1970s rural new-build development.

This finding was duplicated in further studies (e.g. Phillips, 2002; 2004; 2005a) that identified extensive conversion and rebuilds of a series of properties including barns, stables, other agricultural outbuildings; schools, chapels, shops, railway stations, telephone exchanges, tea-rooms, workshops, dovecots and numerous forms of residential property. The extent of the conversion/rebuild was often such that it was hard to discern the degree to which there was much retention of an original structure, at least from visual inspection. Examples included the building of extensions to completely envelope existing buildings (see Phillips, 2002, p. 300) and the transformation of properties from the outside in, as in the 'conversion' of a 1970s detached house into a 'rustic farmhouse' (see Figure 2).

Such instances raise questions about the degree to which it is possible to clearly differentiate new-build and conversion in specific instances, although it is also evident that in general new-build development has figured more prominently than conversions within popular and academic consciousness over rural change, in large part because they clearly challenge development control ethics which have dominated countryside policies in 'developed'/'urbanised' countries such as Britain (e.g. see Curry & Owen, 1996; Spencer, 1997) as well as promote NIMBY-like reactions (see Phillips, 2002). Having said this, it is clear that in many rural areas of Britain, the number of conversions and substantial house extensions far exceed the number of new-build constructions and can exert a significant, and not necessarily expansive, impact on the availability of household units within rural settlements (see Phillips, 2002, 2005a; Spencer, 1995).

In the discussion of urban studies of new-build gentrification it was noted that there have been suggestions that there may be a temporal dimension to its significance, it being identified as one element of a third-wave of gentrification. Whilst new-build, and policies seeking to restrict its construction, have a long history in rural areas (e.g. see Spencer, 1997), there have been some studies which suggest that there may be a temporal/stage dimension to the employment of refurbishment and new-build in the countryside. Phillips (2005a), for example, suggests that in the village of Thornage, Norfolk, conversions dominated in early periods of gentrification, but these came to be increasingly marginalized in form in the 1980s, leading to the conversion of very small properties, which in effect were often virtually complete new-builds, prior to a 'massification' of gentrification, whereby professional development companies engaged in larger scale, complete new-builds (see Table 1).

This temporal interpretation, like many of those advanced in urban studies, implicitly incorporates many of the lines of distinction outlined in Figure 1. It includes, for instance, the distinction between householder and corporate developers, and small- and large-scale developments, suggesting that the new-build developments of the 1990s were both larger in scale and undertaken by corporate rather than household gentrifiers. Similar arguments are also enacted in Smith and Phillips (2001, p. 459) who identify a shift in gentrification in the Hebden Bridge area of Yorkshire from the situation in the 1960s where individual householders bought up "cheap, decaying properties, often in remote areas", to renovate and develop themselves and the 1970s when there was also commercial production of "ready-made 'rural' commodities". They add that the latter were "aimed at attracting managerial and professional inhabitants" from the "the surrounding metropolitan areas and beyond" (ibid), whilst many of the earlier 'pioneer gentrifiers' were counter-cultural in orientation, seeking to escape, or at least 'step-out' from aspects of mainstream commodified society, and therefore making much use of their own-labour, or sweat equity.

Table 1. Property relation, demographic change and gentrification phases in Thornage

Period	Property relations	Demographic character	Agents and forms of gentrification
Early C20th–1950s	Closed village	Depopulation	–
1950s–1970s	Increase in 'closure' as large landowners produce prosperous agriculture by substituting capital for labour	Depopulation	Devalorisation of agricultural and residential properties within the village; heightened need for capital encourages further sale of property; some properties bought and converted
1970s–1980s	Devalorisation of service properties as a result of declining population	Further contraction in population	Devalorisation of service properties within the village; some properties bought and converted; restricted scope for development acts to channel colonisation towards the retired and second-home owners
1980s	Devalorised properties and small parcels of land sold for residential conversion and new build	Increasing population	Increasing marginal gentrification as major properties have become gentrified
1990s	Pressures for development conflict with conservation values and policies	Increasing population	Massification of gentrification—local property companies seek to new-build gentrifiable properties to over-come shortage of properties

Source: Table 6 - Phillips, M. (2005). Differential productions of rural gentrification: illustrations from North and South Norfolk. *Geoforum, 36*, p. 490.

Source: http://www.lowermillestate.com/gallery.html.

Figure 3. New-Build developments at the Lower Mill 'residential nature reserve.'

In Smith and Phillips' arguments there is, as with urban new-build studies, an incorporation of different mechanisms of change along with arguments of scale and agents of change, as well as, as is outlined in Phillips (2004), clear parallels with the urban studies of Ley (1996) which itself bears connections with the asset-based theorisations of gentrification by the like of Butler and Robson (see Phillips, forthcoming-b). As a result it appears that the analysis of Smith and Phillips could quite easily be interpreted as suggesting that the pioneer gentrifiers of Hebden Bridge were cultural capital-rich/economic capital-poor householders, whilst the inhabitants of the subsequent professionally produced ready-made gentrified housing made more use of economic capital and less use of cultural assets (see also D. Smith, forthcoming). Connections might also be made with Phillips' (1993, p. 134) suggestion that some 'sweat-equity' gentrifiers in Gower were 'marginal gentrifiers' undertaking this activity in response to "issues of need and the experience of constraint" rather than as an expression of affluence and choice. These gentrifiers might, hence, be characterised as having limited economic capital, although this does not mean that they might not have been exercising cultural capital (see Cloke, et al., 1998).

It could clearly be objected that the scale of new-build, even at the point of massification identified in Table 1 whereby there were attempts to construct not only new-build housing but also associated retail and leisure investments, was still well below the scale implied within the urban new-build studies discussed earlier. However, large-scale housing new-build construction is far from unknown in the countryside, even if one excludes from consideration suburban extensions to existing urban settlements. Murdoch and Marsden (1991) and Marsden et al (2003), for example, discuss the construction of Watermead, an entirely new settlement of 800 houses designed to resemble "a 'village' with a 'timeless' quality" (Murdoch, et al., 2003, p. 78; see also Phillips & Mighall, 2000). They note that the settlement's developers had come to specialise in 'concept communities', many of which took the form of 'gated communities' which, as Atkinson et al (2004) have observed, expanded rapidly in number in England from mid-1990s, including within rural areas: they suggest that 45 percent of the new gated communities built in Britain between 1995 and 2003 were located in rural or rural fringe locations. Whilst many of these developments are small in scale - 35 percent of the developments identified by Atkinson et al were under 50 houses – there are instances of much larger developments, including several within rural areas: the Burton Waters development in Lincolnshire (Figure 3), for instance, is planned to contain some 200 houses, while the Lower Mill Estate development in the Cotswolds (Figure 4) is aiming for around 575 houses. On a larger scale still are 'eco-town' and 'sustainable urban extension projects', such as the proposed developments of 3,000 homes at Lawley Village, near Telford; 6,100 homes at Weldon Park and Prior Halls near Corby; 3,400 homes at Rackheath, Norfolk; 5,000 homes around china clay pits near St Austell; and 5,500 houses at Whitehill Bordon, in East Hampshire (Charles Church development Limited, 2009; Department for Communities and Local Government, 2009; English Partnerships, n.d.).

Developments such as gated- and concept-communities, eco-towns and suburban extensions raise complex questions about the differentiation of urban and rural space. They might, for instance, be viewed as representing urban developments supplanting pre-existing ruralities, at least in instances where they occur in areas of open countryside separated from existing urban settlements. On the other hand, many of these clearly embody cultural signifiers of rurality in the manner identified by Marsden et al (2003) and might well be characterised as instances of 'post-rurality' as identified by Murdoch (1993, p. 425) in that they are locations "in which the 'rural' may be practiced" even if they themselves may not, or may no longer, be officially characterised as rural. With respect to this, it is relevant to note that Davidson and Lees' (2005) discussion of new-build gentrification along the River Thames, although largely framed as a discussion of an urban phenomena, included reference to the presentation of environmental features which are commonly viewed as constitutive of rural spaces:

> "marketing material ... plays of a particular lifestyle – one where the city (London) and nature (the river front) complement each other: .... A sweeping tree-lined boulevard and riverside piazza gives the development as exciting metropolitan focus. The ambience is enhanced by two community gardens, and a wealth of stylish landscaping ... leafy riverside towpath watching herons dip and rowing boats gently gliding by ... Lazy summer days by the river, kicking leaves in the crisp autumn air. Everything feels so natural. A sense of village life just minutes from central London" (Davidson & Lees, 2005, p. 1182)

Likewise the relationship between these developments and gentrification are complex. Álvarez-Rivadulla (2007, p. 60), for example, has noted a series of parallels between notions of gentrification and studies of gated communities, although she is hesitant to suggest incorporation of the latter into the former as it might be viewed as "yet another stretching of the meaning of gentrification". There clearly may be features of such developments that are not present in other gentrification developments, not least the presence of overt fortification and restrictions on public access, although, on the other hand, these developments exhibit many of the features of new-build gentrification as identified in the dualism outlined in Table 1 and studies have questioned the significance and distinctiveness of some of the physical, and indeed symbolic, barriers associated with gated communities (Blakely & Snyder, 1997; Flusty, 1994; Sternlieb, 1990). Whilst gentrification may be an inappropriate term for some of these developments, in many instances there would seem to be grounds for considering others as instances of new-build gentrification, presuming one is more generally prepared to accept this concept.

Figure 4. New-build development at Burton Water, Lincolnshire.

Reference has been made to the inclusion of signifiers of rurality in many of the larger new-build developments in the countryside such as Watermead, and such features can also be seen in many smaller-scale new-build developments in the countryside which may be seen to employ what might be characterised as a rural 'gentrification aesthetic', employing designs which simulate rural historicity and conversion. These are exemplified in the creation of so-called 'neo-barns' (see Bradbury, 2004; Luscombe-Whyte & Bradbury, 2004), although may also be identified across a wide range of rustic new-builds (e.g. see Phillips, 2000; 2002;

2004; 2005a; 2002; D. Smith & Phillips, 2001). There are clear parallels in this aesthetic to features which have been identified as constitutive of an 'urban gentrification aesthetic' (e.g. see Bridge, 2001a; 2001b; Podmore, 1998), including the blending of historicity and modernity. This can be viewed as illustrative of neglected lines of similarity in urban and rural gentrified space (see Phillips, 2004), although there are also clear senses that rural and urban identities are often played off against each other. The Watermead development, for example, contained not only simulated rural features such as "country cottages" and "the warmth and charm of a traditional ... village" (Murdoch and Marsden, 1991, p. 28), but also symbolically urban features such as town houses, shopping-mall and piazza, while Davidson and Lees' description of the marketing of the Thames riverside quoted earlier also highlights metropolitan as well as village and natural lifestyle elements. Furthermore, Smith (forthcoming) has returned to examine the gentrification of Hebden Bridge, suggesting that there might be a new wave of gentrification which involves the 'cascading' of urban gentrifiers, agents of gentrification and housing design styles from the surrounding urban centres of Leeds and Manchester, such that you can see "modern 'steel and glass'" being "stirred into 'the timber beams and mullion windows'" of earlier gentrification, as "urban-based, large-scale private sector developers increasingly manufacture and supply metropolitan style urban-living spaces within the rural context for young professional workers". He notes how former industrial premises, such as textile mills, as well as agriculture related properties, have become subject to conversions and marketing that, in his view, involve "the superimposition of the ideals of loft-living, apartments and so-called cutting-edge urban design and style" (see Figure 5). Areas of rural, and indeed urban, Britain may hence be sites of conversions and new-builds that incorporate hybrid historic/modern and rural/urban identities.

The incorporation of contrasting, and indeed from modernistic perspectives contradictory, perspectives can be identified quite clearly in the Lower Mill development which not only incorporates simulated historical building design but also explicitly modernist and futuristic designs such as the Orchid House (see Figure 6), which was bought for £7.2 million prior to construction, a figured described as "a world record ... for a country home" (Pierce, 2008). The overall development is promoted as a 'residential nature reserve', a term that itself can be seem to encompass notions of residential development and nature conservations which are widely viewed as incompatible to each other, but which here are marketed as being fully integrated:

> "homes at Lower Mill Estate have been designed to sit comfortably within their setting, with crisp, modern lines complementing the soft natural environment. Inside, extensive use of glass and open plan areas create a feeling of light and space while the highest quality fittings ensure the most luxurious of lifestyles" (Luxury Holidays.uk.com, n.d.)

This development, like many of the larger-scale rural new-build developments, and indeed urban developments, occurs on what might be described as 'brownfield sites', with the Lower Mill Estate being located in an area previously used for gravel extraction. Davidson and Lees (2005, p. 1169) argue that the location of new-build development on "brownfield sites or on vacant and/or abandoned land" is of theoretical significance because "as such they do not displace a preexisting residential population in the same way as classical gentrification has done". As previously discussed, the presence/absence of displacement has been a key issue within discussions over the value or otherwise of the concept of new-build gentrification, and the suggestion that new-build development may not induce displacement is clearly of importance. Care, however, needs to be exercised before the interpretation of brown-field development provided by Davidson and Lees is accepted. Davidson and Lees themselves note that the work of Cameron (2003) highlights that some brown-field sites were themselves previously areas of working class housing which have been demolished, while Raco, more generally, argues:

Source: images from www.mangohomes.co.uk (accessed 7/4/07)

Figure 5. The promotion of urban living in the rural market town of Hebden Bridge.

Source: www. Lowermillestate.com

Figure 6. Historical simulation and futuristic design at Lower Mill Estate.

"policy statements promoting the widespread reuse of brownfield land often overlook the variety of non-commercial meanings and values that are associated with such sites. It is easy for whole areas to be labelled as 'derelict', 'contaminated' and 'vacant', even though they may possess a variety of social, cultural and environmental assets ... local communities tend to value the recreational and leisure uses of brownfield sites ... At the same time, such sites often represent the most significant sources of biodiversity that exist in urban areas, something that national governments have, paradoxically, promoted at the same time as they call for more brownfield redevelopment" (Raco & Henderson, 2006, p. 501)

These arguments suggest that even when the areas of land being used for new-build developments were not previously sites of residential use, there may well be being used by non-middle class residents for recreational, leisure and, indeed, one can add, employment use, and the users of these spaces will hence experience displacement of quite a direct kind (see also Harrison & Davies, 2002; Whatmore & Hinchliffe, 2003). Furthermore, Raco comments highlight that it may not be simply people who experience displacement, but there may also be displacement of non-human actants, such as plants, animals and other forms of biodiversity. If this argument is accepted, it would seem to strengthen claims about the relevance of using the term gentrification in association with new-build developments, although it arguably extends the application of this term to encompass virtually all instances of new-build, a development which is likely to provoke yet further complaints about an over-extension of the concept such that it arguably looses all significance. It may also, as discussed in the next section, raise questions not only about gentrification but also about the concept of the built environment more generally.

RURAL GENTRIFICATION AND THE BUILT ENVIRONMENT: GOING BEYOND BUILDING PERSPECTIVES

As discussed in Phillips (2001, 2007) and Phillips et al (2008) gentrification studies, despite their myriad definitions of the concept and claims of an ever-expanding focus, have retained a highly human, and also building, centred focus. Gentrification studies indeed might be characterised as exhibiting quite direct, as well as broader but perhaps less immediately apparent, connections to the notion of a 'building perspective' as advanced by Ingold (1995; 2000).

With reference to general arguments, Wylie (2007, p. 156) comments that, for Ingold, the building perspective is an "often implicit presence ... at the heart of a series of intellectual dilemmas facing academic disciplines such as human geography, anthropology, and archaeology", all of which, one could add, have made considerable use of the concept of the built environment (e.g. see Rapoport (1994), as well as comments made earlier in this article). Ingold (1995, p. 59) indeed notes that the focus of many studies has been on "the meaning of the built environment", an issue which has been of clear significance to gentrification studies where, as illustrated in some of the earlier discussion, there has been interest in the way particular building forms and styles may convey various social and cultural meanings. Ingold suggests, however, that attention also needs to be paid "on what it means to say an environment is built" (ibid), suggesting that answers to this question often draw upon some notion that a built environment is one that is humanly produced or constructed. This is a viewpoint that can be identified within conceptions of the built environment discussed earlier in this paper, with production-side theorisation by the likes of Harvey and Smith viewing the built environment as the product of the investment of human labour power. More generally, the built environment is widely set up in contrast with a so-called natural environment, although numerous researchers, including both Harvey and Smith, have sought to undermine this distinction by highlighting how environments taken to be natural are actually socially produced (e.g. see Harvey, 1996; Smith, 1984). Ingold (1995, p. 59), however, both reverses and subverts such arguments, suggesting that if one accepts the idea that nature is socially produced, "where, in an environment that bears the imprint of human activity, can we draw the line between what is, and is not" part of the built environment, and "why should the products of human activity be any different, in principle, from the constructions of other animals"?

While Ingold uses these questions to raise very general questions about prevailing approaches to studies of people and nature, and their inter-relation, they also can be seen to have quite direct relevance to discussions of third-wave new-build gentrification. The issue of the dividing line between the built and non-built environment is, for example, clearly of relevance given the claims that third-wave gentrification may involve landscape change as well as changes to buildings. In part such claims revolve around use of landscape as simply a scalar term (see Phillips, et al., 2008), with such studies often remaining resolutely focused on humanly constructed built environments, although even here non-human elements or actants make their appearance, as demonstrated by the references to nature in the interpretation of new-build gentrification along the Thames presented by Davidson and Lees (2005), as well as in references to displacement by Raco (2006). Indeed, one as yet not fully realised consequence of discussions of third-wave new-build gentrification may be to bring nature into the landscape of gentrification studies.

Issues of nature have arguably figured rather more prominently in rural gentrification studies. In a several previous publications (e.g. Phillips, 2005c; 2008; Phillips, et al., 2008), for instance, I have pointed out how a series of studies have suggested that rural gentrification may be conditioned by and have an impact on a range of non-human, 'natural' actants and spaces. Studies have, for instance, long suggested that the 'natural environment' of the countryside might be one of the motivational pulls for moving to a rural locality, with Smith and Phillips (2001, p. 457) making direct connection to processes of gentrification, suggesting that rural gentrification be re-titled 'rural greentrification' to reflect the significance of "the demand for, and perception of, 'green' residential space from in-migrant households in … gentrified rural locations". Whilst expressing reservations about this renaming (see Phillips, 2004; 2005b), it is clear that nature, or "actants which are taken to be natural" (Phillips, et al., 2008 p. 584), do play an important role in motivating some migrational movements into the countryside, although such interpretations may themselves be open to criticism for employing a Cartesian distinction of thought and action, a dualism which Ingold sees as a general characteristic of building perspectives (see Wylie, 2007).

Actants taken to be natural do not, however, figure only in motivation studies of gentrification but can also be identified in some production-focused accounts. Phillips et al (2008, p. 55), for instance, note that the gentrification of properties may well involve a series of intentional and non-intentional re-landscapings, with building work both destroying or damaging "existing vegetation and habitats", some of which may be subsequently restored or transformed, while the landscape surrounding buildings may itself be "an object of investment and restructuring activity by gentrifiers, independent of any work related to the refurbishment or construction of buildings (see also Paquette & Domon, 2001; 2003). Parallels between these arguments and 'domestic' or 'front-yard' landscape studies have been identified (see Phillips et al., forthcoming; Saratsi et al., 2006), whilst more generally they can be seen to support Ingold's questioning of the division between the built and natural environment, both because they indicate that social life goes on not only within but also "around dwellings" (Ingold, 1995, p. 67), and also because these environments which are full of actants taken to be natural are themselves clearly the focus on considerable building and re-building activity.

In the case of rural Britain, these issues have been given further prominence through expressions of concern about 'garden grabbing' that emerged in 2008 in relation to the drafting of a planning policy statement defining brownfield. As outlined in Phillips (forthcoming-a), concern was expressed about the inclusion of gardens in the definition, it being suggested that in the context of sustainable planning policies encouraging development of brownfield sites, this would mean that the green space of gardens would be converted into buildings. It could be objected that such practices have a much longer history in rural areas, with research in the late 1990s, for example, coining the phrase 'windfall gentrification' to refer to situations where residential property owners sell-off parcels of land and outbuildings for residential development (Phillips, 1999). However, the recent publicity given to garden development does provide a further example of how brownfield redevelopment may be seen to involve the displacement of nature, and also highlights how nature, even in the form of areas of greenspace, is often widely present within the supposedly humanly constructed/unnatural space of the built environment. Research examining the 'built environment' of gentrified villages in rural Leicestershire has, for example, suggested that over half of the spatial extent of the area so designated was actually vegetated green space (Phillips, et al., 2008); and studies in urban areas have also identified significant amounts of green spaces within these built environments (e.g. Gaston, et al., 2005; Loram, et al., 2007; Mathieu, et al., 2007). More generally, as Whatmore and Hinchliffe (2003, p. 137) observe, "the cartographic opposition between cities and natures", and also one could add between built and natural environments, has come under challenge, it being widely recognised that there are a range of non-human occupants of the built environment of the city. The built environment of the city is a socially constructed environment that is, to use the words of Murdoch (2003, p. 269, original emphasis), "*more than* social".

Murdoch actually wrote these words in the context of a discussion of the significance of nature in migration to rural space. He argued that this significance had often been rather ignored, with migration to rural areas often being attributed to a range of social influences and processes. He adds, however, that just as rural space may be seen as a socially produced environment that is composed of more than the social, it is also a natural environment which is always "*more than* the natural" (ibid, p. 269). The comments return us once again to the issue of the social production of nature and Ingold's questioning of whether we can actually draw a distinction around the built environment where everywhere, even the seeming natural, can be seen to have been produced through human activity.

This is an issue that has direct relevance to the study of rural gentrification given that the countryside is widely, although often far from logically, viewed as a space of nature, and that this sense of rurality may itself play a role in fostering the colonisation of rural areas (see Phillips, 2008; forthcoming). Such perspectives can be seen to have encouraged the employment of behavioural/consumption focused interpretations of rural gentrification as outlined by Darling (2005), and they may also have acted to deter the employment of production-focused interpretations through fostering an overly naturalistic/passive view of rural spaces. The production-side interpretations advanced in urban studies through concepts such as the 'rent-gap' involve consideration of urban space as a produced/built environment whose form actively constitutes gentrification, enabling or thwarting its initiation and providing the material from which new gentrified constructions come to be made, either directly or through some form of commodified reproduction. There is an explicit recognition of the "making of space for ... gentrification" (Phillips, 2005a, p. 477), which has been significantly absent from many studies of rural gentrification, although the two above mentioned studies do seek to address this, drawing attention to processes of disinvestment in rural space that preceded rural gentrification and in a sense make certain areas ready for gentrification. In the case of Phillips (2005a, p. 477), and also Phillips (2004), the focus was on the devaluation of agricultural land and properties and their revalorization as residential properties, shifts which might be associated with conceptualisation such as 'post-productivism'. Darling (2005), however, focuses of the production of a natural landscape, suggesting that in the case of the Adirondacks in New York State, this has occurred as a result of discouragement of investment in residential property development which had the perverse consequence of preventing the development of a built environment that would have in effect destroyed a resource – an undeveloped landscape or wilderness – that subsequently became highly valued by gentrifiers. Darling suggests that gentrification in this area of wilderness has had little to do with processes of devaluation and revaluation of residential space, but rather reflects the production and valuation of wilderness as a recreational space. These claims further highlight that gentrification may not simply revolve around the built environment of buildings, but may also be driven by the construction of another built environment, namely that of the wild or natural environment.

CONCLUSION

This article has considered the significance of the concept of the built environment to gentrification, focusing particularly on its role within conceptions of third-wave new-build gentrification and rural gentrification. It has suggested that the concept of third-wave new-build gentrification challenges some of the theoretical arguments that were seen to be important within many early discussions of urban gentrification. Distinctions between gentrification and redevelopment, or between redevelopment and urban regeneration, or between refurbishment and new-build, all revolved around claims that there were significance differences between developments that incorporated previous built forms and those that supplanted them. The emergence of notions of third-wave new-build gentrification arguably over-turns this argument, although as has been shown here, discussions of this concept, by both advocates and critics, has often involved many other issues besides the incorporation or supplanting of previous built environments. One might even suggest that the rather dualistic discussions of classical and third-wave new-build gentrification have rather over-shadowed consideration of the theoretical arguments advanced previously in connection with conceptions of the built environment.

The over-shadowing of the earlier concerns is not necessarily problematic, not least given that both advocates and critics of the concept of third-wave new-build gentrification often take the view that the world has changed significantly since the years when gentrification first emerged as a topic of academic study and debate. On the other hand, the 'ontology led modification' and 'ontology led abandonment' lines of argument identified in this article might perhaps usefully incorporate greater consideration of whether some of the features and issues identified as characteristic of this new form/wave of gentrification might have had earlier precursors, and indeed whether overly dualistic arguments might be being advanced by advocates and critics alike.

One of the arguments associated with third-wave new-build gentrification is that there has been a spatial extension in gentrification, which now appears in all manner of spaces other than the inner city. Much of the argument over spatial extension relates to notions of progressive diffusion of gentrification rather than specific linkage to conceptions of new-build gentrification, and indeed there have been few studies of new-build gentrification in places other than cities. The present article seeks to address this issue through consideration of whether ideas associated with third-wave new-build gentrification could be applied to rural Britain. It is suggested that they can, although care is clearly needed, not least because the concept of third-wave new-build gentrification has been shown to include at least eight distinct lines of differentiation. Further work is clearly needed to establish the significance of each of these within particular locations, and to explore their interconnections and degrees of interconstitution. Even if the concept of third-wave new-build gentrification is deemed applicable to rural areas, there clearly may well be differences in its form between different rural, and indeed urban, localities, and perhaps some differences between urban and rural spaces, although notions such as the post-rural, and indeed one might add the post-urban, might suggest caution here as well.

Finally this article has returned to the issue of conceptualizations of gentrification and the built environment, predominately through a reflection on the significance of the latter concept to studies of rural gentrification, although reference was also made to issues of brownfield development, which is clearly an issue of considerable significance to studies of urban regeneration and gentrification. Attention has been drawn to issues of displacement, which are then connected to the more general questioning of the concept of the built environment developed by Ingold as part of a wider critique of what he described as building perspectives. Drawing on these arguments, attention has been drawn to how gentrification may involve changes in built environments other than buildings, with attention being drawn to changes in gardens, residential or domestic landscapes and to the wider landscapes of countryside and wilderness. It is suggested that such landscapes should be seen as built environments and that attention needs to be paid to how these landscapes are produced, and how they might influence the course of subsequent gentrification. Whilst this is an argument that is developed with respect to rural gentrification, it would seem to be an argument that is of relevance to studies of urban gentrification as well. It is also an issue that studies of third-wave new-build gentrification may also need to build into future developments of this concept.

REFERENCES

Álvarez-Rivadulla, M. J. (2007). Golden ghettos: gated communities and class residential segregation in Motevideo, Uruguay. *Environment and planning A, 39*, 47-63.

Atkinson, R., Blandy, S., Flint, J., & Lister, D. (2004). *Gated communities in England.* London: Office of the Deputy Prime Minister.

Badcock, B. (1989). An Australian view of the rent gap hypothesis. *Annals, Association of American Geographers, 79*, 125-145.

Beauregard, R. A. (1986). The chaos and complexity of gentrification. In N. Smith & P. Williams (Eds.) *Gentrification of the city* (35-55). London: Allen and Unwin.

Berg, L. D. (1994). Masculinity, place and a binary discourse of 'theory' and 'empirical investigation' in the human geography of Aotearoa/New Zealand. *Gender, place and culture, 1* , 245-260.

Blakely, E. J. & Snyder, M. G. (1997). *Fortress America: gated communities in the United States.* Washington D. C. and Cambridge, Massachusetts: Brookings Institution Press and Lincoln Institute of Land Policy.

Boddy, M. (2007). Designer neighbourhoods: new build residential development in nonmetropolitan UK cities - the case of Bristol. *Environment and planning A, 39*, 86-105.

Bondi, L. (1999). Between the woof and the weft: a response to Loretta Lees. *Environment and planning D: society and space, 17*, 253–255.

Bourassa, S. (1990). Another Australian view of the rent gap hypothesis. *Annals of the Association of American Geographers, 80*, 458-459.

Bourassa, S. (1993). The rent gap debunked. *Urban studies, 30*, 1731-1744.

Bradbury, D. (2004). The rise of the neo-barn. *The Times.*, 9th May.

Bridge, G. (2001a). Bourdieu, rational action and the time-space strategy of gentrification. *Transactions, Institute of British Geographers, 26*, 205-216.

Bridge, G. (2001b). Estate agents as interpreters of economic and cultural capital: the 'gentrification premium' in the Sydney housing market. *International journal of urban and regional research, 25*, 81-101.

Bridge, G. (2006). It's not just a question of taste: gentrification, the neighbourhood, and cultural capital. *Environment and planning A, 38*, 1965-1978.

Bridge, G. (2007). A global gentrifier class? *Environment and Planning A, 39*, 32-46.

Bridge, G. & Dowling, R. (2001). Microgeographies of retailing and gentrification. *Australian geographer, 32*, 93-107.

Butler, T. & Robson, G. (2001). Social capital, gentrification and neighbourhood change in London: a comparison of three south London neighbourhoods. *Urban studies, 38* , 2145-2162.

Butler, T. & Robson, G. (2003a). *London calling: the middle classes and the re-making of inner London.* Oxford: Berg.

Butler, T. & Robson, G. (2003b). Negotiating their way in: the middle classes, gentrification and the deployment of capital in a globalising metropolis. *Urban studies, 40* , 1791-1809.

Cameron, S. (2003). Gentrification, housing redifferentiation and urban regeneration: 'Going for Growth' in Newcastle upon Tyne. *Urban studies, 40* , 2367–2382.

Caulfield, J. (1994). *City form and everyday life: Toronto's gentrification and critical social practice.* Toronto: University of Toronto Press.

Charles Church development Limited. (2009). *Weldon Park: environmental statement, non-technical summary*. York: Charles Church Developments Limited.

Clark, E. (1988). The rent gap and the transformation of the built environment: case studies of Malmö, 1860-1985. *Geografiska annaler B, 70* , 241-254.

Clark, E. (1991). Rent gaps and value gaps: complementrary or contradictory? In J. van Weesep & S. Musterd (Eds.) *Urban housing for the better off: gentrification in Europe* (17-29). Utrecht: Stedelijke Netwerken.

Clark, E. (1995). The rent gap re-examined. *Urban studies, 32*, 1489-1503.

Clark, E. (2005). The order and simplicity of gentrification - a political challenge. In R. Atkinson & G. Bridge (Eds.) *Gentrification in a global context: the new urban colonialism* (256-264). London: Routledge.

Cloke, P., Phillips, M. & Thrift, N. (1998). Class, colonisation and lifestyle strategies in Gower. In M. Boyle & K. Halfacree (Eds.) *Migration to rural areas*. London: Wiley.

Curry, N. & Owen, S. (1996). Introduction: changing rural policy in Britain. In N. Curry & S. Owen (Eds.) *Changing rural policy in Britain: planning, administration, agriculture and the environment*. Cheltenham: Countryside and Community Press.

Darling, E. (2005). The city in the country: wilderness gentrification and the rent gap. *Environment and planning A, 37*, 1015-1032.

Davidson, M. & Lees, L. (2005). 'New-build 'gentrification' and London's riverside renaissance. *Environment and planning A, 37*, 1165 – 1190.

Department for Communities and Local Government. (2009). *Eco-towns: location decision statement*. London: Communities and Local Government Publications.

Deutsche, R. & Ryan, C. (1984). The fine art of gentrification. *October, 31* (Winter), 91-111.

English Partnerships. (n.d.). *Lawley, Telford*. London: English Partnerships.

Flusty, S. (1994). *Building paranoia: the proliferation of interdirectory space and the erosion of the spatial justice*. West Hollwood, California: Los Angeles Forum for Arch & Urban Design.

Gaston, K. J., Warren, P., Thompson, K., & Smith, R. M. (2005). Urban domestic gardens (IV): the extent of the resource and its associated features. *Biodiversity and conservation, 14*, 3327–3349.

Giddens, A. (1976). *New rules of sociological method*. London: Hutchinson.

Giddens, A. (1984). *The constitution of society*. Cambidge: Polity.

Glass, R. (1964). *London: aspects of change*. London: MacGibbon and Kee.

Hackworth, J. & Smith, N. (2001). The changing state of gentrification. *Tidjschrift voor economische en sociale geografie, 29* , 464-477.

Hammel, D. (1999a). Gentrification and land rent: an historical view of rent gap in Minneoplis. *Urban geography, 20* , 116-145.

Hammel, D. (1999b). Re-establishing the rent gap: an alternative view of capitalized land rent. *Urban studies, 36* , 1283-1293.

Hamnett, C. & Whitelegg, D. (2007). Loft conversion and gentrification in London: from industrial to postindustrial land use. *Environment and planning A, 39*, 106-124.

Harrison, C. & Davies, G. (2002). Conserving biodiversity that matters: practioners' perspectives on brownfield development abd urban conservation in London. *Journal of environmental management, 65*, 95-108.

Harvey, D. (1985). *The urbanisation of capital*. Oxford: Blackwell.

Harvey, D. (1987). Flexible accumulation through urbanisation: reflections of 'postmodernism' in the American city. *Antipode, 19* , 260-286.

Harvey, D. (1989). *The condition of postmodernity*. Oxford: Blackwell.

Harvey, D. (1996). *Justice, nature and the geography of difference*. Oxford: Blackwell.

Ingold, T. (1995). Building, dwelling, living: how animals and people make themselves at home in the world. In M. Strathern (Eds.) *Shifting contexts: transformations in anthropological knowledge* (59-80). London: Routledge.

Ingold, T. (2000). *The perception of the environment: essays in livelihood, dwelling and skills*. London: Routledge.

Jager, M. (1986). Class definition and the esthetics of gentrification: Victoriana in Melbourne. In N. Smith & P. Williams (Eds.), *Gentrification of the city* (78-91). London: Unwin and Hyman.

Lambert, C. & Boddy, M. (2002). *Transforming the city: post-recession gentrification and re-urbanisation*. Bristol and Glasgow: University of Bristol and University of Glasgow.

Lees, L., Slater, T. & Wyly, E. (2008). *Gentrification*. London: Routledge.

Ley, D. (1986). Alternative explanations for inner city gentrification: a Canadian reassessment. *Annals, Association of American Geographers, 76*, 521-535.

Ley, D. (1987). Reply: the rent gap revisited. *Annals, Association of American Geographers, 77*, 465-468.

Ley, D. (1996). *The new middle classes and the remaking of the central city*. Oxford: Oxford University Press.

Loram, A., Tratalos, J., Warren, P. & Gaston, K. J. (2007). Urban domestic gardens (X): the extent and structure of the resource in five major cities. *Landscape ecology, 22*, 601-615.

Lowe, M. (2005). The regional shopping centre in the inner city; a study of retail-led urban regeneration. *Urban studies, 42* , 449-470.

Luscombe-Whyte, M. & Bradbury, D. (2004). *Barns: living in converted and reinvented spaces*. New York: Collins Design.

Luxury Holidays.uk.com. (n.d.). The estate: concept. Retrieved 1/9/09, 2009, from www.luxuryholidays.uk.com/concept

Marsden, T., Murdoch, J., Lowe, P., Munton, R. & Flynn, A. (1993). *Constructing the countryside*. London: UCL Press.

Mathieu, R., Freeman, C. & Aryal, J. (2007). Mapping gardens in urban areas using object-orientated techniques and very high resolution satelitte imagery. *Landscape and urban planning, 81*, 179-192.

May, J. (1996). 'A little taste of something more exotic': the imaginative geographies of everyday life. *Geography, 81*, 57-64.

Mills, C. (1988). 'Life on the upslope': the postmodern landscape of gentrification. *Environment and planning D: Society and space, 6*, 169-189.

Murdoch, J. (2003). Co-constructing the countryside: hybrid networks and the extensive self. In P. Cloke (Ed.), *Country visions* (263-282). Harlow: Pearson.

Murdoch, J., Lowe, P., Ward, N. & Marsden, T. (2003). *The differentiated countryside*. London: Routledge.

Murdoch, J. & Marsden, T. (1991). *Reconstituting the rural in an urban region: new villages for old?* Newcastle: University of Newcastle.

Murdoch, J. & Pratt, A. (1993). Rural studies: modernism, postmodernism and the 'post rural'. *Journal of rural studies, 9* , 411-427.

Paquette, S. & Domon, G. (2001). Rural domestic landscape changes: a survey of the residential practices of local and migrant populations. *Landscape research, 26*, 367-395.

Paquette, S. & Domon, G. (2003). Changing ruralities, changing landscapes: exploring social recomposition using a mullti-scale approach. *Journal of rural studies, 19*, 425-444.

Phillips, M. (1993). Rural gentrification and the processes of class colonisation. *Journal of rural studies, 9*, 123-140.

Phillips, M. (1999). *The processes of rural gentrification.* End of award report to the Economic and Social Research Council.

Phillips, M. (2000). Landscapes of defence, exclusivity and leisure: rural private communities in North Carolina. In J. Gold & G. Revill (Eds.) *Landscape of defence* (130-145). Harlow: Prentice Hall.

Phillips, M. (2002). The production, symbolisation and socialisation of gentrification: a case study of two Berkshire villages. *Transactions, Institute of British Geographers, 27*, 282-308.

Phillips, M. (2004). Other geographies of gentrification. *Progress in human geography, 28*, 5-30.

Phillips, M. (2005a). Differential productions of rural gentrification: illustrations from North and South Norfolk. *Geoforum, 36*, 477-494.

Phillips, M. (2005b). People at the centre? The contested geographies of 'gentrification'. In M. Phillips (Eds.) *Contested worlds: an introduction to human geography* (317-352). Aldershot: Ashgate.

Phillips, M. (2005c). *Rural gentrification and the production of nature: a case study from Middle England.* Paper presented at the 4th International Conference of Critical Geography, Mexico City.

Phillips, M. (2007). *The redevelopment of the Shires, Leicester: initial report.* Leicester: Department of Geography, University of Leicester.

Phillips, M. (2008). Rural gentrification and new human and non-human occupants of the English countryside/Gentrification rurale, nouveaux occupants et "non-humains" dans la campagne anglaise. In J.-P. Diry (Eds.) *Les étrangers dans les campagnes* (581-599). Clermont Ferrand: CERAMAC/Press Universitaires Blaise Pascal.

Phillips, M. (forthcoming-a). Attachment, affect and conflict in the spaces of nature in an English village. submitted to *Transactions of the Institute of British Geographers.*

Phillips, M. (forthcoming-b). Rural gentrifcation: a neglected subject? In M. Phillips (Eds.) *The gentrified countryside.* Aldershot: Ashgate.

Phillips, M. & Mighall, T. (2000). *Society and exploitation through nature.* Harlow: Prentice Hall.

Phillips, M., Page, S., & Saratsi, E. (forthcoming). Rural gentrification and nature. submitted to *Progress in human geography.*

Phillips, M., Page, S., Saratsi, E., Tansey, K. & Moore, K. (2008). Diversity, scale and green landscapes in the gentrification process: traversing ecological and social science perspectives. *Applied geography 28*, 54–76.

Pierce, A. (2008, 24 April). Eco-home in Cotwolds sells for a record £7.2m. *Daily telegraph.* Retrieved from http://www.telegraph.co.uk/earth/ earthnews/3340659/Eco-home-in-Cotswolds-fetches-record-7.2m.html

Podmore, J. (1998). (Re)Reading the 'Loft Living' habitus in Monteal's inner city. *International journal of urban and regional research, 22,* 283-302.

Raco, M. (2003). Assessing the discourses and practises of urban regeneration in a growing region. *Geoforum, 34*, 37-55.

Raco, M. & Henderson, S. (2006). Sustainable urban planning and the brownfield development process in the United Kingdom: lessons from the Thames Gateway. *Local environment, 11*, 499–513.

Rapoport, A. (1994). Spatial organisation and the built environment. In T. Ingold (Eds.) *Companion encyclopedia of anthroplogy: humanity, culture and social life*. (460-502) London: Routledge.

Saratsi, E., Phillips, M. & Page, S. (2006). *Studies of residential green space: Gentrify nature, working paper 1*. Leicester: Department of Geography, University of Leicester.

Savage, M. & Butler, T. (1995). Assets and the middle classes in contemporary Britain. In M. Savage & T. Butler (Eds.) *Social change and the middle classes* (345-357). London: UCL Press.

Sayer, A. (1989). Dualistic thinking and rhetoric in geography. *Area, 21*, 301-305.

Sayer, A. (1991). Behind the locality debate: deconstructing geography's dualisms. *Environment and planning A, 23*, 283-308.

Sayer, A. (2000). *Realism and social science*. London: Sage.

Shaw, K. (2008). Gentrification: what it is, why it is, and what can be done about it. *Geography Compass, 2*, 1697–1728.

Short, J. R. (1989). Yuffies and the new urban order. *Transactions of the Institute of British Geographers, 14* , 173-188.

Slater, T., Curran, W. & Lees, L. (2004). Gentrification research: new directions and critical scholarship. *Environment and Planning A, 36*, 1141-1150.

Smith, D. (2002). Rural gatekeepers and 'greentried' Pennine rurality: opening and closing the access gates? *Journal of social and cultural geography, 3*, 447-463.

Smith, D. (forthcoming). The waves of rural gentrification: the cultural new class and green pro-rural lifestyles? In M. Phillips (Eds.) *Gentification in the countryside*. Aldershot: Ashgate.

Smith, D. & Phillips, D. (2001). Socio-cultural representations of greentrified Pennine rurality. *Journal of rural studies, 17*, 457-469.

Smith, N. (1979a). Gentrification and capital: practice and ideology in Society Hill. *Antipode, 11*, 24-35.

Smith, N. (1979b). Toward a theory of gentrification: a back to the city movement by capital not people. *Journal of the Amercian planners association, 35*, 538-548.

Smith, N. (1982). Gentrification and uneven development. *Economic geography, 58*, 139-155.

Smith, N. (1984). *Uneven development: nature, capital and the production of space*. Oxford: Basil Blackwell.

Smith, N. (1987). Gentrification and the rent gap. *Annals, Association of American Geographers, 77*, 462-478.

Smith, N. (1996). *The new urban frontier: gentrification and the revanchist city*. London: Routledge.

Smith, N. (2002). New globalism, new urbanism: gentrification as global urban strategy. *Antipode, 34* , 428-450.

Smith, N. & DeFilippis, J. (1999). The reassertion of economics: 1990s gentrification in the Lower East Side. *International journal of urban and regional research, 23*, 638- 653.

Smith, N. & Williams, P. (1986). *Gentrification of the city*. London: Allen and Unwin.

Soja, E. (1996). *Third space: journeys to Los Angeles and other real-and-imagined places*. Oxford: Blackwell.

Spencer, D. (1995). Counterurbanisation: the local dimension. *Geoforum, 26*, 153-173.

Spencer, D. (1997). Counterurbanisation and rural depopulation revisited: landowners, planners and the rural development process. *Journal of rural studies, 13*, 75-92.

Sternlieb, G. (1990). Charting the 1990s: thing's aren't what they used to be. *Journal of American Planning Association, 56* , 492-96.

Warde, A. (1991). Gentrification as consumption: issues of class and gender. *Environment and planning D: Society and space, 9*, 223-232.

Whatmore, S. & Hinchliffe, S. (2003). Living cities: making space for urban nature. *Soundings, 22*, 137–150.

Wylie, J. (2007). *Landscape*. London: Routledge.

Zukin, S. (1982). *Loft living: culture and capital in urban change*. Baltimore: John Hopkins University Press.

Zukin, S. (1990). Socio-spatial prototypes of a new organization of consumption: the role of real cultural capital. *Sociology, 24*, 37-56.

Zukin, S. & Kosta, E. (2004). Bourdieu off-Broadway: managing distinction on a shopping block in the East village. *City and community, 3* , 101-114.

Zukin, S., Trujillo, V., Fraser, P., Jackson, D., Recuber, T. & Walker, A. (2009). New retail capital and neighbourhood change: boutiques and gentrification in New York city. *City and community, 8* , 47-64.

In: Built Environment: Design, Management and Applications   ISBN: 978-1-60876-915-5
Editor: Paul S. Geller, pp. 63-88   © 2010 Nova Science Publishers, Inc.

Chapter 3

THE MATHEMATICS OF STYLE IN THE ARCHITECTURE OF FRANK LLOYD WRIGHT: A COMPUTATIONAL, FRACTAL ANALYSIS OF FORMAL COMPLEXITY IN FIFTEEN DOMESTIC DESIGNS

Michael J. Ostwald and Josephine Vaughan
The University of Newcastle, Australia

ABSTRACT

Since the late 1980's a range of applications of fractal geometry, for the analysis of the built environment, have been proposed. One of these, the mathematician Benoit Mandelbrot's "box counting" approach, was developed by Carl Bovill who demonstrated a manual method for determining an approximate fractal dimension of architectural plans and elevations. Since that time, a computational variation of the method has been developed and tested providing a mathematical determination of the characteristic visual complexity of a building.

The present chapter records research into the changing formal strategies employed by Frank Lloyd Wright for the design of houses. Wright designed many houses during his almost 70 year professional career. Historians have defined three distinct stylistic periods in his housing oeuvre; the Prairie style (generally the first decade of the 20th century), the "Textile Block" era (the 1920s) and the Usonian era (the 1930s to the 1950s). This chapter undertakes the first mathematical analysis of visual complexity in the domestic architecture of Wright. The research analyzes five houses from each of these three eras using the computational fractal approach to measuring visual properties of architecture. In this way it is possible to provide a mathematical determination of the relationship between the three major stylistic periods in Wright's work.

INTRODUCTION

American architect Frank Lloyd Wright is one of the world's most famous designers. When he died in 1959, Wright left behind a legacy of over 500 completed buildings. Wright's architecture has been extensively documented by twentieth century historians and his theories, principles and approaches have been exhaustively researched by scholars. Despite this level of attention, Wright's architecture has only rarely been subjected to any type of consistent computational or mathematical analysis. Even though researchers have repeatedly identified Wright's predilection for geometric systems and have carefully quantified and categorized his works into clear stylistic periods, the vast majority of what is known about Wright's architecture remains qualitative in nature. In response to this lacuna, the present chapter records the first detailed computational analysis of formal complexity in Wright's domestic architecture. It undertakes a mathematical analysis of the fractal dimensions of three sets of five houses; each set taken from a distinct stylistic period in Wright's career. This research involves two parallel computational processes and an average of thirteen scales of analysis for each of the 56 elevations of Wright's houses. In total more than 1500 data points are compared in this research to provide a mathematical description of the typical level of visual complexity present in each of Wright's three major stylistic periods for domestic architecture. Moreover, once this information is known it is possible to construct a comparison between each of the periods to quantify the level of difference. This last factor is of interest because historians tend to discount the importance of visual or stylistic differences in Wright's work in favor of identifying consistencies in his planning strategies.

When architectural historians review the career of Frank Lloyd Wright they generally agree that throughout his life, his buildings display a consistent set of design principles even though his works have varied in appearance across a number of distinct stylistic periods. Historians tend to acknowledge such obvious visual and stylistic differences while focusing on similarities in the underlying tactics and theories which shape Wright's work. For example, Robert Sweeny concedes that Wright's ability to "renew himself repeatedly throughout his career" (Sweeney 1994) is a characteristic of his approach, but argues that it does not change his underlying values. David De Long supports this view when he proposes that over time, Wright "was able to retain allegiance to earlier principles while arriving at markedly different conclusions" (De Long 1994, xii). Kenneth Frampton similarly maintains that there is a constant thread throughout Wright's work which is related to a modular system of planning and construction which "varied according to local circumstance" (2005, 178). Frampton defines this continuous thread as a "plaited approach to architectonic space" which "prevailed throughout Wright's long career" (2005, 178). Robert McCarter, who claims that Wright's Usonian houses are derivations of his Prairie house ideals, argues that Wright's "architectural designs were a continuous reinvention or rediscovery of the same fundamental principles" (1999, 249). Finally, Donald Hoffmann supports this argument confirming that "the language of Wright's buildings continued to change, but the logic did not; once he grasped the principles, his work no longer evolved" (1995, 52). To support this assertion Hoffman quotes Wright stating that "I am pleased by the thread of structural consistency I see inspiring the complete texture of the work revealed in my designs and plans, […] from the beginning, 1893, to this time, 1957"(qtd. in 1995, 52).

These five examples are typical of late twentieth century scholarship about Wright. They tend to emphasize the similarities and dismiss the differences as largely stylistic or as surface responses to local conditions and materials. While this may be true, the degree to which these stylistic differences are consistent over many years, and within distinct periods in Wright's lifetime, has rarely been investigated using computational or mathematical methods. Nevertheless, a small number of studies do exist and they confirm that it is not only possible to study Wright's architecture using computational means, but it is highly beneficial because of the size of the body of work he produced.

Importantly, several of the computational or geometric studies of Wright's architecture that have previously been undertaken have been focused on his houses. For example, John Sergeant proposes that Wright's flexible use of planning grids is the common link between the Prairie style, the Textile Block period and the Usonian house period. Sergeant suggests that the Prairie style houses were designed on a "tartan grid" where a "vocabulary of forms was used to translate or express the grid at all points – the solid rather than pierced balconies, planters, bases of flower urns, clustered piers, even built-in seats were evocations of the underlying structure of a house" (2005). The Textile Block houses, typically set in the steep topography of the Los Angeles foothills, necessitated a development in Wright's use of the grid which signals, according to Sergeant, the moment when Wright first extended his planning grid into a vertical axis.

Richard MacCormac uses a different, but equally planar diagrammatic method to compare the forms of 11 of Wright's early buildings with the shapes of the Froebel's Gifts; a three-dimensional educational device which Wright used as a child. MacCormac concludes that the Froebel discipline can be found in all of Wright's early work as a "certain continuity of principle" (2005). Hank Koning and Julie Eizenberg agree that "MacCormac was right in the idea that there is a consistent underlying structure informing Wright's work" however, they find MacCormac's method and approach to analysis to be limited by a sense of order "as if order must in some way be tied up with measurable geometry"(Koning and Eizenberg 1981, 296). Koning and Eizenberg propose that a shape grammar approach is more suitable to the analysis and modeling of Wright's organic design approach. Utilizing the shape grammar approach to analyze six of Wright's prairie houses, Koning and Eizenberg conclude that "the composition of Wright's prairie-style houses is based on a few simple spatial relations between parameterized three-dimensional building blocks of the Froebelean type" (Koning and Eizenberg 1981, 296). They further propose a three-dimensional parametric shape grammar to demonstrate the design of new "Wright-style" buildings. Terry Weissman Knight builds on the work of Koning and Eizenberg with a comparative shape grammar analysis of Wright's Prairie and Usonian styles. Knight believes that the elemental "composition of Prairie houses provide[s] the basis for the basic composition of the later Usonian houses" (1994, 222) . Knight produces a shape grammar for six of Wright's Usonian houses, and compares these with "a simplified version of Koning and Eizenberg's earlier [Prairie] grammar" (1994, 224), concluding that "[a]lthough the outward appearance and spatial organization of Usonian houses seems substantially different from that of the Prairie houses, the underlying composition of Usonian designs is closely related to that of Prairie designs" (1994, 236).

In the largest plan-analysis study undertaken into Wright's works, Paul Laseau and James Tice reject the standard historical focus on "underlying principles" and aim for a balanced approach that employs analytical illustrations to generate typological studies of the plans of 131 of Wright's buildings. Laseau's and Tice's formal analysis demonstrates a categorization of Wright's work into three typological, rather than chronological, thematic groups; the atrium, the hearth and the tower (Laseau and Tice 1992).

Curiously, all of these geometric and typological studies have been focused on Wright's planning, albeit some in three dimensions, and there are no equivalent studies of Wright's elevations using similar methods. Nevertheless, a single suggestion for such an analysis is found in Carl Bovill's demonstration of a version of the "box-counting" fractal analysis technique to determine the characteristic complexity of the main façade of Wright's Robie House. Bovill's work, while manual in nature and including only a single façade, is the progenitor of the present chapter. In 2008 Ostwald, Vaughan and Tucker revisited the case of Wright's Robie house and developed a computational method to automate the fractal analysis of architecture and to propose a protocol, or method, for the consistent analysis of domestic designs. That method has since been applied to around 30 houses designed by Peter Eisenman (Ostwald and Vaughan 2009), Kazuo Sejima (Ostwald, Vaughan and Chalup 2009), Eileen Gray (Ostwald, Vaughan and Chalup 2008) and Le Corbusier (Vaughan and Ostwald 2009). Given its origins in the analysis of Wright's Robie house, it is appropriate that the present chapter returns to Wright to undertake a much more comprehensive analysis of fifteen of his houses using the computational method for analyzing characteristic visual complexity (as measured by fractal dimension).

In this chapter, the present introduction is followed by an overview of fractal geometry and its application in architectural and urban analysis. A particular method of fractal analysis, the box-counting approach, is then described along with its computational variation. The limitations of the method are then described along with minimization strategies to reduce their impact on the results. The section that follows provides a brief historical and theoretical overview of the fifteen houses that are analyzed in this chapter describing their position in Wright's body of work. Thereafter the results of the fractal analysis of these houses are presented and discussed. Finally the paper offers a comparison between the three stylistic periods using cluster analysis of fractal dimensions.

FRACTAL GEOMETRY AND ARCHITECTURE

In the late 1970s Benoit Mandelbrot famously proposed that Euclidean geometry, the traditional tool used in science to describe natural objects, is fundamentally unable to fulfill this purpose. Mandelbrot's revelation can be paraphrased as the idea that mountains are not conical in form, clouds are not spherical and rivers are not orthogonal. Every attempt to abstract a complex natural feature into a pure geometric form loses some of the essential qualities of that feature and renders the abstraction meaningless. In response to this realization, Mandelbrot proposes that a new model of geometry is required which allows scientists to approach "in rigorous and vigorous quantitative fashion," a range of shapes that have previously been defined as too "grainy, hydralike, in between, pimply, pocky, ramified, seaweedy, strange, tangled, tortuous, wiggly, wispy [and] wrinkled" to be seriously considered (Mandelbrot 1982, 5). Mandelbrot's breakthrough was to consider that objects in nature often possess characteristic complexity: this implies that natural objects frequently look alike when they are viewed at different scales. Thus a tree, when viewed from a distance, will often look similar to a branch of that same tree or even a leaf from that tree, when viewed at a closer scale. Thus a tree could be thought of as possessing a form of consistent complexity, or characteristic irregularity, and if this is the case, then that irregularity can be measured.

Mandelbrot suggests that the characteristic irregularity of a natural form may be measured by imagining that the increasingly complicated path of the form's outline is actually somewhere between a one-dimensional line and a two-dimensional surface. The more complicated the line, the closer it comes to being a two-dimensional surface. Therefore irregular lines and forms can be viewed as being fractions of integers, or what Mandelbrot describes as "fractal geometric forms". Thus, fractal geometry describes irregular or complex lines, planes and volumes that exist between whole number integer dimensions. This implies that instead of having a dimension, or D, of 1, 2, or 3, fractals might have a D of 1.51, 1.93 or 2.74.

According to Richard Voss, "Mandelbrot's fractal geometry provides both a description and a mathematical model for many of the seemingly complex forms and patterns in nature and the sciences" (1988, 21). From Mandelbrot's work it has been shown that star systems, geological formations, clouds, plant ecosystems, mountain ranges, coastlines and other natural systems typically possess similar geometric forms that are repeated over multiple scales of observation. Using Euclidean geometry, it is impossible to measure, with any accuracy, such non-linear forms. However, fractal geometry, by iterating measurements over progressive scales, can be used to determine the characteristic visual complexity of natural forms.

Mandelbrot not only pioneered the application of fractal geometry to nature, he also suggests that it may be useful in understanding the properties of human creations, such as art and architecture. Mandelbrot says of fractal geometry that "[i]t describes many of the irregular and fragmented patterns around us" (1982, 1). In this way, buildings can be considered as "irregular and fragmented patterns", as most built forms or urban layouts produce repeated shapes at different scales. According to Bovill, "[w]e experience architecture by observing the overall profile of a building from a distance; as we approach closer, the patterns of window and siding come into attention; as we approach even closer, the details of doors and window frames come into attention, down to what the door knob is like. [...] The fractal characteristic of an architectural composition presents itself in this progression of interesting detail" (1996, 117). This description is reminiscent of Mandelbrot's claim that "in the context of architecture: a Mies van der Rohe building is a scalebound throwback to Euclid, while a high period Beaux Arts building is rich in fractal aspects" (1982, 23-24). Just as Mandelbrot has used mathematical methods to calculate the fractal dimension of coastlines and compare them, so too could architecture or urban space be analyzed for characteristic visual complexity.

Architects adopted fractal geometry as a design tool in the 1980s but, despite some interesting outcomes, it rarely produced an enduring architectural response and has suffered from a lack of rigorous analysis and critique (Jencks 1995; Ostwald 2001; 2003). In contrast, the history of applications of fractal geometry to the analysis of architectural and urban forms is still evolving and is displaying more promising results. For example, at an urban scale Oku (1990), Cooper (2003; 2005) and Chalup, Henderson, Ostwald and Wiklendt (2008) have separately used fractal geometry to provide a quantitative measure of the visual qualities of an urban skyline. Yamagishi, Uchida and Kuga (1988) have sought to determine geometric complexity in street vistas and Kakei and Mizuno (1990) have applied fractal geometry to the analysis of historic street plans; a project that has been extended by Rodin and Rodina (2000). Cartwright (1991) offered an early overview of the importance of fractal geometry and complexity science in town planning and Batty and Longley (1994) and Hillier (1996) have each developed increasingly refined methods for using fractal geometry to understand the visual qualities of macro-scale urban environments. More recently, Batty (2005) has combined fractal geometry with computational automata to produce a further detailed explanation for the visual and formal complexity observed in cities. In a similar way, Cardillo, Scellato, Latora and Porta (2006) have calculated the fractal dimension of the street patterns of 20 different cities.

While not a computational study, an interesting case exists of an attempt to uncover fractalesque qualities of Wright's domestic architecture. Leonard Eaton defines a fractal as "a geometrical figure in which an identical motif repeats itself on an ever diminishing scale" (1998, 33). Eaton suggests that Wright's work became more visually complex, or more fractal, after 1923 or in the years after the design of the Textile block house La Miniatura; a building which Eaton feels has no strong fractal presence. By comparison, Eaton suggests that Wright's work of the 1950's and 1960's, his Usonian period, suggests a "striking anticipation of fractal geometry" (1998, 31). Eaton's method of analysis involves identifying self-similar equilateral triangles in the plan of Wright's Palmer house (Figure 1). Details ranging from the triangular slabs of the cast concrete floors down to the triangular shape of the fire iron rest are noted. Eaton counts "no less than eleven scales of equilateral triangles ascending and descending from the basic triangle" (1998, 32) leading him to conclude that the Palmer house has "a three-dimensional geometry of bewildering complexity" (1998, 35) (Figure 2).

Yannick Joye agrees with Eaton's assertion that the "Palmer House seems to be the culmination point" (2006, 312) of Wright's intuitive use of fractal geometry in architecture. In a similar vein, Daniele Capo (2004) uses Eaton's analysis of Wright's domestic architecture to build on her own fractal analysis of architecture, suggesting that Wright's architecture combined smaller elements to create larger works which show fractal similarities to Palladio's architectural orders. Ferrero, Cotti, Rossi and Tadeschi also accept Eaton's conclusion that fractals can be found in the Palmer house (Ferrero *et al.* 2009). However, before Eaton's work was published, Ostwald and Moore (1996) rhetorically demonstrated that even the most Euclidean of buildings, like Mies Van Der Rohe's Seagram Building, can have more than 12 scales of conscious self-similarity and that this does not make them, or the Palmer house, fractal (Ostwald 2003). James Harris also disagrees with Eaton's simplified analysis of the Palmer House, saying that Eaton's paper "points out the misconception that a repetition of a form, the triangle in this case, constitutes a fractal quality. It is not the repetition of the form or motif but the manner in which it is repeated or its structure and nesting characteristics which are important." (2007, 98) Harris proceeds to apply an iteration method (IFS) to a triangle, in an attempt to produce a similar plan to the Palmer House, finally creating a design with some similarities in plan to the final design by Wright. Harris concludes that Wright's architecture may not be as strongly generated by fractal geometry as Eaton suggests, however he finds the relationship to be "analogous" (2007, 98). The word analogous is appropriate because Eaton proposes the existence of a symbolic or metaphoric relationship between fractal geometry and repetitious forms in the plan of the Palmer House.

In contrast to Eaton's analogical method, the approach for calculating the actual fractal dimension of a building was first demonstrated by Bovill who, adopting Mandelbrot's box-counting technique, showed how it was possible to determine the approximate fractal dimension of architectural plans and elevations. Bovill's method has since been employed to calculate the fractal dimension of a range of buildings, including Mesoamerican pyramids (Burkle-Elizondo 2001; Burkle-Elizondo and Valdéz-Cepeda, 2006) and the fractal dimensions of Doric, Corinthian and Composite orders (Capo 2004). Gozubuyuk, Cagdas and Ediz (2006) use the box-counting method to analyze the urban layout and typical buildings of two historical districts of the Turkish cities; Istanbul and Mardin. Although many scholars refer to Bovill's work it remained largely untested until a computational variation was first considered by Lorenz (2003) and later developed and applied to the analysis of around 40 houses (Ostwald, Vaughan and Tucker 2008; Ostwald and Vaughan 2008; Ostwald and Vaughan 2009; Ostwald, Vaughan and Chalup 2009).

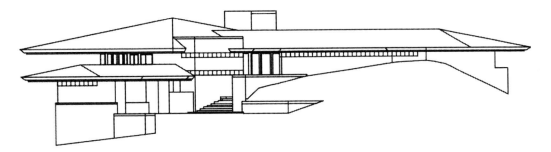

Figure 1. Triangle-Plan Usonian set; Palmer House, Elevation 4 (North).

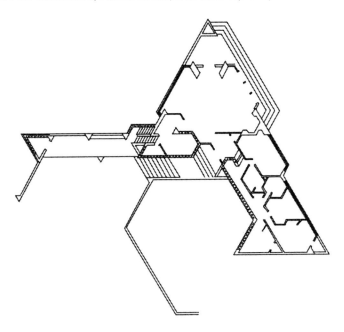

Figure 2. Triangle-Plan Usonian set; Palmer House, Plan.

THE FRACTAL ANALYSIS METHOD

The box-counting method – for calculating approximate fractal dimension – takes as its starting point a line drawing, say the façade of a building. A grid is then placed over the drawing and each square in the grid is analyzed to determine whether any lines from the façade are present in it. Those grid boxes that have some detail in them are then marked. This data is then processed using the following numerical values;

$_{(s)} =$ the size of the grid
$N_{(s)} =$ the number of boxes containing some detail
$1/_s =$ is the number of boxes at the base of the grid

Next, a grid of smaller scale is placed over the same façade and the same determination is made of whether detail is present in the boxes of the grid (Figure 3). A comparison is then made of the number of boxes with detail in the first grid ($N_{(s1)}$) and the number of boxes with detail in the second grid ($N_{(s2)}$). Such a comparison is made by plotting a log-log diagram ($\log[N_{(s)}]$ versus $\log[1/_s]$) for each grid size. When the process is repeated a sufficient number of times, it leads to the production of an estimate of the fractal dimension of the façade; actually it is an estimate of the box-counting dimension (D_b) which is sufficiently similar that most researchers don't differentiate between the two.

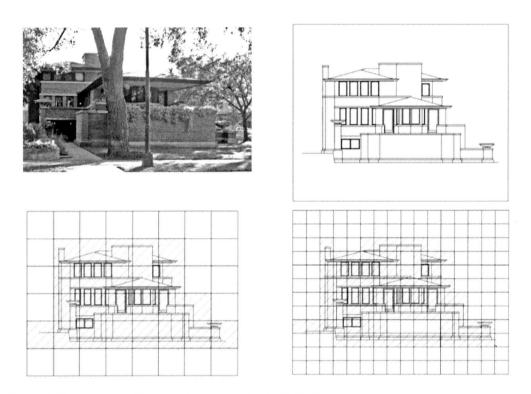

Figure 3. Example of the Box-counting process for the Robie House.

The slope of the line (D_b) is given by the following formula:

$$D_b = \frac{[\log(N_{(s2)}) - \log(N_{(s1)})]}{[\log_{(1/s2)} - \log_{(1/s1)}]}$$

where ($1/_s$) = the number of boxes across the bottom of the grid. (Bovill 1996: 42)

While the box-counting process was traditionally a manual exercise, TruSoft's *Benoit* (vers. 1.3.1) program and *Archimage* (vers. 2.1), a program co-developed by the authors, automate this operation.

There are many variations of the box-counting approach that respond to known deficiencies in the method. These variations in part explain the lack of consistency in the way in which previous studies have both recorded and reported their data. The following five points describe the primary issues and how the present research responds to each.

1. *White space.* The volume and distribution of white or empty space around the source image can alter the result. Prior to Lorenz (2003) identifying this problem with architectural analysis, none of the research had taken this into account. Since that time, to solve the problem, Foroutan-Pour, Dutilleul and Smith (1999) offer an algorithm to optimize the way in which an image is positioned against its background and suggestions on how to derive an ideal analytical grid. In essence, careful sizing of the initial image in combination with solutions to the starting grid problem (described hereafter) limits the impact of this problem.

2. *Image proportion.* If the original image is not pre-sized to produce a clear starting grid, then an additional step must be added to ensure that a divisible grid is determined. The software *Benoit* solves this problem by cropping the image size to achieve a whole-number starting grid. This variation of the process seems especially useful for analyzing complex natural forms. In contrast, *Archimage* software enlarges the source by adding small amounts of empty space to the boundaries of the image. While neither of these variations changes the elevation in the source image, they produce subtle differences in the resultant *D*; meaning that *Benoit* is likely to produce a slightly lower result for architecture than *Archimage*. While the present authors consider the boundary growth strategy superior for abstract, or architectural images, they acknowledge that both variations have merit for different types of images.

3. *Line width.* The wider the lines in the source image, the more chance they have of being counted twice when grid sizes become very small, leading to artificially increased *D* values. This is a common problem in most architectural applications of the method prior to 2008 and at least partially explains many of the inaccurate results that were produced. To counter this situation, *Archimage* software pre-processes images using a line-detection algorithm that reduces all lines to one-pixel-width thereby eliminating any prospect of double-counting. *Benoit* overcomes this problem by allowing the analytical grid to be rotated or resized to provide a point of comparison against which the software can check for double counting and minimize the impact of line weight. However, in *Benoit*'s checking process the software optimizes the result automatically to the lowest *D* result in the range seemingly reducing the impact of double-counting but also artificially lowering the overall result across multiple scales.

4. *Scaling coefficient.* The factor by which successively smaller grids is produced is called the scaling coefficient. Bovill, in his original examples, halved the grid dimension for each comparison (a ratio of 2:1) whereas *Benoit* and *Archimage* use a flexible scaling coefficient allowing a more accurate ratio, typically 1.3:1, to more gradually reduce the grid size and generate a more accurate result.

5. *Statistical divergence.* The average slope of the log-log graph may be the approximate *D* value, but the points generating the line are not always consistent with it. The *D* value is only a reasonable approximation when most of the points in the chart correspond with the resultant line. The question then becomes, how are divergent points handled?

    Bovill, in his manual variation of the method, appears to have intuitively chosen the point when the analysis was yielding less useful results as the limit. This procedure may be viable for a manual variation but it is difficult to replicate and repeat for larger computational studies. Indeed, until 2008 and the first publication of the computational variation, this issue has rarely been mentioned in architectural analysis. The reason that this has become a problem is that the computational variation derives highly accurate data for each grid comparison and this tends to produce slightly more divergent results. While the ultimate level of divergence may be trivial, it is nevertheless more apparent in the computational variation than in any previous version.

    The ideal solution to the issue of statistical divergence is to identify a reasonable level of variation and then exclude values that are above or below this point. However, in practice this is difficult to achieve because the results are so sensitively dependent on the four previous variables (white space, image proportion, line width and scaling coefficient). The first response of the present research to this issue is to limit its significance. This can be achieved by ensuring that the impact of each of the four other factors is consistently minimized. Once this has been achieved, then the consistency of the method is more important than trivial errors caused by statistical divergence. Thus, analytical scales (or grid sizes) are chosen that have less likelihood of being divergent. This is not a solution to the issue but it limits its impact without undermining the consistency and repeatability of the method.

While the box-counting method can be applied to any two dimensional architectural image, the computational technique developed by the present authors has been refined for the testing of sets of designs by an architect. The reason for focusing on sets is that comparable mathematical methods are optimized for considering variations across bodies of data. Although this approach could be used to construct a simple comparison between two elevations, it would at best confirm a person's intuitive reading of the difference. Thus, for the method to provide more meaningful and useful results a larger body of initial data is required. For this reason, the authors have settled on comparisons of sets of five houses by the same architects. A set of five designs is a useful starting point because this implies an analysis of 480 box-counting operations, divided into 40 sets of calculations for 20 elevations; this is a reasonable body of data from which to draw conclusions.

Houses have been chosen as the ideal focus for this method because they typically possess a similar scale, program and materiality. While it is possible to analyze smaller sets of works, or other building types, the focus on a minimum of five houses ensures the production of statistically valid results. In order to maximize the potential consistency of the results, past research using this method has also tended to select completed houses, in preference to unbuilt works, and designs from a similar time frame and geographic distribution. This is, once again, to reduce the possible number of external variables that can have an impact on the data.

The houses of Wright present a unique opportunity for this method to be applied to a single, master architect who produced many houses during his lifetime and in several distinct stylistic periods. The process commences with redrawing the elevations of fifteen of Wright's houses using consistent graphic conventions. For this research, the elevations of Wright's work were all digitally traced from his original working drawings reproduced by Storrer (1993), Pfeiffer and Futagawa (1987). Where the particular house was altered by Wright during construction, or only an incomplete set of working drawings is available, the measured drawings of the *Historic American Buildings Survey* were used to supplement the originals. In total 56 elevations were reproduced; typically four for each house although two of the triangular-plan Usonian houses only have three elevations and scale drawings are not available for two views of La Miniatura. Following Bovill's convention, the primary lines considered in the analysis correspond to changes in form, not changes in surface or texture. Thus, major window reveals, thickened concrete edge beams and steel railings are all considered while brick coursing and control joints are not. Once all of the elevations are prepared for analysis the method is as follows.

1. The elevations of each individual house are grouped together and considered as a sub-set of the data.
2. Each elevation of the house is analyzed using *Archimage* and *Benoit* programs producing $D_{(Elev)}$ results for each program for each elevation $D_{(Elev\ 1-4)}$. The settings for *Archimage* and *Benoit,* including scaling coefficient and scaling limit are preset to be consistent between the programs. The starting image size ($IS_{(Pixels)}$), largest grid size ($LB_{(Pixels)}$), and number of reductions of the analytical grid ($G_{(\#)}$), are recorded so that the results can be tested or verified. *Archimage* results are typically slightly higher than those produced by *Benoit* although the variation is consistent.
3. The $D_{(Elev\ 1-4)}$ results produced by both *Archimage* and *Benoit* are averaged together to produce a composite result, $D_{(Comp)}$, for the house. The composite result is a single D value that best approximates the characteristic visual complexity of the house.
4. This process (steps 2 and 3) is repeated for each house in the set producing five $D_{(Comp)}$ values. These values are averaged together to create an aggregate result $D_{(Agg)}$ which is a reflection of the typical, characteristic visual complexity of the set of the architect's works.
5. The gap ($D_{(Range)}$) between the highest $D_{(Comp)}$ and the lowest $D_{(Comp)}$ result in each set is calculated and reported as a percentage ($\%_{Gap}$). This is a reflection of the degree of diversity within the $D_{(Agg)}$ result. If there is a relatively high degree of diversity (>5%) a determination of the best cluster of three $D_{(Comp)}$, results is made and a cluster result for $D_{(Range)}$ and $\%_{Gap}$ is produced. The "best cluster" is the set of results which are most likely to produce a consistent pattern if there is too much variation on the $D_{(Agg)}$ result for a complete set of five houses.

This five-part process takes the 56 starting elevations and generates around 1500 data sets from which a series of 13 composite results for each house, and three aggregate results for stylistic periods, are produced. Importantly, this method does not produce a D result for the three-dimensional form of the house rather, it generates a series of average D results for the two-dimensional visual qualities of a structure. The list of abbreviations and definitions is in Table 1. All of the data recorded using this method is calculated to three decimal places although, when these are averaged or combined together (in the $D_{(Comp)}$ and $D_{(Agg)}$ data), the resultant number of figures after the decimal point can change.

FIFTEEN HOUSES, THREE STYLISTIC PERIODS

During his lengthy career Wright pioneered many architectural design strategies for housing. The first set of Wright's house designs considered in the present chapter were completed between 1901 and 1910 and are from his Prairie Style period; regarded by some critics as "Wright's greatest invention in this first phase of a long career" (Kostof 1985, 684). The second stylistic set is from the Textile Block period (1923 -1929) which were arguably amongst "the most livable and attractive" (Wright 1960, 209) of the designs. Finally, five triangle-plan Usonian houses, which were completed between 1950 and 1955, are analyzed. These fifteen houses represent a "continuation of [Wright's] lifelong quest to destroy boxlike rooms" (Lind 1994, 9), and transcend the simple spatial qualities of "the familiar ranch house" (Hoffmann 1995, 80). These three distinct periods, spanning 40 years, mark significant early, mid and late eras in Wright's domestic architecture.

Table 1. Abbreviations and definitions

Abbreviation	Meaning
D	Approximate Fractal Dimension.
$D_{(Elev\ 1-4)}$	D result for an elevation of a house as determined by the stated software; either *Archimage* or *Benoit*. The subscript number refers to the elevation.
$D_{(Comp)}$	Composite D is the average of all $D_{(Elev)}$ results for a house.
$D_{(Agg)}$	Aggregated result of five composite values used for producing an overall D for a set of architects' works.
$IS_{(Pix)}$	The size of the starting image measured in pixels.
$LB_{(Pix)}$	The size of the largest box or grid that the analysis commences with, measured in pixels.
$G_{(\#)}$	The number of scaled grids that the software overlays on the image to produce its comparative analysis.
$D_{(Range)}$	The difference between the highest $D_{(Comp)}$ result and the lowest $D_{(Comp)}$ result that have been combined into a $D_{(Agg)}$ result.
$\%_{Gap}$	The $D_{(Range)}$ result expressed as a percentage of the possible maximum range being between $D = 1.0$ and $D = 2.0$. [$\%_{Gap} = 100 \times D_{(range)}$]

The five Prairie Style houses by Wright were constructed between 1901 and 1910. Four of the five are in the state of Illinois and the fifth is in Kentucky. All of the houses are characterized by strong horizontal lines, overhanging eaves, low-pitched roofs, open floor plan and a central hearth. Importantly, the five houses span the period between the first publication of Wright's Prairie Style, in the *Ladies Home Journal* in 1901, and what is widely regarded as the ultimate example of this approach, the Robie house (Figure 4). The first Prairie Style design is the F. B. Henderson House in Elmhurst, Illinois which was completed in 1901. The house is a wooden, two-storey structure with plaster rendered elevations. A range of additions were made to the house in the years following its completion until, in

1975, the house was restored to its original form. The Tomek House, completed in 1907, in Riverside, Illinois, is also a two-storey house although it possesses a basement and is sited on a large city lot (Figure 5). This house is finished with pale, rendered brickwork, dark timber trim and a red tile roof. Storrer notes that, in response to the Tomek family's needs, Wright later allowed posts to be placed beneath the cantilevered roof to heighten the sense of support and enclosure (Storrer 1993, 128). As the posts were not required for structural reasons, and Wright found them personally unnecessary, they have been omitted from the analysis. The Robert W. Evans House, completed in Chicago, Illinois in 1908, features a formal diagram wherein the "basic square" found in earlier Prairie Style houses is "extended into a cruciform plan" (Thomson 1999, 100). The house is set on a sloping site and possesses a plan similar to one Wright proposed in 1907 for a "fireproof house for $5000". The Evans house was later altered to enclose the porch area and the stucco finish on the façade was also cement rendered. The Zeigler House in Frankfort, Kentucky was constructed in 1910 and has a similar plan to the Evans house. Designed as a home for a Presbyterian minister, this two-storey house is sited on a small, city lot and it was constructed while Wright was in Europe. After a decade of development and refinement the quintessential example of the Prairie Style, the Robie House was constructed in Chicago, Illinois in 1910. Designed as a family home, the three storey structure fills most of its corner site. Unlike many of Wright's other houses of the era, the Robie house features a façade of exposed, raked bricks.

Figure 4. Prairie House set; Robie House, Elevation 4 (West).

Figure 5. Prairie House set; Tomek House, Elevation 3 (East).

Wright expanded his practice in California in the early 1920's and during the following decade he designed many buildings although only five houses were built. These five houses, which share some of the character of Wright's famous Hollyhock House, have since become known as the Textile Block homes. Appearing as imposing, ageless structures, these houses were typically constructed from a double skin of pre-cast patterned and plain exposed concrete blocks held together by Wright's patented system of steel rods and concrete grout. The plain square blocks of the houses are generally punctuated by ornamented blocks and for each house a different pattern is employed.

The first of the houses, the Millard Residence or "La Miniatura" was completed in Pasadena in 1923. This house is the only one of the Textile Block works not to feature a secondary structure of steel rods. Reflecting on La Miniatura, Wright wrote that in this project he "would take that despised outcast of the building industry – the concrete block – out from underfoot or from the gutter – find a hitherto unsuspected soul in it – make it live as a thing of beauty-textured like the trees" (1960, 216-217). The second of the Textile Block Houses, the Storer Residence, was completed in Hollywood in 1923 (Figure 6). It is a three-storey residence with views across Los Angeles. The Samuel Freeman Residence in Los Angeles was also completed in 1923. It is regarded as the third Textile Block house and the first to use mitered glass in the corner windows of the house; all of the previous works in this style have solid corners. It is a two-storey, compact, flat roofed house made from both patterned and plain textile block and with eucalyptus timber detailing. The fourth Textile Block house, the Ennis Residence, is probably the most famous of Wright's works of this era (Figure 7). Also completed in 1923 and sited overlooking Los Angeles, it is regarded as the "most monumental" (Storrer 1974, 222) of the houses of this era. It has been described as "looking more like a Mayan temple than any other Wright building except [the] Hollyhock House" (Storrer 1974, 222). The Ennis house is conspicuously sited and is made of neutral colored blocks with teak detailing. Some of the windows feature art glass designed by Wright in an abstraction of wisteria plants. The final design in this sequence, the Lloyd-Jones House is in Tulsa; completed in 1929, it is the only non-Californian Textile Block house. Designed for Wright's cousin, it is a large house with extensive entertaining areas and a four-car garage. The Lloyd-Jones House is notably less ornamental than the others in the sequence with Wright rejecting richly decorated blocks "in favor of an alternating pattern of piers and slots" (Frampton 2005, 170).

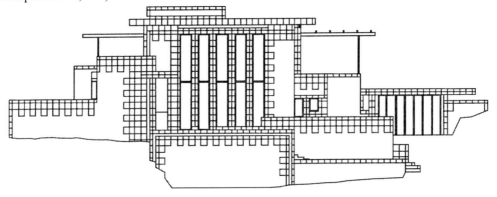

Figure 6. Textile Block set; Storer House, Elevation 4 (South).

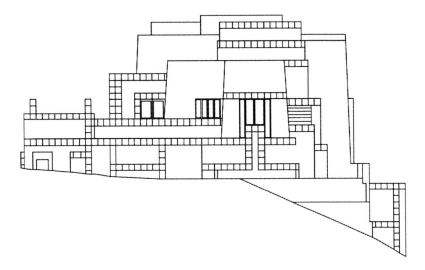

Figure 7. Textile Block set; Ennis House, Elevation 4 (West).

More than twenty years was to pass before Wright developed his third major sequence of domestic works; the Usonian houses. Wright explains that the Usonian house is intended to be "integral to the life of the inhabitants", be truthful in its material expression ("glass is used as glass, stone as stone, wood as wood") and embrace the elements of nature (1954, 353). Hoffman describes the Usonian house as "a simplified and somewhat diluted prairie house characterized by the absence of leaded glass and the presence of [...] very thin wall screens with a striated effect from wide boards spaced by recessed battens" (1995, 80). While there were multiple variations on the Usonian house, the five works featured in the present chapter are all based on an underlying equilateral triangular grid and were constructed between 1950 and 1956.

The first of the triangle-plan Usonian Houses, the William Palmer House is located in Ann Arbour, Michigan and was completed in 1950. The house is a two-storey brick structure, set into a sloping site, with wide, timber-lined eaves, giving the viewer an impression of a low, single level house. The brick walls include bands of patterned, perforated blocks, in the same color as the brickwork. The second house, the Reisley Residence was the last of Wright's Usonian houses built in Pleasantville, New York; it was completed in 1951 (Figure 8). This single level home with a small basement is constructed from local stone with timber paneling and is set on a hillside site. To accommodate the Reisley's expanding family, five years later Wright designed an addition for the children's bedrooms. In contrast in the same year the Chahroudi Cottage was built on an island in Lake Mahopac, New York, and constructed using Wright's desert masonry rubblestone technique with some timber cladding and detailing. The cottage was originally designed by Wright as the guest quarters for the Chahroudi family home however, only the cottage was built and subsequently used as the primary residence (Figure 9). The Dobkins House was built in 1953 for Dr John and Syd Dobkins in Canton, Ohio. This small house is constructed from brick with deeply raked mortar joints. However, unlike the Robie house, the mortar color contrasts with the bricks in the vertical as well as the horizontal joints. Finally, the Fawcett Residence, completed in 1955, had an unusual brief for Wright to design a home for a farming family. The house is set on the large flat expanse of the Fawcett's walnut farm in Los Banos, California. The single storey house is constructed primarily of grey concrete block with a red gravel roof.

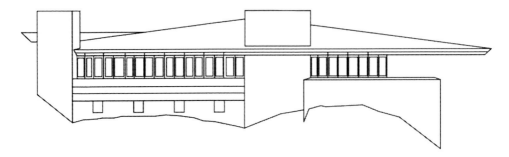

Figure 8: Triangle-Plan Usonian set; Reisley House, Elevation 4 (West).

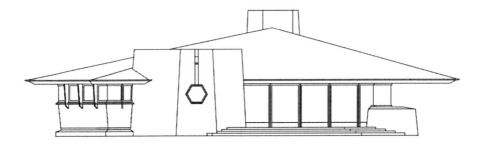

Figure 9: Triangle-Plan Usonian set; Chahroudi House, Elevation 1 (North).

RESULTS

The 15 houses described in the previous section were subjected to the fractal analysis method outlined earlier in the chapter. The numerical results ranged from the low of $D = 1.259$, which indicates an elevation with a low visual complexity, to $D = 1.673$, suggesting a highly complex façade. The remaining results fall within this range and their relative complexity can be judged against these low and high results.

The results of the computational analysis of the Prairie Style Houses are recorded in Table 2. These results indicate that the least visually complex of Wright's Prairie Style works is the Ziegler House: $D_{(Comp)} = 1.503$. Appropriately, the elevations of the Ziegler house are also the simplest of Wright's Prairie Style works with Elevation 1 recording the lowest result for *Archimage* ($D_{(Elev)}1 = 1.500$) and the third lowest result for *Benoit* ($D_{(Elev\ 1)} = 1.471$) (Figure 10). The most visually complex of Wright's Prairie Houses is the Evans house: $D_{(Comp)} = 1.577$. This house features several complex facades including Elevation 1 which has an *Archimage* result of $D_{(Elev\ 1)} = 1.614$ and a *Benoit* result of $D_{(Elev\ 1)} = 1.587$; respectively the second highest and the highest results produced for this set or houses (Figure 11). The results for Wright's most famous Prairie style house, the Robie house, fall neatly between these two extremes with a $D_{(Comp)}$ value of 1.5375. The Robie house is interesting in this regard because, unlike the rest of the houses in the set, its elevations display a relatively high degree of diversity. Most of the other Prairie Houses have some consistency in $D_{(Elev)}$ values across the four facades implying that the houses have a similar visual character when viewed from any direction. The Robie House is on a tight suburban block and two of its façades have little or no outlook. This explains the fact that the Robie house has one of the most complex elevations Wright produced at this time, along with some of the simplest. The complex elevation is number 2: *Archimage* result of $D_{(Elev\ 2)} = 1.620$ and *Benoit* result of $D_{(Elev\ 2)} = 1.526$. When the different results for the Robie House facades are averaged together they reduce the overall complexity of the house to below that of the Evans House.

Table 2. *D* values for the five Prairie Style Houses

House	Elevations analyzed with *Archimage*				Elevations analyzed with *Benoit*				
	$D_{(Elev\ 1)}$	$D_{(Elev\ 2)}$	$D_{(Elev\ 3)}$	$D_{(Elev\ 4)}$	$D_{(Elev\ 1)}$	$D_{(Elev\ 2)}$	$D_{(Elev\ 3)}$	$D_{(Elev\ 4)}$	$D_{(Comp)}$
Henderson	1.589	1.574	1.565	1.534	1.541	1.527	1.534	1.509	1.546625
Tomek	1.530	1.560	1.578	1.600	1.488	1.476	1.502	1.507	1.530
Evans	1.614	1.608	1.586	1.575	1.585	1.587	1.526	1.536	1.577125
Ziegler	1.500	1.530	1.546	1.561	1.471	1.473	1.488	1.456	1.503
Robie	1.558	1.620	1.595	1.565	1.452	1.526	1.530	1.454	1.5375

Typical parameters for the data calculation: $IS_{(Pix)} = 1182 \times 795$; $LB_{(Pix)} = 394$; $G_{(\#)} = 14$.

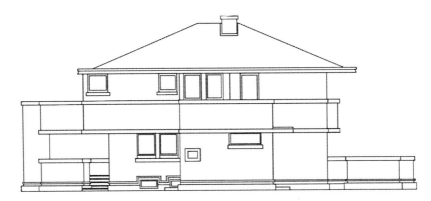

Figure 10. Prairie House set; Ziegler House, Elevation 1 (North), *Archimage* $D_{(Elev\ 1)} = 1.500$ and *Benoit* $D_{(Elev\ 1)} = 1.471$.

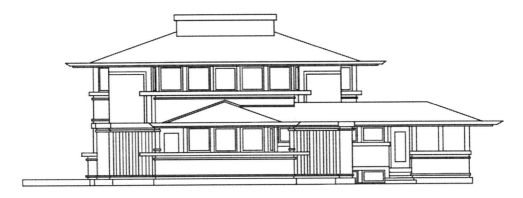

Figure 11. Prairie House set; Evans House, Elevation 1(North), *Archimage* $D_{(Elev\ 1)} = 1.614$ and *Benoit* $D_{(Elev\ 1)} = 1.587$.

The calculated results for the visual complexity of the Textile Block Houses are contained in Table 3. Three houses in the set, the Storer House, the Freeman House and the Lloyd Jones House all have very similar levels of visual complexity. The least visually complex in this set, although only by a narrow margin, is the Freeman house, with a $D_{(Comp)}$ result of 1.5065; this is slightly higher than the lowest result for the Prairie Style set, the Zeigler House. While the Freeman house has the lowest $D_{(Comp)}$ result of the Textile Block Houses, it, along with the Storer House and the Lloyd Jones House all have a similar spread of results across the complete set of elevations. Elevation 1 of the Freeman House is amongst the least complex in the set of Textile Block Houses with an *Archimage* result of $D_{(Elev\ 1)}$ = 1.497 and a *Benoit* result of $D_{(Elev\ 1)}$ = 1.391 (Figure 12).

While only two elevations of La Miniatura are available for analysis, this house has the highest overall result for the set: $D_{(Comp)}$ = 1.61425. With only two results it is also anticipated that La Miniatura's elevations must be amongst the most complex of any of the Textile Block houses. The *Archimage* results show the two elevations are the most visually complex of this set with $D_{(Elev\ 1)}$ = 1.673 and $D_{(Elev\ 2)}$ = 1.640. The *Benoit* results concur with La Miniatura having the most complex facades using this method of calculation as well: $D_{(Elev\ 1)}$ = 1.580 and $D_{(Elev\ 2)}$ = 1.564 (Figure 13). The results for La Miniatura are also the highest for any of the 15 houses considered in the present chapter. The Ennis House is only slightly less visually complex than La Miniatura and all four Ennis House elevations have a relatively consistent degree of complexity: for *Archimage* results 1.558 < D < 1.629 and for *Benoit*, 1.480 < D < 1.550.

Table 3: *D* values for the five Textile Block Houses

House	Elevations analyzed with *Archimage*				Elevations analyzed with *Benoit*				
	$D_{(Elev\ 1)}$	$D_{(Elev\ 2)}$	$D_{(Elev\ 3)}$	$D_{(Elev\ 4)}$	$D_{(Elev\ 1)}$	$D_{(Elev\ 2)}$	$D_{(Elev\ 3)}$	$D_{(Elev\ 4)}$	$D_{(Comp)}$
La Miniatura	1.673	1.640	–	–	1.580	1.564	–	–	1.61425
Storer	1.569	1.579	1.565	1.485	1.466	1.530	1.480	1.393	1.508375
Ennis	1.615	1.629	1.565	1.558	1.513	1.522	1.550	1.480	1.554
Freeman	1.497	1.621	1.544	1.534	1.391	1.520	1.491	1.454	1.5065
Lloyd Jones	1.513	1.560	1.585	1.583	1.4675	1.511	1.538	1.526	1.510625

Typical parameters for the data calculation: $IS_{(Pix)}$ = 1182x795; $LB_{(Pix)}$ = 394; $G_{(\#)}$ = 14.

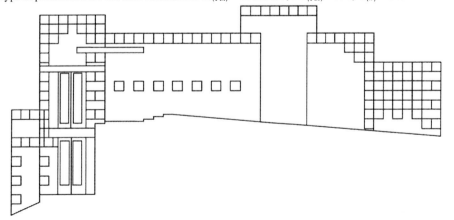

Figure 12. Textile Block House set; Freeman House, Elevation 1 (North), *Archimage* $D_{(Elev\ 1)}$ = 1.497 and *Benoit* $D_{(Elev\ 1)}$ = 1.391.

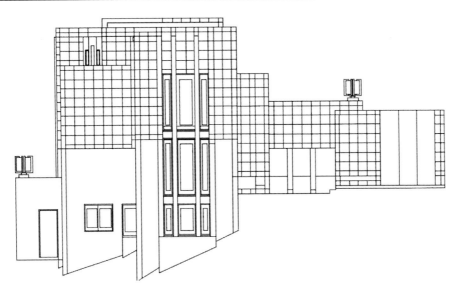

Figure 13: Textile Block House set; La Miniatura, Elevation 1 (East), *Archimage* $D_{(Elev\ 1)}$ = 1.614 and *Benoit* $D_{(Elev\ 1)}$ = 1.587.

Table 4. *D* values for the five Triangle-Plan Usonian Houses

House	Elevations analyzed with *Archimage*				Elevations analyzed with *Benoit*				
	$D_{(Elev\ 1)}$	$D_{(Elev\ 2)}$	$D_{(Elev\ 3)}$	$D_{(Elev\ 4)}$	$D_{(Elev\ 1)}$	$D_{(Elev\ 2)}$	$D_{(Elev\ 3)}$	$D_{(Elev\ 4)}$	$D_{(Comp)}$
Palmer	1.543	1.520	1.542	1.559	1.418	1.419	1.428	1.452	1.486
Reisley	1.446	1.446	1.441	1.477	1.435	1.432	1.429	1.440	1.434
Chahroudi	1.536	1.560	1.486	-	1.396	1.422	1.435	-	1.403
Dobkins	1.426	1.582	1.449	1.500	1.322	1.460	1.351	1.446	1.442
Fawcett	1.379	1.494	1.290	-	1.314	1.362	1.259	-	1.350

Typical parameters for the data calculation: $IS_{(Pix)}$ = 1182x795; $LB_{(Pix)}$ = 394; $G_{(\#)}$ = 14.

The results for Wright's Triangle-Plan Usonian houses are contained in Table 4. The composite results for the five Triangle-Plan houses produce values ranging from a low of $D_{(Comp)}$ = 1.350 for the Fawcett House, to a high of $D_{(Comp)}$ = 1.486 for the Palmer House. The low results for visual complexity in the Fawcett House are influenced by Elevations 1 and 3; the former of which is relatively low in comparison to the other five houses in the set and the latter which is the lowest for any of these five houses. Elevation 3 of the Fawcett House has a *Archimage* result of $D_{(Elev\ 3)}$ = 1.290 and a *Benoit* result of $D_{(Elev\ 3)}$ = 1.259. These are also the lowest results for any of the 15 houses considered in the present chapter (Figure 14).

The Palmer house, the most complex of Wright's Usonian Triangle-plan houses in this set, was the one singled out by Eaton as "an excellent illustration of the concept" (1998, 33) of a fractal architecture. However, the Palmer House result is not a high in the context of Wright's oeuvre. The visual complexity of the Palmer House elevations range between, for *Archimage*, 1.520 < D <1.559 and, for *Benoit*, 1.418 < D <1.452. Despite this apparent consistency in the Palmer House visual character, the most complex elevation in the complete Triangle-plan set is Elevation 2 of the Dobkins house – *Archimage* $D_{(Elev\ 2)}$ = 1.582 and *Benoit* $D_{(Elev\ 2)}$ = 1.460 (Figure 15).

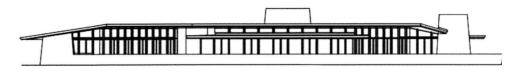

Figure 14: Triangle-Plan Usonian set; Fawcett House, Elevation 3 (South), *Archimage* $D_{(Elev\ 3)}$ = 1.290 and *Benoit* $D_{(Elev\ 3)}$ = 1.259.

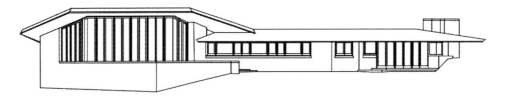

Figure 15: Triangle-Plan Usonian set; Dobkins House, Elevation 2 (South), *Archimage* $D_{(Elev\ 2)}$ = 1.582 and *Benoit* $D_{(Elev\ 2)}$ = 1.460.

Table 5. $D_{(Comp)}$ and $D_{(Agg)}$ values for all Frank Lloyd Wright houses analyzed

15 Houses	Prairie Style Houses	$D_{(comp)}$	Textile Block Houses	$D_{(comp)}$	Triangle-Plan Houses	$D_{(comp)}$
	Henderson	1.546625	La Miniatura	1.61425	Palmer	1.486
	Tomek	1.530	Storer	1.508375	Reisley	1.443
	Evans	1.577125	Ennis	1.554	Chahroudi	1.403
	Ziegler	1.503	Freeman	1.5065	Dobkins	1.442
	Robie	1.5375	Lloyd Jones	1.510625	Fawcett	1.350
$D_{(Range)}$	0.074125		0.10775		0.136	
%$_{Gap}$	7.4%		10.7%		13.6%	
$D_{(Agg)}$	1.53885		1.538745		1.425	

Table 5 records the aggregate results for the three sets of five houses along with information about the stylistic and formal consistency in each set of houses. The $D_{(Agg)}$ result for the complete set of Prairie Houses is 1.53885 and for the Textile Block Houses the $D_{(Agg)}$ result is 1.538745. This appears to be an almost identical result and not what might be anticipated from a cursory review of photographs of the buildings. Taken in isolation, the $D_{(Agg)}$ results imply that the 20 Prairie style elevations are mathematically similar to the 18 Textile Block elevations, from the point of view of average characteristic complexity. This may be true when these results are averaged together, but it is much more informative to consider the degree of consistency in each set. In order to analyze this consistency, the difference ($D_{(Range)}$) between the $D_{(Comp)}$ results in each set is reviewed and the magnitude of the difference is determined as a percentage (%$_{Gap}$); this is a calculation of the degree of variability across the set.

For the Prairie House set, the difference in visual complexity between the most complex and the least complex houses is 7.4%. For the Textile Block houses the difference is 10.7%. This suggests that across the five houses, the Prairie House set is slightly more stylistically consistent. However, in both sets of five designs there are houses which are inconsistent and in order to optimize the mathematical fit for the set, the "best cluster" of three $D_{(Comp)}$ results is isolated for comparison in Table 6.

Table 6. $D_{(Comp)}$ and $D_{(Agg)}$ values for "best cluster" of three houses in each set

9 Houses	Prairie Style Houses	$D_{(comp)}$	Textile Block Houses	$D_{(comp)}$	Triangle-Plan Houses	$D_{(comp)}$
	Tomek	1.530	Freeman	1.5065	Reisley	1.443
	Robie	1.5375	Lloyd Jones	1.510625	Dobkins	1.442
	Henderson	1.546625	Storer	1.508375	Chahroudi	1.403
$D_{(Range)}$	0.016625		0.004125			0.04
$\%_{Gap}$	1.6%		0.4%			4%
$D_{(Agg\ for\ 3\ cluster)}$	1.53804		1.5085			1.42933

The "best cluster" method for analyzing the Prairie House set of results produces a range of visual complexity of $D_{(Range)} = 0.016625$ or 1.6%; around six times less than the result for all five houses in the set. This means that three of the five houses have much more in common, formally, than the other two. While this is a better result, for the best cluster of the Textile Block houses, the $D_{(Range)} = 0.004125$ or 0.4%; more than 26 times less than the range for the complete set of five houses. This result means that three of Wright's Textile Block houses are incredibly consistent across all facades. A comparison of the best cluster results, when aggregated, reveals that the Prairie Houses are still slightly more visually complex than the Textile Block Houses. This slightly counterintuitive result occurs in part because the most famous the Textile Block houses, La Miniatura and Ennis House are the most complex in the set, but they are also least typical of the set and are thus, not part of the best cluster.

The $D_{(Agg)}$ result for the set of Triangle-Plan Houses is 1.425. This is around 11.4% less visually complex than either the Prairie set or the Textile Block set. The $D_{(Range)}$ result for this set of 0.136 is a $\%_{Gap}$ of 13.6%; meaning that these are the least consistent set of five designs in terms of their visual complexity. Thus, not only are the Triangle-Plan Houses the least complex of the works, they are also the most variable within that low range. This is also evident in the best cluster analysis of the three most statistically similar houses in the set. The best cluster for the Triangle-Plan Houses reveals a $D_{(Range)}$ of 0.04 or 4% and a $D_{(Agg)}$ of 1.42933 confirming that, even for the best possible set of results, the Triangle-Plan Usonian houses are both less complex and less consistent than their Prairie and Textile Block counterparts.

What then, does a $\%_{Gap}$ result imply about how a person perceives architecture? Past research using this method has focused on the consistent analysis of twentieth century houses; this has tended to limit the possible range of visual complexity results that have been recorded. While two variations have been used in past research to determine the $\%_{Gap}$, if the method outlined in this paper is applied to past data, the lowest result recorded using this method is for a Minimalist house by Kazuo Sejima with a $D_{(Comp)}$ result of 1.192 (Ostwald, Vaughan, Chalup 2009). To clarify what this result implies; a house with three blank elevations and a fourth with a single door in it, might produce a result of $D_{(Comp)} = 1.08$. The highest result produced so far is recorded in this paper for La Miniatura: $D_{(Comp)}$ result of 1.61425 although multiple of Peter Eisenman's elevations have similar results (Ostwald and Vaughan 2009). All of this implies that a practical maximum difference for the visual complexity of house designs along the scale $1.0 < D < 2.0$, is in the order of 50% (the $\%_{Gap}$ between Sejima and Eisenman). Using the same scale, past research has demonstrated that houses which have a high degree of visual similarity, and which might be regarded as being in the same style, will have a $\%_{Gap}$ of 2% or less. Conversely, a $\%_{Gap}$ of 15% or greater tends to signal a lack of similar formal qualities.

When considered in the context of these past results, two of the three groups of Wright's houses have less than a 10% difference across their respective sets and the third, the triangle-plan set, is slightly higher at 13.6%. Thus, each of the sets falls within the mathematical range that might define a consistent formal language or style. When a gap comparison is made across the three groups, comparing the aggregate results for each set, the overall gap is 11.38%; implying that there are also similarities across Wright's works. Finally, when the best cluster group is analyzed both the Prairie houses and the Textile Block works display a very high degree of similarity for characteristic visual complexity; respectively 1.6% and 0.4%.

CONCLUSION

The mathematical analysis of visual complexity in 15 of Wright's house designs supports the following four conclusions and observations.

First, despite some stylistic differences, there is a notable degree of consistency across Wright's housing design (<12% difference across the $D_{(Agg)}$ results for the three sets of five houses). This supports the arguments of a range of historians referred to in the introduction to this chapter including those of Frampton (2005), McCarter (1999) and Hoffman (1995). Moreover, within each of his stylistic periods, Wright's works display a very high degree of consistency (between 0.4% and 4% difference within the best cluster of results for each set).

Secondly, Wright's Triangular-plan Usonian Houses have the least formally complex elevations of those being considered ($D_{(Agg)}$ = 1.425). This result matches the historical analysis of these buildings which suggests that, as Wright's architectural skills became more refined his houses became less formally expressive. For example, Hoffman argues that "the Usonians lacked the intensity of discipline and detail in the best of the Prairie houses" (1995, 67). Similarly, Sergeant proposes that the "Usonian Houses develop and simplify the methods of grid planning that were first used in the Prairie Houses" (2005, 190).

Next, La Miniatura and the Ennis House are amongst Wright's most visually compelling compositions. This supports the common contention that Wright's architecture reached a degree of tactile complexity at this point in his career. De Long argues that it was in the designs of the Textile Block Houses that "Wright conceived bold visions for a new architecture. It was an architecture new in scale, new in form, and new in means" (1994, xii). However, when taken in their totality, the Textile Block works are marginally less complex than Wright's earlier house designs. This directly supports Sergeant's claim that the Textile Block houses "marked the transition from the mature and complex geometric organization of the Prairie house to the freer yet more rational ordering of space in the Usonian houses" (2005, 195).

The fourth finding is that the Palmer House is less visually complex, or less fractal, than all of the Prairie Houses and Textile Block Houses analyzed in this chapter. This is contrary to Eaton's (1998) argument that the Palmer House is the most fractal of Wright's houses. While acknowledging that Eaton is talking about symbolic or metaphorical fractal properties in plans, and this chapter is concerned with the actual properties of elevations, the result still suggests a lack of support for Eaton's argument. Furthermore, Eaton maintains that there is nothing fractal in the Millard house, La Miniatura, even though this paper has mathematically determined that it is the most visually complex of Wright's 15 houses that are tested here and that this complexity operates across the widest possible range of scales.

Finally, it is hoped that the mathematical methods presented in this chapter, along with the comprehensive data for 15 houses, can be used to assist historians and design scholars to ground their analysis of Wright's architecture and to promote future, more detailed computational analysis of historic buildings.

ACKNOWLEDGMENTS

This research was supported by an Australian Research Council, Discovery Project grant (ARCDP): DP0770106.

REFERENCES

Batty, Michael. (2005). *Cities and complexity: understanding cities with cellular automata, agent-based models, and fractals.* Cambridge Mass: MIT Press.

Batty, Michael, & Paul Longley. (1994). *Fractal Cities: A Geometry of Form and Function.* San Diego: Academic Press.

Bovill, Carl. (1996). *Fractal Geometry in Architecture and Design. Design science collection.* Boston: Birkhäuser.

Burkle-Elizondo, Gerardo. (2001). Fractal Geometry in Mesoamerica. *Symmetry: Culture and Science, 12*, no. 1-2, 201-214.

Burkle-Elizondo, Gerardo, & Ricardo David Valdez-Cepeda. (2006). Fractal analysis of Mesoamerican pyramids. *Nonlinear Dynamics, Psychology, and Life Sciences, 10, no.* 1 (January): 105-122.

Capo, Daniele. (2004). The Fractal Nature of the Architectural Orders. *Nexus Network Journal, 6, no.* 1 (April 13), 30-40. doi:10.1007/s00004-004-0004-9.

Cardillo, Alessio, Salvatore Scellato, Vito Latora, & Sergio Porta. (2006). Structural properties of planar graphs of urban street patterns. *Physical Review. E, Statistical, Nonlinear, and Soft Matter Physics, 73*, no. 6 Pt 2 (June): 066107.

Cartwright, T. J. (1991). Planning and Chaos Theory. *Journal of the American Planning Association, 57*, no. 1, 44. doi:10.1080/01944369108975471.

Chalup, Stephan, Naomi Henderson, Michael J Ostwald, & Lukasz Wiklendt. (2008). A method for cityscape analysis by determining the fractal dimension of its skyline. In *Innovation Inspiration and Instruction: New Knowledge in the Architectural Sciences.*, (Ed). Ning Gu, Lehman Figen Gul, Michael J Ostwald, & Anthony Williams, 337-344. Newcastle, NSW, Australia.

Cooper, Jon. (2003). Fractal Assessment of Street-level Skylines: A Possible Means of Assessing and Comparing Character. *Urban Morphology, 7*, no. 2, 73-82.

Cooper, Jon. (2005). Assessing Urban Character: the use of fractal analysis of street edges. *Urban Morphology, 9*, no. 2, 95-107.

De Long, David G. (1994). Foreword. In *Wright in Hollywood: visions of a new architecture.* New York, N. Y: Architectural History Foundation.

Eaton, Leonard. (1998). Fractal Geometry in the Late Work of Frank Lloyd Wright. In *Nexus II: Architecture and Mathematics*, ed. Kim Williams. Fucecchio (Florence): Edizioni dell'Erba.

Ferrero, Giovanni, Celestina Cotti, Michela Rossi, & Cecilia Tedeschi. (2009). Geometries of Imaginary Space: Architectural Developments of the Ideas of M. C. Escher and Buckminster Fuller. *Nexus Network Journal, 11*, no. 2 (July 1), 305-316. doi:10.1007/s00004-008-0090-1.

Foroutan-pour, K., Dutilleul, P. & Smith. D. L. (2000). Advances in the implementation of the box-counting method of fractal dimension estimation. *Appl. Math. Comput, 105*, no. 2-3, 195-210.

Frampton, Kenneth. (2005). The Text-Tile Tectonic. In *On and by Frank Lloyd Wright: A Primer of Architectural Principles*. London: Phaidon.

Gozubuyuk, Gaye, Gulen Cagdas, & Ozgur Ediz. (2006). Fractal Based Design Model for Different Architectural Languages. In *The architecture co-laboratory: GameSetandMatch II: on computer games, advanced geometries, and digital technologies* , (Ed). *Kas Oosterhuis and Lukas Feireiss*, 280-286. episode publishers.

Harris, James. (2007). Integrated Function Systems and Organic Architecture from Wright to Mondrian. *Nexus Network Journal, 9*, no. 1 (March 21), 93-102. doi:10.1007/s00004-006-0031-9.

Hillier, Bill. (2007). *Space Is the Machine: A Configurational Theory of Architecture*. Cambridge: Cambridge University Press.

Hoffmann, Donald. (1995). *Understanding Frank Lloyd Wright's Architecture*. New York: Dover Publications.

Jencks, Charles. (1995). *The Architecture of the Jumping Universe: A Polemic: How Complexity Science Is Changing Architecture and Culture*. London: Academy Editions.

Joye, Yannick. (2006). An interdisciplinary argument for natural morphologies in architectural design. *Environment and Planning B: Planning and Design, 33(2)*, 239-252.

Knight, Terry Weissman. (1994). *Transformations in Design: A Formal Approach to Stylistic Change and Innovation in the Visual Arts*. Cambridge: Cambridge University Press.

Koning, Hank, and Julie Eizenberg. (1981). The language of the prairie: Frank Lloyd Wright's prairie houses. *Environment and Planning B: Planning and Design, 8*, no. 3, 295, 323. doi:10.1068/b080295.

Kostof, Spiro. (1985). *A History of Architecture: Settings and Rituals*. New York: Oxford University Press.

Laseau, Paul, & James Tice. (1992). *Frank Lloyd Wright: Between Principle and Form*. New York: Van Nostrand Reinhold.

Lind, Carla. (1994). *Frank Lloyd Wright's Usonian houses*. San Francisco: Promegranate Artbooks.

Lorenz, Wolfgang. (2003). Fractals and Fractal Architecture. Masters Diss. *Vienna University of Technology*.

MacCormac, Richard. (2005). Form and philosophy : Froebel's kindergarten training and Wright's early work . In *On and by Frank Lloyd Wright: A Primer of Architectural Principles*, (Ed). Robert McCarter. Phaidon.

Mandelbrot, Benoit B. (1982). *The Fractal Geometry of Nature*. Updated and augmented. San Francisco: Freeman.

McCarter, Robert. (1999). *Frank Lloyd Wright*. London: Phaidon.

Mizuno, Setsuko, & Hidekazu Kakei. (1990). Fractal analysis of street forms. *Journal of architecture, planning and environmental engineering*, *8*, no. 414, 103-108.

Oku, Toshinobu. (1990). On Visual Complexity On the Urban Skyline. *Journal of Planning, Architecture and Environmental Engineering*, *6*, no. 412 (June), 61-71.

Ostwald, Michael J. (2001). "Fractal Architecture": Late Twentieth Century Connections Between Architecture and Fractal Geometry. *Nexus Network Journal, Architecture and Mathematics*, *3*, no. 1 (Winter-Spring): 73-84.

Ostwald, Michael J. (2003). Fractal Architecture: The Philosophical Implications of an Iterative Design Process. *Communciation and Cognition*, *36*, no. 3&4, 263-295.

Ostwald, Michael J. & John Moore, R. (1996). Fractalesque Architecture: An Analysis of the Grounds for Excluding Mies Van Der Rohe from the Oeuvre. In *Traditions and Modernity*, A., Kelly, K., Bieda, J. F. Zhu, & W. Dewanto, (Edited by) 437,453 Jakarta: Mercu Buana University.

Ostwald, Michael J, & Josephine Vaughan. (2008). Determining the fractal dimension of the architecture of Eileen Gray. In *Innovation, Inspiration and Instruction*, 9-16. Newcastle, NSW, Australia: ANZAScA.

Ostwald, Michael J, & Josephine Vaughan. (2009). Calculating visual complexity in Peter Eisenman's architecture: A computational fractal analysis of five houses . In *CAADRIA, 2009*, 75-84. Taiwan.

Ostwald, Michael J, Josephine Vaughan, & Stephan Chalup. (2008). A Computational Analysis of Fractal Dimensions in the Architecture of Eileen Gray. In *ACADIA 08: Silicon + Skin: Biological Process and Computation*, 256-263. Minneapolis: Acadia.

Ostwald, Michael J, Josephine Vaughan, & Stephan Chalup. (2009). A Computational Investigation into the Fractal Dimensions of the Architecture of Kazuyo Sejima. *Design Principles and Practices: An International Journal*, *3*, no. 1, 231-244.

Ostwald, Michael J, Josephine Vaughan, & Christopher Tucker. (2008). Characteristic visual complexity: Fractal dimensions in the architecture of Frank Lloyd Wright and Le Corbusier. In *Nexus VII: Architecture and Mathematics*. Turin: Kim Williams Books.

Pfeiffer, Bruce Brooks, & Yukio Futagawa. (1987). *Frank Lloyd Wright Monograph* 1907-1913. Vol. *3*. Tokyo, A.D.A. Edita.

Rodin, Vladimir, & Elena Rodina. (2000). The fractal dimension of Tokyo's streets. *Fractals*, *8*, no. 4, 413-418.

Sergeant, John. (2005). Warp and woof : a spatial analysis of Wright's Usonian houses. In *On and by Frank Lloyd Wright: A Primer of Architectural Principles*, Ed. Robert McCarter. London: Phaidon.

Storrer, William Allin. (1974). *The Architecture of Frank Lloyd Wright, a Complete Catalog*. Cambridge, Mass: MIT Press.

Sweeney, Robert L. (1994). *Wright in Hollywood: Visions of a New Architecture*. New York, N.Y: Architectural History Foundation.

Thomson, Iain. (1999). *Frank Lloyd Wright: A Visual Encyclopedia*. London: PRC.

Vaughan, Josephine, & Michael J Ostwald. (2009). A Quantitative Comparison between the Formal Complexity of Le Corbusier's Pre-Modern (1905-1912) and Early Modern (1922-1928) Architecture. *Design Principles and Practices: An international journal*, *3*, no. 4, 359-372.

Voss, Richard. (1988). Fractals in Nature: From Characterization to Simulation. In *The Science of fractal images* , ed. Heinz-Otto Peitgen, Dietmar Saupe, and Michael F Barnsley. New York: Springer-Verlag.

Wright, Frank Lloyd. (1960). *Frank Lloyd Wright: Writings and Buildings*. Cleveland, Ohio: World Publishing Co. 1984. *Frank Lloyd Wright*. Vol. 5. Tokyo: A.D.A. Edita.

Yamagishi, R., Uchida S. & Kuga. S. (1988). An Experimental Study of Complexity and Order of Street-Vista. *Journal of Architecture, Planning and Environmental Engineering*, *2*, 384, 27,35.

In: Built Environment: Design, Management and Applications   ISBN: 978-1-60876-915-5
Editor: Paul S. Geller, pp. 89-120   © 2010 Nova Science Publishers, Inc.

Chapter 4

THE USE OF THE 3D LASER SCANNER IN THE BUILT ENVIRONMENT

Yusuf Arayici

School of the Built Environment, University of Salford,
Greater Manchester, UK

EXECUTIVE SUMMARY

Capturing and modelling 3D information of the built environment is a big challenge. A number of techniques and technologies are now in use. These include EDM (Electronic Distance Measurement), GPS (Global Positioning System), and photogrammetric application, remote sensing and traditional building surveying applications. However, the use of these technologies cannot be practical and efficient in regard to time, cost and accuracy. Furthermore, a multi disciplinary knowledge base, created from the studies and research about the regeneration aspects is fundamental: historical, architectural, archeologically, environmental, social, economic, etc. In order to have an adequate diagnosis of regeneration, it is necessary to describe buildings and surroundings by means of documentation and plans. However, at this point in time the foregoing is considerably far removed from the real situation, since more often than not it is extremely difficult to obtain full documentation and cartography, of an acceptable quality, since the material, constructive pathologies and systems are often insufficient or deficient (flat that simply reflects levels, isolated photographs,..). Sometimes the information in reality exists, but this fact is not known, or it is not easily accessible, leading to the unnecessary duplication of efforts and resources.

Systems that measure range from the time-of-flight of a laser pulse have been available for about 25 years, so this does not constitute new technology. However, the development of fast measurement (up to 10000 measurements per second) and a scanning mechanism (using rotating mirrors) has only occurred in this decade or so. Packaging these components into a robust and reliable instrument has resulted in the innovation of a 3D laser scanner.

In this chapter, we discussed 3D laser scanning technology, which can acquire high density point data in an accurate, fast way. Besides, the scanner can digitize all the 3D information concerned with a real world object such as buildings, trees and terrain down to millimetre detail Therefore, it can provide benefits for refurbishment process in regeneration in the Built Environment and it can be the potential solution to overcome the challenges above. The chapter introduces an approach for scanning buildings, processing the point cloud raw data, and a modelling approach for CAD extraction and building objects classification in IFC (Industry Foundation Classes) format. The approach presented in this section can lead to parametric design and Building Information Modelling (BIM) for existing structures. In this chapter, while use of laser scanners are explained, the integration of it with various technologies and systems are also explored for professionals in the Built Environment.

1. INTRODUCTION

Advanced digital mapping tools and technologies so-called 3D laser scanner are enablers for effective e-planning, consultation and communication of users' views during the planning, design, construction and lifecycle process of built and human environments. The regeneration and transformation of cities from the industrial age (unsustainable) to the knowledge age (sustainable) is essentially a 'whole life cycle' process consisting of; planning, development, operation, reuse and renewal (Arayici and Hamilton, 2005a). In order to enhance the implementation of build and human environment solutions during the regeneration and transformation of cities, advanced digital applications can have a significant impact (Arayici and Hamilton, 2005b).

This innovation is significant because it has potential to solve the problems that are always been associated with design and construction of existing buildings. It can provide faster, better quality and more precise analysis and feature detection for building survey (Arayici et al, 2004). Within the built environment, the use of the 3D laser scanner enables digital documentation of buildings, sites and physical objects for reconstruction and restoration including cultural heritage. It also enables the creation of educational resources within the built environment, as well as the reconstruction of the built environment. Besides, it has also potential to accurately record inaccessible and potentially hazardous areas. As a result, it can facilitate "virtual refurbishment" of the buildings and allows the existing structure and proposed new services to be seen in an effective manner. The use of the 3D scanner in combination with building and city systems is key enablers to the creation of new approaches to the 'Whole Life Cycle' process within the built and human environment for the 21st century.

Besides, the 3D Laser Scanners in the natural environment also allow constructing Digital Elevation Models (DEMs) that accurately represent landform surface variability and offer an excellent opportunity to measure and monitor morphological change across a variety of spatial scales such as flooding scenarios and the causes of potential flooding in the natural environment (Brasington *et al.* 2000; Fuller *et al.* 2005) (Hetherington et al, 2005). Although many studies have been conducted in the natural environment to collect spatial data by means of sophisticated surveying equipment such as aerial LIDAR, EDM theodolites, GPS, photogrammetry, many of these studies continue to suffer area or resolution limitations due to a trade off between spatial coverage and morphologic detail captured (Hetherington et al, 2005). For example, techniques such as terrestrial photogrammetry produce dense accurate morphometric data but aerial coverage is restricted, aerial photogrammetry offers increased spatial coverage but reduced elevation accuracy, EDM surveys suffer from long collecting times resulting in reduced data density if large areas are surveyed (Hetherington et al, 2005).

Traditional building surveying techniques also suffer for accurate and complete data capture due to accessibility and manual measurement and laborious activities on field. For example, due to various hazards building surveyors may have accessibility problems on site, which leads them to estimation. As a result, that causes inconsistency in accurate data capture.

In this chapter, the use of laser scanner and its potential benefits for the built environment is explained based on the research studies undertaken in the built and natural environment fields. In the following section, background information and literature search are presented in the research studies in regard to 3D laser scanner. This includes technical description of 3D laser scanner concept, use of this technology in the literature, the vision for the use of this technology and the limitations experienced in the initial experiments of this tool in the research studies in the built and natural environment.

2. THE 3D LASER SCANNER TECHNOLOGY

2.1 Concept and Definitions

These instruments, based on laser technology, are commonly known as 3D laser scanners or LIDAR (Light Detection and Ranging). While laser scanner instruments based on the triangulation principle and high degrees of precision have been widely used since 80's, TOF (Time of Flight) instruments have been developed for metric survey applications only in this decade. These types of laser scanners can be considered as highly automated total stations. They are usually made up of a laser, which has been optimised for high speed surveying, and a set of mechanisms that allows the laser beam to be directed in space in a range that varies according to the instrument that is being used.

Figure 1: Specification of LMS Z210 scanner used in the research.

Range:	
r³80%	300m
r³10%	
Minimum	2m
Spot size/ beamwidth	25mm @ 100m
Precision	25mm
Max resolution	25mm
Capture	6.000 pts/sec
SCAN	
Vertical	0°-80°
angular resolution	0.002°
horizontal	0°-333°
angular resolution	0.025°
Weight	
Software	RiSCAN Pro

For each acquired points, a distance is measured on a known direction: X, Y, and Z coordinates of a point can be computed for each recorded distance direction. Laser scanners allow millions of points to be recorded in a few minutes. Because of their practicality and versatility, these kinds of instruments are today widely used in the field of architectural, archaeological and environmental surveying.

There are airborne and terrestrial scanners. Airborne laser scanner is mounted beneath a plane to scan the earth while flying, whilst terrestrial laser scanner is mounted on a tripod and uses a laser to measure the three dimensional coordinates of a given region of an objects surface automatically, in a systematic order at a high rate in near real time (Bryan, 2003). The focus in this section will be on the 3D terrestrial laser scanners.

The scanner provides a combination of wide field-of-view, high accuracy, and fast data acquisition. The scanner is connected to a 12V battery and a ruggedised laptop. Terrestrial laser scanning offers fast 3D terrain data acquisition. It has advantages over current survey techniques including EDM, GPS and photogrammetric applications obtaining high density point data without the need for a reflector system. Merged data clouds have sufficient points to negate the need for DEM interpolation techniques potentially providing the optimum representation of any scanned surface. The advantages of speed and high data point density must be viewed against the data point accuracy which may reduce instrument performance below that achievable using EDM techniques (such techniques are, however, much more time consuming). Any Improvement to measurement range, resolution, field of view and error/accuracy would further fill the research gap relating to spatial and temporal measurement of space and change in the natural environment and resolve the accuracy issue with regard to EDM techniques.

The scanner is targeted to the physical objects to be scanned and the laser beam is directed over the object in a closely spaced grid of points. By measuring the time of laser flight, which is the time of travel of the laser from the scanner to the physical objects and back to the scanner, the position in three-dimensional space of each scanned point on the object is established. The result is a "cloud of points" which consists in thousands of points in

3-dimensional space that are a dimensionally accurate representation of the existing object (Schofield, 2001). This information can then be converted in a 3D CAD model that can be manipulated using CAD software, and to which the design of new equipment can be added. This innovation is significant because it has potential to solve the problems that are always associated with design and construction of existing buildings for reuse goals. For example, it can provide faster, better quality and more precise analysis and feature detection for building survey. The table below shows a list of riegl scanners and their suitability for various built environment applications.

2.1.1 Point clouds

The XYZ co-ordinates of points in a common co-ordinate system portray an understanding of the spatial distribution of objects on the earth. It also includes additional information such as intensity or RGB value. Generally a point cloud contains a relatively large number of co-ordinates according to the volume of the point cloud occupies, rather than a few widely distributed points. For example, the images in figure 2 and 3 shows point cloud data in 2D interface and 3D interface respectively in true colour mode.

Table 1. Comparison of Different Riegl Scanners in regard to their specification and their use in different applications www.riegl.com

	3D-Imaging Scanner Systems			3D-Laser Profile Measuring Systems	
Type	**LMS-Z420i**	**LMS-Z390**	**LMS-Z210ii**	**LPM-i800HA**	**LPM-2K**
Main Specs	High-Accuracy, Long Range & High-Speed	High-Accuracy & High Resolution	General Purpose	High-Accuracy & Long-Range	Long-Range
eye safety	Laser Class 1	Laser Class 1	Laser Class 1	Laser Class 1M	Laser Class 1M
min range	2m	1m	4m	10m	10m
max range $r^3 80\%$ $r^3 80\%$	1000m 350m	300m 100m	650m 200m	800m 250m	2500m 800m
Repeatability Averaged Single shot	4mm 8mm	2mm 4mm	10mm 15mm	- 15mm	- 50mm
Accuracy	10mm	6mm	15mm	-	-
Field of view Line scan Frame scan	80deg 360deg	80 deg 360 deg	80 deg 360 deg	-20/+130deg 360 deg	- 60/+130deg 360 deg
Angular resolution	0.0025deg	0.001deg	0.005deg	0.009deg	0.009deg
Measurement per seconds Rotating mirror Oscillating mirror	Up to 8000 Up to 12000	Up to 8000 Up to 12000	Up to 8000 Up to 12000	typically 1000	up to 4
laser wavelength	near IR	near IR	near IR	near IR	near IR
beam divergence	0.25 mrad	0.25 mrad	2.7 mrad	1.3 mrad	1.2 mrad
data interface	TCP/IP	TCP/IP	TCP/IP	TCP/IP &RS 422	TCP/IP&RS 422

Table 1. (Continued)

3D Projects	LMS-Z420i	LMS-Z390	LMS-Z210ii	LPM-i800HA	LPM-2K
Archaeology & Cultural Heritage Documentation	●●●	●●●	-	●	-
Architecture&Façade Measurements	●●●	●●●	-	-	-
City Modelling	●●●	●●●	●	-	-
Civil Engineering	●●●	●●●	-	●●	-
Monitoring	●●●	-	●●●	●●●	●●●
Process Automation & Robotics	-	●	●●●	-	-
As-Built Surveying	●●●	●●●	-	-	-
Topography&Mining	●●●	●●	●●●	●●●	●●●
Tunnel Surveying	-	●●●	-	-	-
Virtual Reality	●●●	●●●	●	-	-

●●● Excellent for the selected applications, ●● very good ● good

Point clouds are suitable for use on a wide variety of subjects including small objects, details of architectural design, building facades and whole sites. The variety of laser scanning systems is available and their differing design and operation of the exact procedures used to perform a survey will differ from system to system.

Figure 2: An image of field of view of an individual scan.

Figure 3: Point cloud data of an individual scan.

Figure 4: Interior wall scans and tie points with outside scans through door spaces in gray colors.

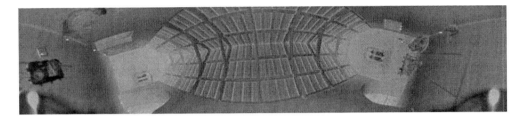

Figure 5: Roof Structure Scan in gray color.

Figure 6: Outside scan at the rear door step for linkages with interior scans through door space in gray colors.

2.1.2 Overview and detail scans

A combination of overview and detail scans provides a clear description of the interested area. In addition, they enable the acquisition of detailed information of selected parts of the area or buildings. It is likely that scanning process on field requires the combination of a mixture of scans. Figure 4, 5 and 6 shows some detail interior scans in gray colour mode, the small white spots are the tie points to register the consecutive scans

2.1.3. Overlapping Scans

Overlapping scans are generally required to ensure a full record of an object is collected. Overlapping data can also provide users with confirmation that the registration process has been successful (Bryan, 2003). It is possible to filter overlapping scan data in order to reduce the point density in the final registered point cloud and hence reduce file sizes and strain on software systems during processing.

2.1.4 Data voids

Since data, which is collected during the scanning survey, will form part of the record of that site or the building, it is necessary to make sure that the data collected is complete. Data voids should therefore be minimised during the scanning process on field through selection of appropriate scanning positions (overlapping if necessary) and minimising temporary obstructions to the scanner during operation such as vehicles and pedestrians. Data voids due to occlusions to the line of sight can normally only be eliminated by using multiple scans. Data voids due to temporary obstructions should be limited by appropriate positioning of the scanner or restriction of such obstructions.

2.1.5 Survey control, co-ordinate systems and registration

In the majority of surveys dealing with fixed objects, such as buildings and monuments, a site co-ordinate system will be available to which to reference the data. Where a previous co-ordinate system does not exist therefore a new system may need to be established. In order to ease the processing of point cloud in CAD packages, the Z axis should be defined as the vertical axis. Although a single scan may be sufficient to fully record certain scenes, multiple scans are likely to be required, especially when dealing with a large site or buildings. It is likely to be necessary to transform the collected point clouds to the local site co-ordinate system. The process of registration can be performed using four methods (Bryan, 2003):

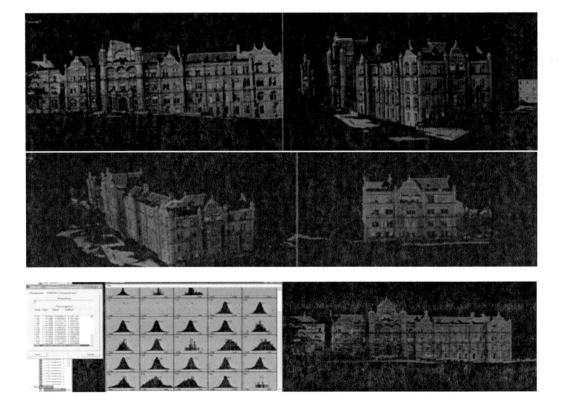

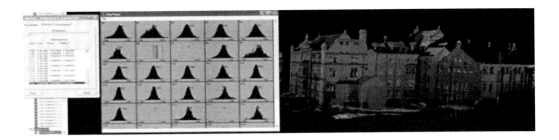

Figure 4: Overleaping Scans are registered one by one using the tie points in the overlapping areas.

- Data collection can be performed from a station with known co-ordinates and known orientation
- Targets such as cylindrical reflectors can be used to transform the data onto common co-ordinate systems
- Surface matching algorithms for alignment of the individual scans can be used to transform data onto common co-ordinate systems. The use of surface matching algorithms alone would normally involve the use of an arbitrary co-ordinate system
- A combination of the previous methods may be used to improve the speed and ease of the registration process

Figure 5. Lack of data, due to people and cars in front of the building scanned.

Generally speaking, each method will provide an indication on the quality of the registration process. Data collection from a known station and known orientation will be reliant upon the precision to which the control information is known. In the case of using reflectors as control points the quality of the registration process is best indicated by the residuals of the transformation process and the estimated precision for each transformation parameter. In the case of surface matching alone the quality of the registration would be indicated by the residuals of surface matching algorithms, along with the estimated precision of the transformation parameters.

Figure 6. shows a number scan registered into the same coordinate system to form a holistic point cloud model.

It is unlikely that surface matching algorithms alone would be suitable for metric survey applications and that some targeted points should be defined during the post-processing.

2.1.6   Intensity / colour information and additional data

The majority of systems provide intensity information, in addition to XYZ position as well as a colour value for each point. Additional image data such as those obtained by camera mounted on top of the scanner can be collected to provide an overview of the subject being scanned, in addition to providing imagery for narrative purposes. This imagery should be of a high resolution and clearly portray the building or objects in the interested area.

This additional data will help to reduce scan artefacts caused by poor framing of the scanner. For example, it can be used for interpretation of actual geometric edges, even though scans may simply have been created by poor positioning of the scanner. Furthermore, additional image data can help to interpret the effect of different materials on the scanning measurement.

2.1.7   Weather

Weather can have an impact on the quality of data collected. For example, scanning in heavy rain may give rise to data voids due to falling raindrops or erroneous data points due to return from airborne raindrops or erroneous range measurement due to refraction of the measurement beam.

2.1.8   Data format

Currently there are no standard formats for the distribution of scan data. This leads to issues relating to compatibility, exchange of information and data archiving and preservation of data access. It is likely that the most desired or appropriate data formats relates to the software system to be used. In these cases, it is possible that a particular format can be specified. However, in cases where the data is to be preserved as an archive data source, the required data format may not be well defined.

To assist in the future management of scan data, all data is required to be delivered in a pre-specified format with emphasis on the transferability of data between software systems.

3. The Main Barriers and the Limitations in the Real World Data Modelling

Currently, there are no standard formats for the distribution of terrestrial scan data. This leads to issues relating to data archiving and preservation of data access. It is likely that the most desired or appropriate data format relates to the software system to be used. In these cases, it is possible that a particular format can be specified. However, in cases where the data is to be preserved as an archive data source the required data format may not be well defined. Through the attempt for integration with those systems in Figure 12, the main limitation is the extracting the information from the 3D model produced through 3D laser scanning system. This is because the laser scanner acquires millions of point data of the existing objects. As a result, it brings about the difficulty in extracting significant information accurately and fast from "point clouds" data such as feature lines for 2D and 3D CAD plans and models. Otherwise, use of the post-processed model of the laser scanning system is not easy to handle for the other systems. The process is illustrated in Figure 13 below to address the limitations and data issues at certain stages of the process.

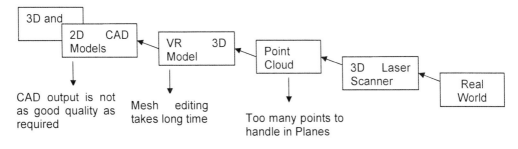

Figure 7. The development process in the 3D laser scanner system.

In the use of the 3D laser scanner system according to the above process, the followings advantages and disadvantages of the system are explored.

Table 2: The Advantages and Disadvantages of This Technology is Shown in the Below Table

3D Laser Scanning	
Advantages	Disadvantages
Applicable to all 2D and 3D surfaces	Some systems do not work in sun or rain
Rapid 3D data collection-near real time- requiring substantially less site time	Large, high resolution 3D data sets require post-processing to produce a useable output
Very effective due to large volumes of data collected at a predictable precision	Difficulty in extracting the edges examples from indistinct data clouds
Ideal for all 3D modelling and visualisation purposes	Output requires manipulation to achieve acceptable recording quality
Both 3D position and surface reflectance generated which, when processed, can be viewed as an image	No common data exchange format-such as DXF-currently in use to allow ease of processing by third parts
Rapidly developing survey technology	Difficult to stay up-to-date with developments
Extensive world-wide research and development currently undertaken on both the hardware and software tools	With hardware still expensive and sophisticated software required to process data, cost is prohibited for many projects

4. THE POST-PROCESSING THE SCAN DATA

The PolyWorks software (including only IMAlign-IMMerge-IMEdit-IMCompress Modules) for processing the point data into a 3D model is the main tool for the post processing the point cloud data. During data acquisition, the 3D-RiSCAN software is used. It allows the operator to perform a large number of tasks including sensor configuration, data acquisition, data visualization, data manipulation, and data archiving. Numerous export functions allow the scanned data to be passed to post-processing data packages for, e.g., feature extraction or volume estimation (www.riegl.com).

A field methodology has been developed to rapidly capture data for a building survey. For each scanner position, first a scan of the full field of view (333° horizontally) is made to aid merging of data from subsequent scan positions and to aid the definition of detailed scans. This scan takes about 4 minutes. Next, one or more detailed scans are made with a finer resolution of the area of interest (typically the building). This scan typically takes 10 minutes depending on resolution and field of view. Once back in the office, PolyWorks software (developed by Innovmetric Software Inc.) provides comprehensive set of tools for quickly processing 3D scanned data (www.innovmetric.com). This software has traditionally been used in the manufacturing industries with very short range scanners, but the advent of longer range laser scanners, it has seen widespread use in surveying and architecture, especially within North America. The software can handle many millions of data points while still retaining the ability to model very fine details very accurately.

4.1. 3D Point Cloud Data, Scan Registration and 3D Mesh Modeling

After 3D point cloud data is captured using the laser scanner together with the RiSCAN Pro software, Scan registration can be comfortably done in either RiSCAN Pro software or the Polyworks IMAlign software module based on project requirements and scanning strategy. Once all the scans are registered, a point cloud model of the scanned object is obtained. A 3D mesh model can be generated using the IMMerge Module of Polyworks software. Meshing parameters such as surface sampling step, reduction tolerance, smoothing and maximum distance, are important to create smooth and filtered high quality mesh model with high resolution. To achieve this, it is important to take into account the scanning parameters. The accuracy and resolution of the model will be dependent on the scanning resolution and the laser scanner accuracy. The output from IMMerge is a polygonal mesh model of the building, scene or object that has been scanned and merged.

Figure 8. 3D models of building after registration of the different scans taken from different perspectives.

The final merged model can then be exported from IMMerge and imported into IMEdit for refining. This will allow scans from each scanner position to be merged together in space. The software will compare geometric features in overlapping areas of adjacent scans to calculate the correct alignment of each scan. This is a very accurate method of aligning laser scan data as it is using many thousands of measurements to make a comparison rather than a few observations that would be made in a conventional surveying scenario. The created 3D mesh model (sometimes termed a wireframe) is created uses all the measured data points without applying any dangerous point sampling techniques. In its approach, it uses tolerance-based smoothing and tolerance-based adaptive meshing which provide a robust industrial strength process (Gilbert et al, 1995).

The 3D model can then be viewed, analysed and edited as necessary. It is also possible to use an intelligent simplification algorithm on the model to reduce the number of triangle vertices while retaining as much information as possible (especially on edges).

Figure 8 shows images of a complete holistic model of interior and exterior of the scanned buildings after registering all the scans into a common coordinate system using the tie points.

4.2. Editing the 3D Model and CAD Extraction:

The editing process is crucial to produce a neat CAD model. This is done by means of the Polyworks IMEdit software module. Several phases exist in stages 4 and 5 in order to complete the editing process to a high standard: The phases are explained below (Arayici, 2007)

1. Phase, Lining up Model: The first activity in the IMEdit module of Polyworks software is to orientate the 3D Mesh model into an XYZ coordinate system. Because the model could be oblique in the space, the first step is to line up the model according to the IMEdit XYZ coordinate system. As a result, width, height and length of the model can be viewed horizontally and vertically when the model is viewed from X, Y or Z perspectives of the coordinate system. To do that, a plane object is inserted through a surface of the model by selecting 3 vertices distributed almost equally on the surface. This plane is situated through the surface. Another plane object is inserted aligned with any of XY, XZ, YZ dimensions. The second plane is situated in the space as parallel to one of the XY, XZ, YZ dimensions. Now the first plane inserted into the model surface is converted to the second plane. This also allows the model to be aligned to an intended vertical dimension such as XY, XZ, and YZ. Once this lining up process is done three times, the whole model will be converted to the XYZ coordinate system of the IMEdit module of Polyworks software. This can be done automatically by developing scripts below. The script below is to align a mesh model in XZ plane

```
version "4.0"
VIEW POSE Y_NEG
TREEVIEW MODEL VIEW DEFAULT_STATIC COLOR VERTEX_COLOR
EDIT PLANE CREATE XZ_PLANE
SELECT ELEMENTS
EDIT PLANE CREATE FROM_3_VERTICES
TREEVIEW SELECT NONE
TREEVIEW MODEL SELECT ( 1, "On" )
DECLARE I
```

```
DECLARE J
DECLARE MYPLANEONE
DECLARE MYPLANETWO
TREEVIEW PRIMITIVE PLANE GET_NB (I)
set j expr($i-1)
TREEVIEW PRIMITIVE PLANE NAME GET ($I, MYPLANEONE)
TREEVIEW PRIMITIVE PLANE NAME GET ($J, MYPLANETWO)
ECHO ("$MYPLANEONE")
ECHO ("$MYPLANETWO")
TREEVIEW PRIMITIVE PLANE SELECT ( $I, "On" )
ALIGN ROTATE_PLANE_A_TO_PLANE_B ( $MYPLANEONE, $MYPLANETWO )
TREEVIEW SELECT NONE
TREEVIEW PRIMITIVE PLANE SELECT ( $J, "On" )
TREEVIEW PRIMITIVE PLANE SELECT ( $I, "On" )
VIEW VISIBILITY OBJECTS HIDE ( )
```

With the same logic of coding the model can also be aligned in XY and YZ planes. This activity is important to conduct the next phase effectively and correctly. However, this phase also includes some tedious work. Therefore to speed up the process, macro programming can be developed in Polyworks programming editor. These macro programs within Polyworks editor can make this phase relatively straightforward.

2. Phase, Plane insertion: After lining up the model to the coordinate system, it is much easier to insert planes to the model surfaces. There are a number of ways for inserting planes such as

- Inserting planes by picking three points
- Inserting planes normal to a Curve
- Inserting planes parallel to another plane
- Inserting planes from two used picked points
- Inserting planes coplanar with three vertices
- Inserting planes coplanar with a triangle
- Fitting a plane on a set of vertices / triangles
- Inserting planes aligned with the coordinate system
- Inserting a Plane by specifying its equation

In our case, *inserting planes coplanar with three vertices, inserting planes aligned with the coordinate system* options are mostly used options because in the previous phase, the model is aligned according to the coordinate system. There are three ways of inserting planes aligned with the coordinate system of the polygonal model.

(a) *Insert planes aligned with the X-Y Plane:* Select the X-Y plane item of the "Plane" submenu of the "Edit" menu, or press the "Create a plane co-planar with the x-y plane" button in the Plane section of the creation/editing toolbar, to create a plane parallel to the one formed by the x and y axes. The plane is translated to the middle of the model's bounding box.

(b) *Insert planes aligned with the X-Z Plane:* Select the X-Z Plane item of the "Plane" submenu of the "Edit" menu, or press the "Create a plane co-planar with the x-z plane" button in the Plane section of the creation/editing toolbar, to create a plane parallel to the one formed by the *x* and *z* axes. The plane is translated to the middle of the model's bounding box.

(c) *Insert planes aligned with the Y-Z Plane:* Select the Y-Z Plane item of the "Plane" submenu of the "Edit" menu, or press the "Create a plane co-planar with the y-z plane" button in the Plane section of the creation/editing toolbar, to create a plane parallel to the one formed by the y and z axes. The plane is translated to the middle of the model's bounding box.

Planes aligned with the coordinate system can be inserted accurately and quickly by selecting only one vertex on the model surface. Once a vertex is selected, an aligned plane is inserted automatically. The inserted plane is parallel or perpendicular to XY, YZ or XZ dimensions. Plane insertion is important because any points diverted from the average surface can be projected using plane insertion. Without this it is not possible to produce straight lines for CAD modeling.

3. Phase, Point Selection: Polygon editing is all about selecting and editing triangles and vertices. To enter selection mode, the SPACEBAR must be pressed. The cursor then becomes a cross and the mouse is used to make individual, surface, volumetric or pre-programmed menu selections. Point selection can be done by using different but complex key combinations and a variety of options can be used in the Polyworks IMEdit software module. This can be a tedious but needs full attention to select the corresponding points on the surface because it is not easy task to select the right amount of vertices in one or two attempts. When the number of surfaces is too many in a model, it is easy to lose concentration and make wrong selection. Extra care is needed when selecting points towards the edges because it could be risky selecting points on the adjacent surfaces of the model. Because vertices selection is a laborious and time intensive activity, improvement is necessary to reduce the amount of time and labor in this phase. This Improvement can be made by defining a selection strategy to select all the relevant vertices on a surface in one click. For example, select a good point on the average surface and project this point on the plane inserted before. After this allow the software to select the points away from the plane at a certain distance such as 1mm or 5 mm. As a result, all the relevant points can be selected at a certain accuracy based on project requirements.

4. Phase, Project Points onto Plane: After carefully selecting points in phase 3, it is time to project them onto the planes inserted and in turn into the corresponding surfaces of the model. Besides projecting points onto planes, points can also be projected on the curve surfaces when necessary as well as the plane surfaces. This is a straightforward process although it is necessary to pay full attention in order to avoid any clashes while making up the edges between surfaces. In case of any clash, phases 3 and 4 can be repeated iteratively. If there are any arc edges on the model, curves can be used to construct sharp edges at defined deviation parameters from the curve. As an example, the script below can help to achieve the phases from 2 to 4 automatically for any lined up surface.

```
version "4.0"
EDIT PLANE CREATE PARALLEL_TO_PLANE
DECLARE I
TREEVIEW PRIMITIVE PLANE GET_NB (I)
TREEVIEW PRIMITIVE PLANE SELECT NONE
TREEVIEW PRIMITIVE PLANE SELECT ($I, "On" )
VIEW VISIBILITY OBJECTS HIDE ( )
SELECT VERTICES USING_PLANES ABOVE_AND_BELOW ( 4.5e-002, "Off" )
EDIT VERTICES PROJECT ONTO_PLANE
SELECT ELEMENTS
```

This script above can select and project vertices away from the surface at a maximum distance of 45mm. if necessary, this threshold distance for selecting and projecting vertices can be customized,

5. Phase, Optimize the Mesh Model: it is necessary to optimize the mesh model to make the model consistent and if necessary reduce the number of points at some regions in the model to reduce the file size and avoid point intensity and heterogenic point scatter. A mesh model is optimized when triangle edges are parallel to the flow of a contour. An optimized mesh bends to conform to the flow of the shape, providing a smoother surface. IMEdit uses angle criteria to determine if there should be an edge swap for each triangle. Mesh optimization may be applied to a selection, or to the entire model. There are a number of parameters that need to be adjusted for mesh optimization. These are (i) sensitivity, (ii) minimum number of triangles per vertex, (iii) max number of triangles per vertex, (iv) min inner angle, (v) max dihedral angle

The mesh optimization procedure is applied to pairs of triangles. For each processed pair of triangles, IMEdit checks the pair's neighborhood to see if the region is concave or convex. If the neighborhood is convex and the pair of triangles is concave, the common edge will be permuted by means of the two adjacent triangles to make the pair of triangles convex. The same operation is performed if the neighborhood is concave and the pair of triangles is convex. After mesh optimization, the triangulated mesh is more consistent, and the surface curvature is better described. Mesh optimization works best if the polygonal mesh is relatively smooth.

6. Phase, Create Cross-sections: After completion of phase 5, a regular mesh model with planes and defined edges should be in place. At phase 6, cross-sections are created through the planes inserted into the corresponding surfaces. Since the main tool in the cad extraction process is the planes inserted in the previous phases, the cross-sections will be created from the planes. Each cross-section will create a CAD line on the edges of the corresponding surfaces. This CAD lines describes the characteristics features of the model. For example, the images in figure 9 shows the transition of CAD lines from the mesh to the CAD system. The images on the right are the mesh models from the scan data. Many different planes are inserted and edited and cross-sectioned and exported to CAD as shown on the left.

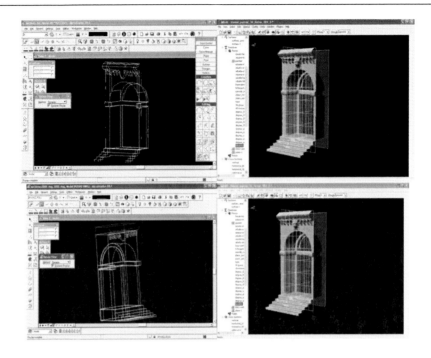

Figure 9. 3D mesh model of a door is being edited and CAD models are generated via cross-sections.

7. Phase, Export Cross-section: Cross-sections can be exported in various formats such as DXF, IGES, and so on. Generally, exporting to DXF is the preferred option because it is CAD oriented format. The export of cross-sections can be done in a variety of combinations such as individually selections all cross-sections selections or certain cross-section selections. The ideal way of exporting cross sections is to export the only cross-sections that have relationships among themselves to describe a building element. With this in mind the export process should be done for each building element separately. For example, if a window in a building model were to be exported then only the cross-sections that define the edges of this window should be selected and exported. Once this has been done for each element of the model, a CAD model will result that includes all the building's elements geometry in an object-oriented manner. The output is a 3D CAD model, which can be manipulated or edited if necessary in CAD software such as Microstation Triforma. This output is ready to use for building surveyors or engineers.

Figure 10. 3D mesh models produced from point cloud data.

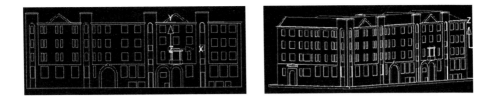

Figure 11. 3D CAD models generated from the mesh models in figure 10.

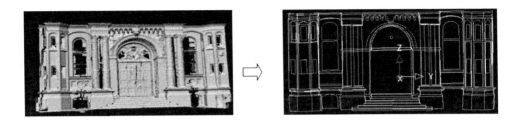

Figure 12. 3D CAD model on the right is produced from the 3D mesh model on the left.

5. THE SCOPE FOR THE 3D LASER SCANNING TECHNOLOGY

The scope is to help in thinking about real world problems. The spatial data can be obtained by using 3D scanner. By post-processing the captured spatial data, outputs for different purposes can be obtained such as CAD modeling, physical modeling by prototyping and visualization in different platforms (Arayici et al, 2006) as illustrated in figure 13. The use of these technologies is a key facilitator in the creation of an integrated system to capture, process, and display 3D information. For example, it can be used for civil engineering and environmental analysis. It permits the user to acquire irregular point clouds of land areas, rivers and infrastructure in a fast and cheap way.

The use of raw laser scanner data requires orientation and filtering procedures to generate a 3D model of the surveyed object. It is possible to automatically derive a set of geometric information from this 3D model that is useful for a variety of particular engineering and environmental applications such as dense Digital Terrain Model (DTM) generation, sections and profiles, contour maps, and volumes (Bornaz, et al, 2002), (Arayici, et al, 2004). One of the major problems that building surveyors have encountered is the inaccessibility. That is to say, on some occasions, they have to measure the details of building and relevant architectural details on the walls for example and in particular when developing elevation plans. In these circumstances, the building surveyors are required to estimate the measurements of these details, which is very much likely to involve a variety of errors regarding the accuracy and the actual design of the building due to wrong assumption and judgement.

The use of the 3D laser scanner has enormous potential to benefit building surveyors and their clients in terms of accuracy, speed and productivity in plan preparation and then to extend the range of services offered through modeling applications. The output of feasibility studies would improve immeasurably with modeling in areas such as assessing disabled access and fire safety. With current demand for whole life costing, asset management planning and database application, building surveyors are called upon to link information in CAD files to database files. The possibility of linking this information to 3D models would suggest all sorts of additional benefits. For example, producing a computer model of historical building with 3D laser scanner technology can only take a week, which includes scanning interior and exterior of the building in two days and post processing the point cloud data in three days. For this work only two people would be required for both completing the initial scans and the post-processing the raw data. Consequently, this technology is significantly faster and more effective than any other surveying applications in use. However, the post-processing may continue depending on the end product aimed. For example, if it is aimed at rapid prototyping.

5.1. Rapid Prototyping

The term rapid prototyping refers to a class of technologies that automatically construct physical models from Computer Aided Design (CAD) data. These "three dimensional printers" allow designers to quickly create tangible prototypes of their designs, rather than just two dimensional pictures. Such models have numerous uses. They make excellent visual aids for communicating ideas with co-workers or clients. In addition, prototypes can be used for design testing. There are research activities in rapid prototyping for the reverse engineering and design of buildings in particular historic buildings (Arayici et al, 2006).

The laser scanner can provide reverse engineering in construction for the reuse of the existing facilities. Producing building design and CAD models and VR (Virtual Reality) models from an existing facility or groups of facilities, by means of the laser scanner, will facilitate an analysis of the latest conditions of the buildings taking into account the original drawings of the same buildings, if they still exist. Besides, they even have the potential to accurately record inaccessible and potentially hazardous areas such as pitched rooftops. Consequently, it facilitates "virtual refurbishment" of the buildings and allows the existing structure and proposed new services to be seen in an effective manner (Arayici, 2007). Rapid prototyping for buildings is conducted based on the ground based 3D laser scanner data. Although several rapid prototyping techniques exist, all employ the same basic five step process. These steps are as follows;

1. Create a CAD model of the design: First, the object to be built is modelled using a Computer Aided Design (CAD) software package such as Microstation, which is used together with Polywork point cloud modeller in order to create a CAD model from the laser scanned data of real objects such as buildings.

Firstly, the model is lined up to the reference the coordinate system of Polywork software and co-planar surfaces with different distances and depths are defined according to model's façade, some planes are parallels to the model and others perpendiculars. Next, vertices on the corresponding surfaces are selected and projected to the co-planar surfaces previously defined. With this process of projection of points onto the co-planar surfaces, the edges are also defined. Last but not least, the mesh model is optimised to rectify the imperfection in the triangulation that existed in the initial model. Cross-sections are inserted through the co-planar surfaces to obtain the feature lines of the model. Lastly, the cross-sections are exported in dxf and then imported into Microstation. The result in Microstation is a regular 3D CAD model of the building.

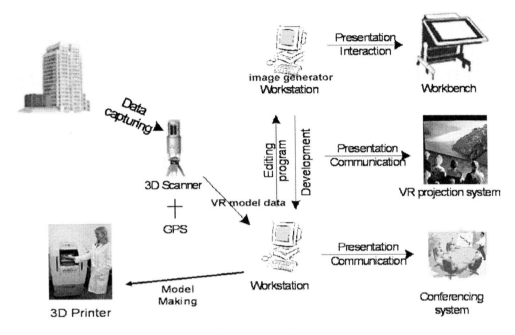

Figure 13. Integration of spatial data with 3D printer and the virtual environments.

2. Convert the CAD model to STL format: To establish consistency, the STL (Stereolithography) format has been adopted as the standard of the rapid prototyping industry. This format represents a three dimensional surface as an assembly of planar triangles. The STL file contains the coordinates of the vertices and the direction of the outward normal of each triangle. Because STL files use planar elements, they cannot represent curved surfaces exactly. Increasing the number of triangles improves the approximation, but at the cost of bigger file size. Large, complicated files require more time to pre-process and build, so the designer must balance accuracy with manageability to produce a useful STL file. Since the .stl format is universal, this process is identical for all of the rapid prototyping techniques.

3. Slice the STL file into thin cross-sectional layers: in the third step, a pre-processing program prepares the STL file to be built. Several programs are available, a most allow the user to adjust the size, location and orientation of the model. The pre-processing software slices the STL model into a number of layers from 0.001 mm to 07 mm thick, depending on the build technique. The program may also generate an auxiliary structure to support the model during the build. Supports are useful for delicate features such as overhangs, internal cavities and thin-walled sections.

4. Construct the model one layer on top another: The forth step is the actual construction of the part. Using one of several techniques, rapid prototyping machines build one layer at a time from polymers, gypsy or powdered metal.

5. Clean and finish the model: The final step is post processing. This involved removing the prototype from the machine and detaching any supports. Some photosensitive materials need to be fully cured before use. Prototypes may also require minor cleaning and surface treatment.

This innovation is significant because it has potential to solve the problems that are always been associated with design and construction of existing buildings for reuse goals. For instance, in regard to cultural heritage, many historical sites are slowly deteriorating due to exposure to the elements. Although remediation efforts can reduce the rate of destruction, a digital model of the site will preserve it indefinitely. Models of historical artefacts also allow scientists to study the objects in new ways. For example, archaeologists can measure artefacts in non contact fashion, which is useful for fragile objects. Also digital models allow objects to be studied remotely, saving time and travel expenses and enabling more archaeologists to study them.

Figures below show some examples of rapid prototyping for buildings using laser scanner data.

In the above figure, the first picture labeled 1 is the point cloud raw data captured by the laser scanner. The second image is the polygonal mesh VR model produced from the raw scan data in the Polyworks modeler, the third image is the CAD model extracted from the VR model in the second image. The fourth image is the solid model in STL format (Standard Template Library) for 3D printing to produce the physical model. Figure 15 shows the picture of physical model produced.

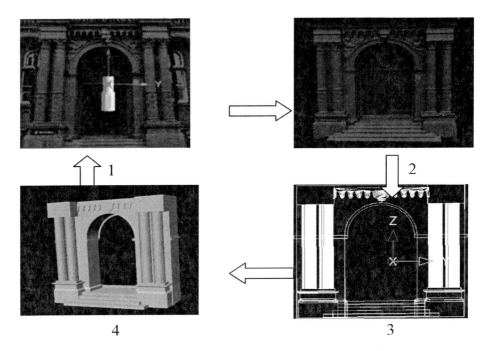

Figure 14. the process of creating physical models from digital laser scanned data.

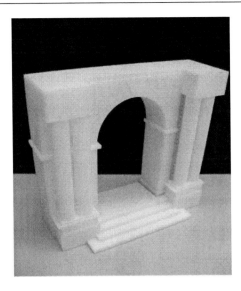

Figure 15. Printed physical model of the peel building in figure 14.

In regard to civil engineering, modelling from reality offers an efficient alternative to surveying. For example, bridges need to be surveyed periodically to determine whether significant settling or other movement has occurred. A 3D model of the site allows engineers to make the same measurements with equivalent accuracy in a fraction of the time. With regard to reverse and rapid prototyping, modelling from reality can be used to reverse engineer manufactured parts for which a CAD model is unavailable or never existed. Modelling from reality "imports" the prototype into a computer, creating a digital model that can be edited with a CAD program. Regarding architecture and construction, architects frequently need to determine the "as built" plans of building or other structures such as industrial plants, bridges, and tunnels. A 3D model of a site can be used to verify that it was constructed according to specifications. When preparing for a renovation or a plant upgrade, the original architectural plans might be inaccurate, if they exist at all. A 3D model allows architect to plan a renovation and to test out various construction options.

Overall, this advanced digital mapping tools and technologies are enablers for effective e-planning, consultation and communication of users' views during the planning, design, construction and lifecycle process of built and human environments. The regeneration and transformation of cities from the industrial age (unsustainable) to the knowledge age (sustainable) is essentially a 'whole life cycle' process consisting of; planning, development, operation, reuse and renewal. In order to enhance the implementation of build and human environment solutions during the regeneration and transformation of cities, advanced digital applications can have a significant impact.

Within the built environment, the use of the 3D laser scanner enables digital documentation of buildings, sites and physical objects for reconstruction and restoration. The use of the 3D scanner in combination with the 3D printer provides the transformation of digital data from the captured CAD model back to a physical model at an appropriate scale – reverse prototyping. The use of these technologies is key enablers to the creation of new approaches to the 'Whole Life Cycle' process within the built and human environment for the 21st century.

The VR environment also is developed using the laser scanner technology which also provides data models in different formats including the Virtual Reality Modelling Language (VRML). Figure 8 below shows demonstration in the VR projection system as an example of VR environment or alternatively it can be made available online for distant users to interact with the models. The research is of potential benefits and practical applications to the construction industry and professions. It will provide a better support for evaluation and visualisation of building maintenance works so that informed policies can be effectively targeted. It will benefit construction companies, facility and estate managers, and all those concerned with building maintenance issues.

The ultimate beneficiaries of this work will be professionals and stakeholders of the construction industry involved in the stages depicted in the picture below:

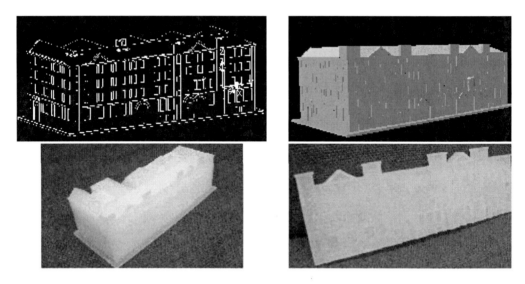

Figure 16. Polygonal VR mesh models developed from the raw scan data and CAD Models extracted from the VR models.

Figure 17. Peel building data capture by laser scanner is presented in VR Projection System.

Figure 18. in the image on the left, a view from the rear of the historical building and in the image on the right a view from the front of the same historical building (http://vp.salford.ac.uk/stteilos/).

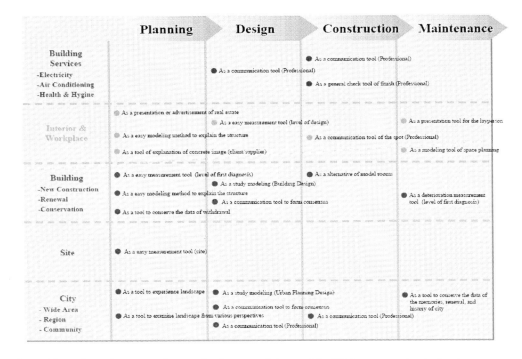

Figure 19. envisaged stages and areas for the potential use of the 3D laser scanning technology.

6. A WAY FORWARD TO BETTER IMPLEMENTATION TOWARDS BIM (BUILDING INFORMATION MODELLING)

Building Information Modeling (BIM) is the term used to describe a range of discipline-specific software applications that support all phases of the project lifecycle from conceptual design and construction documentation, to coordination and construction, and throughout ongoing facility management, maintenance, and operations. BIM is an integrated 3D digital description of a building, its site and related geographic information system (GIS) context. A BIM comprises individual building, site or GIS objects with attributes that define their detailed description and relationships that specify the nature of the context with other objects. BIM is called a rich model because all objects in it have properties and relationships and this information can be for data mining to develop simulations or calculations using the model data (Ballesty, 2007).

The principal difference between BIM and 2D CAD is that the latter describes a building by 2D drawings such as plans, sections, and elevations. Editing one of these views requires that all other views must be checked an updated, an error-prone process that is one of the major causes of poor documentation today. In addition, the data in these 2D drawings are graphical entities only such as lines, arcs and circles, in contrast to the intelligent contextual semantic of BIM models, elements and systems such as spaces, walls, beams and piles (Ballesty, 2007). The generic attributes of BIM are listed below:

- Robust geometry: objects are described by faithful and accurate geometry that is measurable.
- Comprehensive and extensible object properties that expand the meaning of the object. Objects in the model either have some predefined properties or the IFC specification allows for the assignment of any number of user or project specific properties are richly described with items such as a manufacturer's product code or cost or date of last service.
- Semantic richness: the model provides for many types of relationships that can be accessed for analysis and simulation.
- Integrated information: the model holds all information in a single repository ensuring consistency accuracy and accessibility of data.
- Lifecycle support: the model definition supports data over the complete facility lifecycle from conception to demolition, for example, client requirements data such as room areas or environmental performance can be compared with as designed, as built or as performing data

The key benefits of BIM is its accurate geometrical representation of the parts of a building in an integrated data environment are listed below (Ballesty, 2007)

- Faster and more effective processes - information is more easily shared can be value added and reused.
- Better design – building proposals can be rigorously analysed, simulations can be performed quickly and performance benchmarked, enabling improved and innovative solutions
- Controlled whole life costs and environmental data – environmental performance is more predictable, lifecycle costs are understood.
- Better production quality - documentation output is flexible and exploits automation.
- Automated assembly – digital product data can be exploited in downstream processes and manufacturing
- Better customer service – proposals are understood through accurate visualisation
- Lifecycle data – requirements, design, construction and operational information can be used for, for example, facilities management.
- Integration of planning and implementation processes – government, industry, and manufacturers have a common data protocol
- Ultimately, a more effective and competitive industry and long term sustainable regeneration projects

Interoperability is defined as the seamless sharing of building data between multiple applications over any or all applications (or disciplines) over any or all lifecycle phases of a building's development. Although BIM may be considered as an independent concept, in practice, the business benefits of BIM are dependent on the shared utilisation and value added creation of integrated model data.

To access the model data therefore requires an information protocol, and although several vendors have their own proprietary database formats, the only open global standard are IFC (Industry Foundation Classes) that published by the international Alliance for interoperability.

There have been many researches on shape representation and retrieval at lower level, and semantic retrieval on colour image recently (Zhang and Lu, 2004). To semantically annotate a shape database can be very difficult. However, if the objects are the CAD graphics as illustrated in figure 16, it is possible to annotate it automatically. There are efficient techniques available in the literature. However, a simply way is to design some templates representing each type of objects, and then mapping those new objects into templates. If there is already an annotated object database, they can be used to approximate new objects to be created.

The complexity of annotation depends on how large it is and the nature of projects. If a project is new and not large, newly created objects can be manually annotated and stored into database, so that the annotated and stored objects may be reused in the project. That is, whenever an object is necessary later in the project, it can be simply found in the database and manipulated for new situation. If a required object cannot be found in the database, it can be created and annotated before storing into the database. However, if there is a large amount of objects to be defined and annotated, which also need to be stored into the database, these objects can be defined with some object recognition techniques such as the contour based shape descriptors, like elongation, compactness, Fourier descriptors etc, normally those descriptors are size, translation and rotation invariant (Zhang and Lu, 2004).

The object-oriented CAD modelling approach in the research utilised the Microstation Triforma software, which employs the building information modelling concept. All information about a building (or at least as much as possible) is recorded in a 3 dimensional model. Traditionally a given door in a building would be drawn in at least three or four places (plan, building elevation, building section, interior elevation, etc). In the Triforma building information modelling, it is constructed once and these various drawings are later extracted automatically. It requires building objects that are defined, edited and stored in the triforma library.

Since the entire process is not fully automated, manual interaction with the model is needed for the time being. For example, individual entities from the Polywork software are achieved by many subsequent exports of group of cross-section that represent a building entity. Furthermore, instead of only assigning part attributes for the sample entities, the former is assigned to each entity manually since there is no search engine embedded into the process yet. However, once a search engine is adopted, the whole process can be fully automated for OO CAD modelling for the entire process in figure 20.

The building objects are called "parts" in the Triforma software. Attribute information is available in addition to geometrical information for all elements. The individual entities in a CAD model are manually selected and defined as triforma part elements, which are already available in triforma. However, this triforma part library can also be enriched with new definitions of part elements and stored in the relevant category of the triforma library.

TriForma Parts are stored in Families, which are in turn stored in one or more .xml files. The location of these .xml files is determined by the configuration variable TFDIR_PART, as can be seen under Workspace > Configuration > TriForma > Parts Libraries.

The Parts libraries can be accessed via TriForma > Dataset > Parts, in the Dataset Explorer. Specific information about using the Dataset Explorer can be found in TriForma's help system. This Part .xml files can also be edited in an XML editor. A Triforma Part isn't a physical 3D element, a Part is a collection of attributes or database entries, which can be attached to any 3D element in a file. The Part information can define symbology, such as colour, line weight and level, which can be set when an element is placed in the file with that part attached. The Part also contains information that is used when 2D section drawings are generated from the model, for example symbology to be used for elements in the forward view, the section cut, and the reverse view. Pattern, hatching and centerline symbology for the section cut is also defined by the Part definition. Rendering materials and Quantities information is also stored in the Part definition.

Contour shape techniques only exploit shape boundary information. There are generally two types of very different approaches for contour shape modeling: continuous approach and discrete approach. Continuous approaches do not divide shape into subparts; usually a feature vector derived from the integral boundary is used to describe the shape. The measure of shape similarity is usually a metric distance between the acquired feature vectors. Discrete approaches break the shape boundary into segments; called primitives using a particular criterion (Zhang and Lu, 2004). Discrete approaches differ in the selection of primitives and the organisation of the primitives for shape representation. Common methods of boundary decomposition are based on polygonal approximation, curvature decomposition and curve fitting (Pavlidis, 1982). The result is encoded into a string of the general form:

$S=s_1, s_2, s_3....s_n,$ where s_i may be an element of a chain code, a side of a polygon, a quadratic arc, a spline, etc. s_i may contain a number of attributes like length, average curvature, maximal curvature, bending energy, orientation, etc. The string can be used directly for description (Pavlidis, 1982).

According to the approach in figure 20, a 3D element will be attached with a suitable part attributes in Triforma as a sample. The Pattern matching software to be developed will search the whole model with the defined constraints and thresholds for the sample to find the similar ones. Once found, the same part attributes will be assigned these new copies of the sample.

Shape representation and description techniques can be classified into two class of method: contour based methods and region based methods. The classification is based on whether shape features are extracted from the contour only or extracted from the whole shape region. Under each class, the different methods and further divided into structural and global approaches. This sub-class is based on whether the shape is represented as a whole or represented by segments/sections (primitives).

Since contour based techniques only exploit shape boundary information, applying a contour based pattern matching technique is more suitable for this case. The contour based techniques encompass a variety of different shape descriptors such as area, circularity (perimeter^2 /area), eccentricity (length of major axis/length of minor axis), major axis orientation, and bending energy, convexity, ratio, of principle axis, circular variance and elliptic variance (Yong et al, 1974), (Peura et al, 1997). These descriptors can be used in combination or individually depending on the pattern matching condition. Added to these, simple contour based shape descriptors, more complex shape descriptors can also be implemented such as correspondence-based shape matching (Hausdorff distance, and Euclidean distance), shape signature (centroidal profile, complex coordinates, centroid distance, tangent angle, cumulative angle, curvature, area and chord-length), Boundary moments, Elastic matching, stochastic method, scale space method, spectral transform (Fourier descriptors, wavelet descriptors) (Zhang and Lu, 2003).

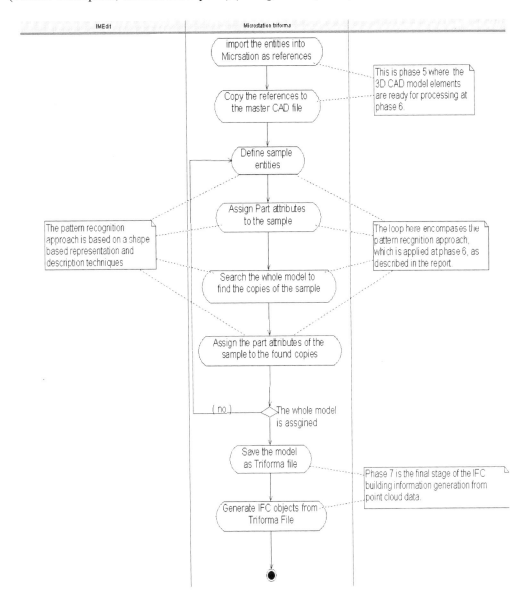

Figure 20. object oriented CAD modelling.

Global contour shape techniques take the whole shape contour as the shape representation. For the shape description, there is always a trade-off between accuracy and efficiency. On the one hand, shape should be described as accurately as possible; on the other hand, a shape description should be as compact as possible to simplify indexing and retrieval. Simple global shape descriptors are compact, however, they are very inaccurate shape descriptors to create practical shape descriptors. They need to be combined with other shape descriptors to create practical shape descriptors. Elastic matching and wavelet methods are complex to implement and match. Autoregressive methods involve matrix operations which are expensive and it is difficult to associate autoregressive descriptors with any physical meaning. The implementation and matching of curvature scale space (CSS) are also complex. However, perceptually meaningful and compact features are appealing for shape description. Fourier descriptor is simple to implement and involves less computation by either using fast Fourier transform or using truncated Fourier transform computation. The resulting descriptor is also compact and the matching is very simple. Compared with CSS, Fourier descriptor is simpler to compute and more robust. Boundary moment descriptor is similar to Fourier descriptor and is east to acquire. However, unlike Fourier descriptor, only the few lower order moment descriptors have physical interpretation. After all the entities are defined as triforma part objects, the file can be saved as a triforma file and it is now ready for IFC generation from triforma parts.

Based on the logic above, the pattern matcher in figure 20 will access the triforma library with the criteria in hand to do search and match. Two different type of matching can be done such as exact pattern matching and approximate pattern matching. Exact pattern matching consists of finding the exact pattern looked for. In the case of approximate pattern matching, it is generalisation of the pattern looked for and a determined number of differences between the pattern looked for and the objects found in the library is allowed.

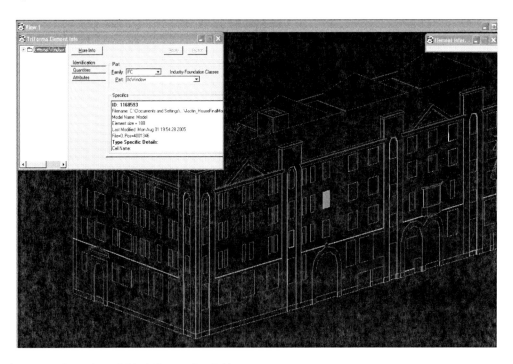

Figure 21. The IFC model in Microstation Triforma.

As a result of various matching processes, object recognition can be worked out for the interested building frames in the 3D CAD model. Attributes of the objects matched in the library will be assigned to the building frame in the CAD model, which result in the building frames to be building objects defined. Subsequently, object-oriented (OO) CAD model will be obtained. This OO CAD model will be mapped into IFC schema to save and store the model in IFC data model into the nD modelling database. After all the entities are defined as triforma part objects, the file can be saved as a triforma file and it is now ready for IFC generation from triforma parts. The image in figure 13 shows the IFC model of the Jactin house in the Microstation triforma environment that has IFC 2X plug-in installed.

9. REFERENCES

Arayici, Y., (2007). "An Approach for Real World Data Modelling with the 3D Terrestrial Laser Scanner for Built Environment" *The International Journal of Automation in Construction, Vol 16*, Issue 6, 816-829, September 2007.

Arayici Y., Hamilton, A., Gamito, P. (2006). "Modelling 3D Scanned data to Visualise and Analyse the Built Environment for Regeneration", *Surveying and Built Environment, Vol 17(2)*, 7-28 December 2006, ISSN1816-9554.

Arayici Y., Hamilton A. (2005a). *"Modelling 3D Scanned data to Visualise and Analyse the Built Environment"*, 9th International Conference of Information Visualisation, London, 6,7,8 July 2005.

Arayici, Y., Hamilton, A. (2005b). "Built Environment Reverse and Forward Prototyping", SCRI Symposium, *University of Salford, Greater Manchester, April 12th 2005*.

Arayici, Y., Hamilton, A., Gamito, P., Albergaria, G., (2004). The Scope in the INTELCITIES Project for the Use of the 3D Laser Scanner, in the Proceeding of ECT2004: *The Fourth International Conference on Engineering Computational Technology*, 7-9 September 2004, Lisbon, Portugal. ISBN 0948749962.

Ballesty S, (2007). "Building Information Modelling for Facilities Management", project report by Co-operative Research Centre (CRC) for Construction Innovation, Queensland, Australia, 2007, www.consruction-innovation.info, ISBN 978-0-9775282-2-6.

Bornaz L., Lingua A., Rinaudo F. (2002). *Engineering and environmental applications of laser scanner techniques*, International Archives of Photogrammetry and Remote Sensing. ISSN: 1682-1777. Volume XXXIV part 3A+BCD-ROM.

Brasington, J., Rumsby, B.T., McVey, R.A. 2000. Monitoring and modelling morphological change in a braided gravel-bed river using high-resolution GPS-based survey. *Earth Surface Processes and Landforms*, 25, 973-990

Bryan, P., (2003), An Addendum to the Metric Survey Specifications for English Heritage, The Metric Survey Team, YORK.

Fuller, I. C., Large, A. R. G., Heritage, G. L., Milan, D. J., Charlton, M. E. (2005). Derivation of reach-scale sediment transfers in the River Coquet, Northumberland, UK. In M. Blum., S. Marriott, S. Leclair, (eds) Fluvial Sedimentology VII, *Special Publication, International Association of Sedimentologists*, 35, 61-74.

Gilbert, R., Stevenson, D., Giradet, H., and Stren, R. (1996). Making cities work: *The role of Local Authorities in the Urban Environment*, Earthscan, London.

Hetherington D, Heritage G. L., and Milan D. J. (2005). Daily fine sediment dynamics on an active Alpine glacier outwash plain, Proceedings of symposium S1 (sediment budgets) held during the Seventh IAHS Scientific Assembly at Foz do Iguaçu, Brazil, April 2005. IAHS Publ. 291.

Pavlidis, T., Algorithms for Graphiques and Image Processing, Computer Science Press, Rockville, MD, 1982, p143.

Peura et al. (1997). Polywork User Guide, www.innovmetric.com RiSCAN PRO User Guide, www.riegl.com.

Schofield, W. (2001). "Engineering Surveying 5th Edition: *Theory and Examination Problems for Students*".Yong et al, 1974, Zhang and Lu, 2003

Zhang, D., Lu G., "Review of shape representation and description techniques", Pattern Recognition Journal 37 (2004) 1-19.

In: Built Environment: Design, Management and Applications   ISBN: 978-1-60876-915-5
Editor: Paul S. Geller, pp. 121-145   © 2010 Nova Science Publishers, Inc.

Chapter 5

ANTHROPOCENTRIC BIOCYBERNETIC APPROACHES TO ARCHITECTURAL ANALYSIS: NEW METHODS FOR INVESTIGATING THE BUILT ENVIRONMENT

Stephan K. Chalup and Michael J. Ostwald
The University of Newcastle, Australia

ABSTRACT

Anthropocentric biocybernetic computing uses machine learning to provide a heightened level of understanding of human perceptions of complex systems. In this chapter, such processes are the catalyst for the development of two new methods for architectural and urban analysis and the revision and expansion of a third, existing method. The two new methods rely on, respectively, the 'Hough transform' algorithm to model line segmentation in images of buildings and a newly authored program to investigate facial pareidolia in façades. The third method is a computational variation of the fractal analysis technique for the classification and review of urban skylines.

Anthropocentric biocybernetic methods are ideal for the analysis of the built environment because architecture evokes a complex emotional and perceptual response in viewers and building users. While architectural and urban spaces may be readily measured and understood in terms of their formal characteristics, the phenomenal and semiotic qualities of the built environment have rarely been investigated using computational means. This chapter responds to this situation by providing examples of new techniques that are ideal for expanding the computational analysis of architecture beyond the simple analysis of form.

INTRODUCTION

The built environment has traditionally posed a particular challenge to scientific researchers and to scholars seeking objective or quantifiable information about the physical and temporal characteristics of constructed space. This is because urban spaces and buildings are not only physical, measurable forms, but they are also complex cultural and social artifacts that have both phenomenological and symbolic qualities. Only the first of these, form, has traditionally been the subject of mathematical or computational analysis. However, recent developments in anthropocentric biocybernetic computing (ABC) have begun to suggest new ways of expanding the study of architectural form to encompass more perceptual and experiential analytical possibilities (Chalup and Ostwald, 2009).

This chapter describes two new methods for the computational analysis of architectural and urban form and a new variation on an existing method. The new methods are focused on geometric perception and facial pareidolia and the revised method is for the fractal analysis of urban boundaries. All three are significant because they expand the traditional focus of architectural computational research away from studies of form and spatial hierarchy and towards an improved modeling of how humans perceive the constructed world. In order to understand the significance of this research and these new methods it is first necessary to describe the traditional division that has existed in architectural scholarship between studies of formal, phenomenal and semiotic qualities in buildings.

Architects use the word 'form' to describe the shape or volume of a building or space. Form is one of the few characteristics of architecture or urban space that is relatively straightforward to record. For this reason, several computational methods for the analysis of architecture, including space syntax (Hillier and Hanson, 1984) and shape grammars (Stiny, 1975; Steadman, 1983), have been developed from studies of exterior form and interior planning. However, the relationship that exists between the form of a building and the individual user or viewer is more difficult to assess. Architecture is necessarily experienced through the senses or sight, touch, hearing and smell. This is why the concept of phenomenology has become so important in architectural analysis.

Derived from the philosophies of Husserl and Heidegger, phenomenology in architecture is concerned with the sensorial properties of buildings; the experience of form, texture and materiality (Norberg-Shultz, 1980). Phenomenological readings of buildings have, in the past, typically relied on descriptive or qualitative analytical methods to suggest the way in which a building may shape a person's behaviour, memory or sense of place (Bachelard, 1958). Such works seek to explain or replicate the way in which people react to buildings. While phenomenological responses to architecture tend to be either subconscious or emotional, the final analytical dimension is more social and linguistic in nature.

The symbolic properties of a building or space are a product of a shared understanding of the meaning of certain forms and materials. This way of viewing architecture has its origins in semiotics and the concept that the city is a coded system of communication that people with a similar social or cultural background can understand (Eco, 1980). Thus, for example, in European society the silhouette of a tall spire surmounted by a cross will typically communicate to the viewer the idea of a place of Christian worship. Throughout history, architecture has served as a rich medium for communication and the messages conveyed in built form can be literal and illustrative (through, say, pictographic ornamentation) or symbolic and inferred (often by way of iconic representation or use of materials) (Jencks, 1980). The study of buildings and spaces from a semiotic perspective has a long tradition and has, most recently, begun to adopt social sciences methods to support the analysis (Krampen, 1979; Stamps, 2000).

While all three of these dimensions in architectural analysis – the formal, experiential and symbolic – are interconnected, only the first of these has been successfully and repeatedly investigated using computational methods. However, recent advances in machine learning have begun to produce software tools and methods that have the potential to expand the analysis of form in such a way that it may support a heightened understanding of its social and cultural impact. In particular, a range of anthropocentric biocybernetic techniques have recently been developed for the purpose of providing a greater understanding of the way in which people perceive complex environments. While these techniques are still being refined, several concepts and programs developed in this research have just begun to be applied to the analysis and evaluation of buildings and urban spaces.

There are two broad areas in which anthropocentric biocybernetic approaches suggest new ways of understanding the built environment. The first is concerned with improved understanding of visual material. The computer is ideally suited to examining images and rapidly developing a detailed understanding of the information contained in them. New techniques for improved data extraction and analysis include line and area segmentation, perimeter analysis and centroid mapping. The second area where advances have been made is in perceptual modeling. Humans do not process data in the same way that a machine does. Instead, the human mind seeks patterns to assist way-finding and differentiate spatial structures or building forms. This chapter presents three computational techniques that seek to model the way humans process and understand information about the built environment. While each of these methods is derived from studies of architectural and urban form, the general aim of ABC is to gain a better understanding of the mechanisms underlying human information processing, including those that are relevant to the perception of aesthetics in the built environment.

There are several important dimensions to ABC including that it is human centered, it is concerned with complete systems and it is reliant on computational techniques. In the first instance, the word anthropocentric in ABC does not imply that its purpose is to explain the world from a human-centered perspective. Rather, the purpose is to gain new insights into human functionality, development and survival. ABC regards humans as complex information processing systems and investigates this information on different levels, including the cell, body, language and interaction with the environment. Anthropocentric concerns include modelling and analyzing the impact of design perception on social and emotional factors. The second component of ABC, biocybernetics, is a part of systems biology; a field which focuses on living systems and their environment. Like cybernetics (Wiener, 1948) the biocybernetic approach is holistic and integrates a broad scale of information processing, from sub-cellular levels to ecosystems. The aim of biocybernetics is to gain a new and deeper understanding of how complex organisms function by analyzing the whole organism and its interaction with the environment. Finally, ABC is intricately reliant on concepts derived from machine learning to allow a computational system to be iteratively refined until it can model almost human levels of performance (Mitchell, 1997; Bishop, 2006).

In order to simulate the way in which humans process visual sensory signals associated with design, a possible ABC strategy is to implement relevant physiologically grounded perceptual mechanisms in an artificial system and to analyze how this system performs on selected examples. This approach is associated with several relatively new research areas in artificial intelligence (AI) (Russell and Norvig, 2003; Luger, 2009). These areas include, affective computing (Picard, 1997), artificial life (Bedau, 2003; Johnston, 2008), bionics (Passino, 2005), computational models of emotion (Palensky and Barnard, 2009), human robot-interaction (Dautenhahn, 2007), and intelligent autonomous agent design (Braitenberg, 1984; Brooks, 1990; Brooks, 1999; Albus and Meystel, 2001; Doya, 2002; Siegwart and

Nourbakhsh, 2004; Choset *et al.*, 2005; Thrun, Burgard and Fox, 2005; Chalup, Murch and Quinlan, 2007; Siciliano and Khatib, 2008; Dietrich, Fodor, Zucker and Bruckner, 2009).

For the present research, when characterizing relevant computational concepts for design perception within the framework of ABC, physiologically grounded mechanisms were sought. This means that the visual design features of the built environment that are either believed to, or are proven to, cause measurable physiological impact on the human emotional system are the focus. Of particular interest are those mechanisms that appear to occur automatically or unconsciously. The following section briefly reviews selected results from the broader field of cognitive neuroscience that have motivated the implementation of the computational techniques presented in this chapter. Thereafter, the remainder of this chapter consists of three parts each dedicated to a new analytical method or a novel variation of an existing method. The first of these addresses Hough transform (HT)-based image analysis and the perception of geometry. The second and third parts describe, respectively, facial expression in architectural façades and fractal dimension perception in relation to urban image analysis.

PHYSIOLOGICALLY GROUNDED MEASURES IN VISUAL PERCEPTION

The present chapter's approach is motivated by three physiologically grounded properties of human visual perception. The first property is linked to mechanisms of geometric processing that occur at several levels of the human visual system. The second property is that the recognition of facial expressions is one of the most highly developed perceptual skills in humans and one for which designated brain regions have evolved or developed (Kanwisher, McDermott and Chun, 1997; Farah and Aguirre, 1999; George, Driver and Dolan, 2001; Pelphrey, Singerman, Allison and McCarthy, 2003; Engell and Haxby, 2007). The third property is based on the hypothesis that the fractal dimension (Mandelbrot, 1983; Barnsley, 1988) of a perceived image has physiologically measurable effects on the human body (Taylor *et al.*, 2005; Taylor, 2006; Hagerhall *et al.*, 2008). The following three paragraphs briefly review key results in cognitive neuroscience that motivate the computational approaches to design analysis which are described in the remaining sections of this chapter.

Line Detection

The human visual cortex is organised in such a way that local signals received at the retina are processed through a hierarchy of neuron layers into higher-level representations that are multimodal, abstract and holistic (Grill-Spector and Malach, 2004). Retinal ganglion cells are able to report the direction of movement (Barlow, Hill and Levick, 1964; Barlow and Levick, 1965; Masland, 2004). An important discovery by Hubel and Wiesel (1962; 1968) was that so-called simple cells in the striate cortex (visual area V1) of cats and monkeys respond to oriented bars/edges in their receptive fields. Later it was found that these simple cells are not only sensitive to the orientation and location of bars, but also to the spatial frequency of visual stimuli (R. L. De Valois and K. K. De Valois, 1988). Engel (2005) extrapolated from this research to suggest the presence of orientation-selective adaptation in human visual area V1. These results are supported by Larsson, Landy and Heeger (2005), who also found that second-order stimulus orientation can emerge through continued processing in multiple visual areas, including V1 and V2.

These discoveries of modern neuroscience have parallels in Gestalt psychology (Wertheimer, 1923; Koffka, 1935; Westheimer, 1999) where one of the fundamental assumptions is that visual perception prefers continuous over broken transitions or that there are neurons sensitive to collinearity (Spillmann and Ehrenstein, 2004), and that the human perceptual system automatically supplements missing parts of structures, such as lines, to connect them into coherent wholes. The perception and processing of line directions is, therefore, a fundamental concept of the human brain's visual information processing system and it is regarded as essential for form recognition.

Face Processing

Research in cognitive and brain science has found that large areas of the brain are concerned with recognising faces and with communicating the emotional content of facial expressions (Vuilleumier, Armony, Driver and Dolan, 2003; Engell and Haxby, 2007; Vuilleumier, 2007). Recent results show that the processing of facial information is a special task that, in contrast to information about other objects, can be processed subcortically, non-consciously and independently of visual attention (Johnson, 2005; Finkbeiner and Palermo, 2009). The human visual system is therefore optimised for associating abstract, face-like patterns, such as 'smileys', with emotions of corresponding human facial expressions. This phenomenon is also responsible for the pareidolia effect wherein the human mind is drawn to intuitively find face-like shapes and forms in a wide range of natural and synthetic systems including the façades of buildings. This may explain the reason why some architectural façades trigger emotional responses in observers and further suggests that it may be possible to determine whether or not a building will promote pareidolia and, if so, what the response will be (Chalup, Hong and Ostwald, 2008; 2009).

Fractal Dimension Perception

The final method being proposed in the present chapter is reliant on the growing awareness that certain levels of characteristic complexity that occur in nature evoke emotional responses. Taylor *et al.* (2005) report experiments using skin-conductance measurements (Wise and Rosenberg, 1986) in which twenty-four subjects seated in isolated cabins performed a sequence of stress-inducing mental tasks while being exposed to three images showing patterns of low, medium and high fractal dimension. The results indicate that the physiological stress-levels were lowest when patterns of mid-range fractal dimensions were shown (Taylor *et al.*, 2005; Taylor, 2006). A recent study of Hagerhall *et al.* (2008) corroborated these findings; 31 subjects' quantitative electroencephalograms (qEEGs) were continuously monitored while the subjects viewed silhouette patterns of different fractal dimensions on a computer screen. A statistical analysis of the subjects' qEEG recordings indicated that visual stimuli with a fractal dimension of $D=1.3$ are unique. They induced the highest alpha response in the frontal areas of the subjects' brains and at the same time, they had maximal effect on beta waves in the parietal areas. Hagerhall *et al.* (2008) suggest that the observed alpha response indicates a relaxing and restorative effect while its interaction with the beta response would still require further investigation. In summary, these studies support the hypothesis that perception of patterns within a mid-range fractal dimension (say $D=1.3$) causes the least stress for the observer. This result suggests that the fractal dimension of the built environment may be used to determine the level of perceptual wellbeing a person may experience as a result of the visual properties of a space.

HOUGH TRANSFORM-BASED CLUSTERING OF FAÇADES

Horizontal and vertical lines play an important role in the way in which people perceive and evaluate house façades and streetscapes. Our hypothesis is that the visual experience gained through perception of thousands of line distributions can be represented by a non-linear geometric structure which we previously introduced and called a 'streetmanifold' (Chalup, Clement, Ostwald and Tucker, 2006; Chalup, Clement, Marshall, Tucker and Ostwald, 2007; Chalup, Clement, Tucker and Ostwald, 2007).

Geometrically, a line is a set of points $x = (x_1, x_2) \in \mathbf{R}^2$ which can formally be described by $\{x \in \mathbf{R}^2; \ [\cos\phi, \sin\phi] \cdot x - b = 0\}$, where $\phi \in [0, 360^\circ)$ controls the slope of the line's normal vector and $b \in \mathbf{R}$ is its perpendicular distance from the origin. The Hough transform (HT) for lines takes a global view of an image to determine edge directions or lines, including interrupted or virtual lines (Hough, 1962; Illingworth and Kittler, 1988; Shapiro and Stockman, 2001; Gonzalez and Woods, 2002; Forsyth and Ponce, 2003; Shapiro, 2006). The HT associates each image with an array of parameters $(\phi, b) \in [0, 360^\circ) \times \mathbf{R}$ where each point corresponds to a line in the image. This array is often called 'accumulator array' or 'Hough array'. An example of a Hough array showing the line distribution extracted from a digital image of a house façade is shown in Figure 1.

 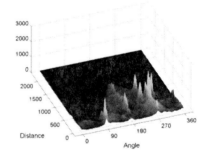

Figure 1. *Left:* lines extracted by Hough transform overlaid on an image of a house façade. *Right:* corresponding Hough array. The height of the dis-played peaks is the intensity of the corresponding lines in the image on the left. In this example peaks at angles 90°, 180°, 270°, and 360° represent dominant lines (Chalup and Ostwald, 2009).

The concept of manifolds which has developed in mathematics over more than 100 years is now central to geometry and topology (e.g. Spivac, 1979; Ito, 1987; James, 1999). Simple two-dimensional examples of manifolds are the surface of a sphere or a torus. Manifolds can represent high-dimensional, non-linear data concepts in a general and precise way. The mathematical definition and properties of topological, differentiable or Riemannian manifolds appear to reflect some of the terminology, for example continuity, smoothness, coherence or connectivity, which was used by Gestalt theorists in the context of vision processing (Wertheimer, 1923; Koffka, 1935; Westheimer, 1999; Spillmann and Ehrenstein, 2004). They also proposed the occurrence of similarity-based clustering, driven by features such as colour, texture, size and form (Kanizsa, 1979; Palmer and Rock, 1994).

Manifold learning describes a set of methods for non-linear dimensionality reduction (NLDR). Burges (2005), and Lee and Verleysen (2007) give an overview of NLDR proposing that its aim is to take a set of high-dimensional data as input and return a low-dimensional representation of the essential geometric structure. This essential structure could be a low-dimensional manifold that was 'hidden in' or 'intrinsic' to the high-dimensional input data space. In the present study, the authors applied isometric feature mapping (Isomap), by Tenenbaum, de Silva and Langford (2000). Isomap can be regarded as a generalisation of classical multi-dimensional scaling (T. F. Cox and M. A. A. Cox, 2001) where the Euclidean distance between input data points is replaced by a graph approximation of their geodesic distances on the intrinsic manifold.

Chalup et al., (2006; 2007a; 2007b) introduced the concept of streetmanifolds to provide a holistic geometrical representation of the visual experience that can be gained through evaluation of a large set of house façades. Navigation within a streetmanifold corresponds to continuous non-linear morphing or interpolating between façade designs represented by the data set. Streetmanifolds are based on 'pairwise' distances between digital images of house façades and can be calculated as follows:

1. Take a large set of images of house façades.
2. Calculate their Hough arrays.
3. Calculate a distance matrix consisting of pairwise distances between the Hough arrays.
4. Apply Isomap (Tenenbaum et al., 2000).

Figure 2 shows an example of a streetmanifold. The displayed manifold is three-dimensional and was calculated with Isomap following the procedure outlined above. The third dimension is encoded in colour; details of the calculation are discussed in Chalup et al. (2007a). On the streetmanifold, similar houses were assembled in clusters. Figure 2 shows four example clusters. Clusters A and B are located at two extreme ends of the manifold; cluster A consists of narrow, two-storey houses and cluster B contains wide one-storey houses. Clusters C and D are more centrally located on the manifold and show houses with a relatively homogeneous distribution of horizontal and vertical lines. Cluster C is characterised by the fact that at least one tree is standing in front of all the houses. More results, including how to obtain proto-plans for new houses by interpolating between points on the streetmanifold, were discussed in Chalup et al. (2007b).

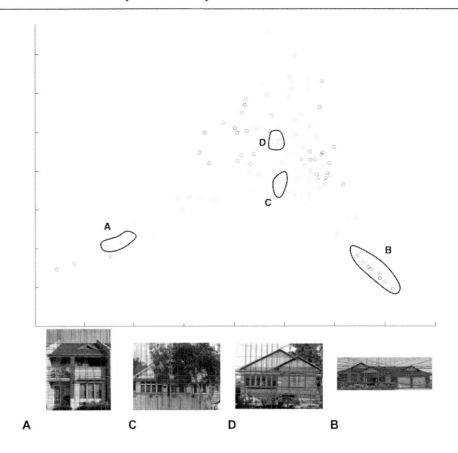

Figure 2. Four clusters of house façades and representative example images (A-D) found in the streetmanifold. Each small coloured circle represents one house and the colour encodes the third dimension (Chalup and Ostwald, 2009).

Biological evidence that local features are integrated by the brain into global shapes through processing in multiple visual pathways comes from fMRI studies in humans and monkeys (Kourtzi, Tolias, Altmann, Augath and Logothetis, 2003). In a similar way, the local features of line distributions, together with some additional features that are described further below, may lead to a sophisticated internal world-model of architectural visual experience.

FACIAL EXPRESSION CLASSIFICATION FOR FAÇADE ANALYSIS

In order to model aspects of the extraordinary face perception abilities of the human visual system for automated architectural design analysis, a two-stage approach was implemented in software. In the first stage, a test image of a house façade was scanned to detect face-like patterns. If a face-like pattern was detected, a box was fitted around it. In the second stage the detected face patterns were classified into eight different classes of facial-expressions of emotions.

Both stages of classification employed ν-support vector machines (ν-SVMs) with radial basis function kernel (Schölkopf, Smola, Williamson and Bartlett, 2000; Schölkopf, Platt, Shawe-Taylor, Smola and Williamson, 2001; Schölkopf and Smola, 2002) as implemented in the libsvm library (Chang and Lin, 2001). Platt's posterior probabilities for class membership output by the one-class SVM for face detection in stage 1 were interpreted as decision values that indicated the 'goodness' of a face-like pattern (Platt, 2000; H.-T. Lin, C.-J. Lin and Weng, 2003). After several series of pilot experiments, ν was set to 0.1 and the width parameter of the RBF kernel was left at libsvm's default value of 0.0025.

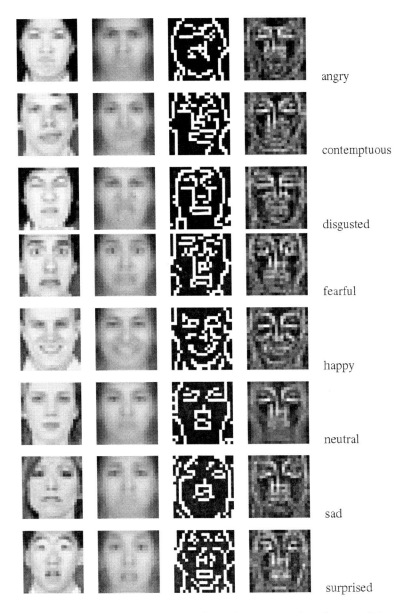

angry

contemptuous

disgusted

fearful

happy

neutral

sad

surprised

Figure 3. Training data in 22×22 pixel resolution; *first column:* examples of greyscale images; *second column:* for each emotion category, averages of all associated greyscale images in the training set; *third column:* Canny edge images; *fourth column:* averages of Canny edge images. The underlying images stem from the JACFEE and JACNeuF image data sets (© Ekman and Matsumoto, 1993).

Figure 4. Application of face detection and emotion classification to the image of Lenna; small (violet) boxes = 'surprised'; white boxes = 'happy'; larger (green) boxes = 'disgusted' (note Lenna's concave shoulder edge).

The face detection stage has five steps and is followed by facial expression classification in stage 2:

Stage 1: Face detection.
(a) Randomly select a box centre point c within the image.
(b) Select a random box size.
(c) Crop the image to extract the interior of the box generated with the random point c as centre. Re-scale the interior of the box to a 22 × 22 pixel resolution.
(d) Perform histogram equalisation and/or edge detection.
(e) Apply a one-class v-SVM classifier to decide if the box contains a face. Return the decision value.

Stage 2: *Facial expression classification using 8-class SVMs.*
(f) Apply an eight-class v-SVM classifier to determine to which of eight different facial expression classes the face in the box should be assigned.
(g) Assign a distinctive colour to the box frame to indicate the class label.

The face detection stage used one-class SVM classification for face detection where the output of the classifier, that is, the decision value indicated the probability that the tested box contains a face. Based on the decision values, a cloud of candidate solutions (highlighted yellow in Figures 4 to 7), which consisted of the centre points of boxes with the highest decision values, was generated.

Once a box that contained a face-like pattern was detected, stage 2 of the procedure was applied to decide to which facial expression class the face pattern belongs. Coloured pixels displayed within the yellow face clouds indicate faces associated with local peaks of high decision values. The colours were selected to indicate the associated facial expression classes, as explained further below.

Research into affect recognition is growing and a recent overview can be found in the work of Zeng, Pantic, Roisman and Huang (2008). Excellent results have also been obtained in multi-modal approaches that use visual and acoustic data (Wang and Guan, 2008). Due to the planned application in architectural image analysis the present chapter relied only on visual data.

The digital images for training the classifiers for face detection and facial expression classification consisted of 280 images of human faces which were taken from the research image data sets of Japanese and Caucasian Facial Expressions of Emotion (JACFEE) and Japanese and Caucasian Neutral Faces (JACNeuF) (© Ekman and Matsumoto, 1993). All images in the training set were cropped and resized to 22 × 22 pixels so that each of them showed a full individual frontal face in such a way that the inner eye corners of all faces appeared in exactly the same position (Figure 3). This normalisation step helped to reduce the false positive rate. Profiles and rotated views of faces were not taken into account.

For training of the multiclass v-SVM in stage 2, Ekman, Friesen and Hager's (2002) facial expression classification system (FACS)'s eight emotional states were used: anger, contempt, disgust, fear, happiness, neutral, sadness and surprise. Facial expression classes were colour coded via the face-enclosing boxes, which were detected in the first stage. The following colours were assigned to the eight emotional states:

- angry = red;
- contemptuous = orange;
- disgusted = green;
- fearful = black;
- happy = white/yellow;
- neutral = grey;
- sad = blue; and
- surprised = violet.

It should be noted that, in science, there is still an ongoing debate about the fundamental nature and understanding of emotions (Damasio, 1994; Panksepp, 1998; Barrett, 2006). Therefore, the representation of emotions by eight classes of facial expressions may be regarded as a compromise.

One example of how the face detection and facial expression classification system was applied is shown in Figure 4. It displays faceboxes overlaid on the well-known image processing test image of 'Lenna'. Boxes that most tightly fit Lenna's face were classified as 'surprised' (violet) or 'happy' (white). Some of the other boxes with high decision values were larger and were classified as 'disgusted' (green).

Another example, shown in Figure 5, demonstrates that the trained classifiers perform as desired on a test image composed of four face images from the Cohn–Kanade data set (Kanade, Cohn and Tian, 2000). Classification accuracies for facial expression classification in the training set were determined by ten-fold cross-validation. In order to determine which preprocessing steps deliver the best classification accuracies we compared the results obtained for greyscale-, Sobel- and Canny-filtered images, each of them with and without equalisation (Chalup et al., 2009b). The highest level of accuracy for correct classification was about 65% and was achieved when non-equalised greyscale images were used for training the SVMs. All four faces in Figure 5 were detected as dominant face patterns by the trained face detection module, except the bottom left face in the case of equalised greyscale filtering. For the equalised greyscale, Sobel- and Canny-filtered versions of the facial expression classification module consistently assigned the same sensible emotion classes to all detected faces. The top left face was classified as 'disgusted' (green), the top right face as 'angry' (red) and the bottom right face as 'happy' (white). Outcomes of processing of the bottom left face showed some instability between the different filtering options. The face was detected and was classified as 'neutral' with the Sobel and Canny filters. It was not detected with equalised greyscale as the input. In the case of Canny filtering, additional smaller faces were detected for the emotion categories 'sad' (blue), 'disgusted' (green) and 'surprised' (violet) (see Figure 5). The boxes associated with the latter two categories were so small that they only contained the mouth and the bottom part of the nose. This indicates that the classifier interpreted the nose openings as small 'eyes' above the mouth.

The yellow face clouds in Figure 5 also show that the desired face pattern was exactly detected as expected in the case of equalised greyscale with the exception of the bottom left image. For the result with Canny filtering the yellow face cloud was large and several local peaks were detected. Many of these can be regarded as false positives.

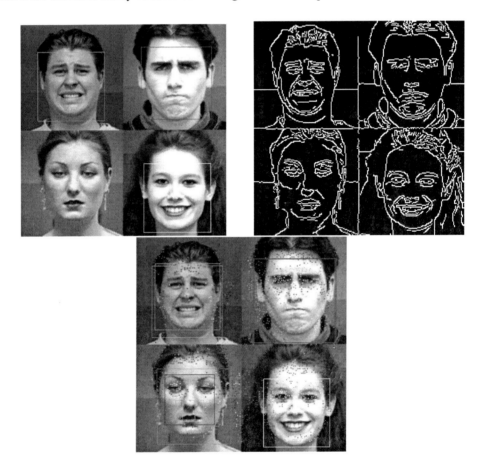

Figure 5. The trained SVMs for face detection and expression classification were applied to a squared test image which was assembled from four images that were taken from standard database of face images (Kanade *et al*, 2000) (© Jeffrey Cohn). *Top:* greyscale image with detected face boxes; *middle:* Canny edge image; *bottom:* Overlaid yellow face clouds and face boxes detected after Canny filtering (Chalup *et al.*, 2009b).

Figure 6. House with face-like pattern. Original (top), Canny edge image (middle) and detected face boxes (bottom). Grey ('neutral'), violet ('surprised') and black ('fearful') boxes were detected with high decision values (Chalup *et al.*, 2009b).

After the system was tuned and trained on an image database of 280 images of human faces, it was applied to images of selected house façades. It was then tested to see if the trained system would perform a kind of artificial pareidolia effect (Chalup *et al.*, 2009b) and recognise, as humans would, face-like patterns in house façades similar to the ones shown in Figures 6 and 7.

Using preprocessing and parameter settings (Canny edge filtering on non-equalised greyscale images and $v = 0.1$ for the SVMs) that had a slight tendency to generate false positives in face detection on human test images (bottom of Figure 5), the authors demonstrated that the system was also able to detect face-expression patterns within images of selected house façades. Most 'faces' detected in houses were very abstract or incomplete and often allowed the assignment of several different emotion categories depending on the choice of the centre point, and the box size.

The violet box in Figure 6 indicates that the section of the façade included in this box corresponds to a 'surprised' face. The back box, which contains a smaller fraction of the garage door, as a 'mouth', contains a 'fearful' face. In this example, all of the displayed boxes had a similarly high decision value. The large grey boxes indicate a 'neutral' facial expression.

Figure 7 shows two examples of house façades which contain several face-like patterns. Violet ('surprised'), green ('disgusted') and red ('angry') boxes were detected within the shown façades. The yellow face cloud contains several other local peaks of lower decision values. These are typical of our approach using Canny filtering which was applied to equalised face boxes in the examples shown in Figures 6 and 7.

The proposed machine learning approach using SVMs produced a statistical model based on the selected training data of human faces. Given the complexity of the human visual system and the still not fully understood interaction between evolutionary and learning processes that establish the extraordinary face recognition abilities of humans the chosen precise but simple statistical model without further constraints or assumptions seemed to be most appropriate. Pilot results as shown in Figures 6 and 7 reflect satisfactory simulation of the pareidolia effect (Chalup *et al.*, 2009b). Future research may further investigate our hypothesis that unconscious perception of facial expressions of emotions via pareidolia plays a significant role in design perception of general objects such as house façades.

Figure 7. Houses with many faces. *Top row:* original images; *middle row:* Canny edge images; *bottom row:* yellow face clouds and extracted face boxes of emotion categories 'anger' (red), 'surprise' (violet), 'disgust' (green), and in the sky 'sadness' (blue).

FRACTAL DIMENSION OF SKYLINES

The word 'fractal' can be used to describe point sets, for example, curves in space, which are highly irregular (Mandelbrot, 1983; Barnsley, 1988). There are many different ways to define and compute the fractal dimension of a set of points (Edgar, 1993). One possibility is based on the box-dimension,

$$D(S) = \lim_{\varepsilon \to 0}(\log(N_\varepsilon(S))/\log \varepsilon)$$
,

where S is a given set of points and $N_\varepsilon(S)$ is the number of boxes in an overlaid lattice of boxes with edge length ε which intersect with S (Bouligand, 1928). Fractal methods have been employed in image analysis (Soille and Rivest, 1996) and for the analysis of the urban environment (Ostwald, 2001; Ostwald, Vaughan and Chalup, 2009). A discrete approximation of the box-dimension can be achieved via box-counting, a method that has become common in architectural image analysis (Bovill, 1996; Ostwald and Tucker, 2007; Ostwald, Vaughan and Tucker, 2008). Several variations of the box-counting approach exist (Li, Du and Sun, 2009).

Using psychological eye-tracking experiments, it has been demonstrated that contours with high-intensity gradients attract subjects' attention (Rayner and Pollatsek, 1992). The skyline is the contour of the sky segment in an image. Its important role for the aesthetic assessment of distant urban views was emphasised by Heath, Smith and Lim (2000). The fractal dimension of skylines can, therefore, be regarded as an important feature that may be used to characterise cityscapes and natural scenes (Keller, Crownover and Chen, 1987; Hagerhall, Purcell and Taylor, 2004). Fractal analysis of urban skylines has previously been conducted by Oku (1990) and Cooper (2000; 2003). The theory of contextual fractal fit implies that cityscapes look better if the fractal dimension of their skyline matches the fractal dimension of the environment (Bovill, 1996; Stamps, 2002).

Part of the present project was the development of a new method for fractal analysis of a cityscape's skyline (Chalup, Henderson, Ostwald and Wiklendt, 2008; 2009). The basic approach developed by the authors extracts the skyline from standard images if the sky is a homogeneous, connected component which is adjacent to the upper edge of the image and if the sky region is of higher brightness than the other parts of the image. Details about how to exclude narrow objects such as cranes, power lines, and poles which intersect the skyline can be found in (Chalup et al., 2008a; 2009a). Future research may address images of cityscapes that display extreme lighting conditions, distinctive cloud structures, extreme weather conditions, and other non-standard situations.

Pilot experiments showed that calculating a sensible skyline approximation can be an unstable process that depends on various characteristics of the input image. It is a challenge to decide which skyline among many possible approximations of the sky contour is the best. We proposed a method that controls the skyline approximation locally via intensity cut-off values (co). The smoothest approximation is determined by a suitable local minimum of the fractal dimension when plotted as a function in dependency of the intensity cut-off values. The process is visualised in Figures 8 and 9. The arrow in Figure 9 highlights the locally best approximation at a local minimum for $co = 146$. For lower cut-off values, the skyline starts to invade façades. For higher cut-off values, the skyline starts to jump gaps between buildings or to take off into the sky region. The graph in Figure 9 suggests that the fractal dimension is locally stable for cut-off values where the smoothest approximation that appropriately separates the sky region from the cityscape is reached. Locally, therefore, the selection of the skyline can be automated by using the local minima, but, globally, the process, currently, is only semi-automated and requires user input to select the most appropriate local minimum.

Figure 8: *Top*: original; *second row*: $co = 95$; *third row*: $co = 146$; *fourth row*: $co = 155$; *fifth row*: $co = 180$. The best skyline approximation was obtained for cut-off $co = 146$ (image in the third row). Associated fractal dimensions are displayed in Figure 9 (Chalup and Ostwald, 2009).

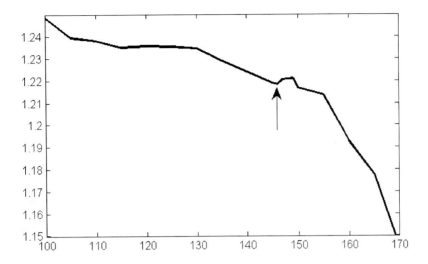

Figure 9. Graph showing the fractal dimension as a function of the intensity cut-off value (horizontal coordinate) for the skyline approximations displayed in Figure 8 (Chalup and Ostwald, 2009).

The described method was further employed in several experiments (Chalup *et al.*, 2008a; 2009a), which support two hypotheses:

1. Intersecting trees typically increase a skyline's fractal dimension. Figure 10 shows an example.

Figure 10. An example of a streetscape's skyline that without trees (top image) has fractal dimension D = 1.0543 and, with trees included (bottom image), D = 1.1567 (Chalup *et al.*, 2009a).

2. Fractal dimension of cityscapes' skylines can be used to distinguish different types of cityscapes, for example, those consisting of historic houses vs. cityscapes consisting of modern apartment and office buildings (Figure 11).

This new variation of the process for the analysis of urban boundaries automates the detection of skylines and the determination of their fractal dimensions; a feature of the visual environment which has a physiological impact on the human emotional system.

Figure 11. *Top*: A skyline in Amsterdam with fractal dimension D = 1.1229. *Bottom*: A skyline in the modern part of Suzhou with D = 1.0591 (Chalup *et al.*, 2009a).

CONCLUSION

Each individual's emotional response to a building, space or place is ultimately unpredictable. A myriad of factors – including cultural background, education, social networks and life experience – shape the way in which a person will perceive and then respond to any given situation. This is precisely the problem with both the phenomenological and semiotic arguments about architecture. In the former case, there is no accurate way of modelling a person's experience of an object or place; it might be possible to record data relating to experience with some degree of objectivity but there is no way of predicting how an individual will respond to this stimuli. Similarly, post-structuralist thinkers have summarily rejected semiotics because there is no 'universal contract' which defines the meaning of signs and symbols; multiple cultures will have different meanings (signifieds) for the same signs (signifiers). However, counter-arguments exist for both of these cases.

From a phenomenological perspective, we cannot predict how a person will react to a space or building, but we can observe many thousands of people entering a Baroque public space in Rome (Bernini's St Peter's square for example) and having demonstrably related experiences. Similarly, a semiotician would argue that many constructed forms can and do carry meaning. For example, a steeple surmounted by a cross and located in a European city will convey differing messages for divergent groups of people but there are similarities too. For many people, the steeple will signify a place of deep spiritual experience; a space of familial celebration and commemoration. For millions of Europeans the steeple conveys the idea that the building is a place of Christian faith. Even those who do not understand the religious significance of the structure will see it as a landmark to navigate their way around a new city; this is a different type of shared meaning. For all of these reasons a balanced approach must be taken to research into human perceptions of the built environment. It must be acknowledged that attempts to replicate or predict human perceptions are necessarily fraught, but with the assistance of new methods much can still be revealed and learnt.

The present chapter is predicated on the view that insights developed in neuroscience, physiology and psychology may inform the design and testing of new methods for the analysis of the built environment. Each of the proposed methods starts with the form of a space or building and then, adopting ABC principles, it develops a way of simulating and analysing possible perceptual reactions to the built environment. While only three perceptual indicators are developed in this chapter – sensitivity to fractal dimension, capacity to identify facial expression and communicate emotion and detection and recognition of linearity – they are, in isolation, critical steps towards a new, biocybernetic way of thinking about the built environment. However, if these methods could be combined together and their results correlated (say in the form of an emotion manifold, like the streetmanifolds presented previously) they would represent a major advance in architectural computing.

ACKNOWLEDGMENTS

This chapter received support from the ARC discovery project grant DP0770106 'Shaping social and cultural spaces: The application of computer visualisation and machine learning techniques to the design of architectural and urban spaces'. We are grateful to all our research assistants and previous co-authors for their contributions to software and experiments in the underlying series of technical papers (Chalup *et al.*, 2006; 2007a; 2007b; 2008a; 2008b; 2009a; 2009b). As indicated in the Figure captions several Figures of the present chapter have previously been published in some of our earlier articles. For Figures from (Chalup and Ostwald, 2009) the publisher is Common Ground. Please note that readers must contact the relevant publisher if they require permission to reproduce any of those Figures.

REFERENCES

Albus, J. S. & Meystel, A. M. (2001). Engineering of mind. *An introduction to the science of intelligent systems*. New York, John Wiley & Sons.

Bachelard, G. (1958). *The poetics of space*. Boston, Beacon Press.

Barlow, H. B., Hill, R. M. & Levick, W. R. (1964). Retinal ganglion cells responding selectively to direction and speed of image motion in the rabbit. *Journal of Physiology, 173(3)*, 377-407.

Barlow, H. B. & Levick, W. R. (1965). The mechanism of directionally selective units in rabbit's retina. *Journal of Physiology, 178(3)*, 477-504.

Barnsley, M. (1988). *Fractals everywhere*. Boston, Academic Press.

Barrett, L. F. (2006). Emotions as natural kinds? *Perspectives on Psychological Science, 1(1)*, 28-58.

Bedau, M. A. (2003). Artificial life: Organization, adaptation and complexity from the bottom up. *Trends in Cognitive Sciences, 7(11)*, 505-512.

Bishop, C. M. (2006). *Pattern recognition and machine learning*. New York, Springer.

Bouligand, G. (1928). Ensembles impropres et nombre dimensionnel. *Bulletin des Sciences Mathématiques France*, *52*, 320-344; 361-376.

Bovill, C. (1996). *Fractal geometry in architecture and design*. Boston, Birkhäuser.

Braitenberg, V. (1984). Vehicles: *Experiments in synthetic psychology*. Cambridge, MA, The MIT Press.

Brooks, R. (1990). Elephants don't play chess. *Robotics and Autonomous Systems*, *6(1,2)*, 3-15.

Brooks, R. (1999). Cambrian intelligence: *The early history of the new AI*. Cambridge, MA, The MIT Press.

Burges, C. J. C. (2005). Geometric methods for feature extraction and dimensional reduction. Chapter 4 in O. Maimon, & L. Rokach, (Eds.), Data mining and knowledge discovery handbook: *A complete guide for researchers and practitioners*. New York, Springer.

Chalup, S. K., Clement, R., Marshall, J., Tucker, C. & Ostwald, M. J. (2007a). Representations of streetscape perceptions through manifold learning in the space of Hough arrays. In: *2007 IEEE Symposium on Artificial Life*, April 1-5, 2007, Honolulu, HI. IEEE.

Chalup, S. K., Clement, R., Ostwald, M. J. & Tucker, C. (2006). Applications of manifold learning in architectural facade and streetscape analysis. *Workshop at NIPS on novel applications of dimensionality reduction*, *Whistler*, CN, extended abstract.

Chalup, S. K., Clement, R., Tucker, C. & Ostwald, M. J. (2007b). Modelling architectural visual experience using non-linear dimensionality reduction. In: M. Randall, H. Abbass, & J. Wiles, (Eds.), Australian Conference on Artificial Life (ACAL 2007). Vol. *4828 of Lecture Notes in Computer Science LNCS*, (84-95). Berlin, Springer.

Chalup, S. K., Murch, C. L. & Quinlan, M. J. (2007c). Machine learning with aibo robots in the four legged league of robocup. *IEEE Transactions on Systems*, Man, and Cybernetics—Part C, *37(3)*, 297-310.

Chalup, S. K., Henderson, N., Ostwald, M. J. & Wiklendt, L. (2008a). A method for cityscape analysis by determining the fractal dimension of its skyline. In: N., Gu, L. F., Gul, M. Ostwald, & A. P. Williams, (Eds.), ANZAScA 2008: Innovation Inspiration and Instruction: *New Knowledge in the Architectural Sciences*, Proceedings of the 42nd Annual Conference of the Australian and New Zealand Architectural Science Association (337-344), November 2008. University of Newcastle, Callaghan, NSW, ANZAScA.

Chalup, S. K., Hong, K. & Ostwald, M. J. (2008b). A face-house paradigm for architectural scene analysis. In: R., Chbeir, Y., Badr, A., Abraham, D. Laurent, & F. Ferri, (Eds.), CSTST 2008, *Proceedings of The Fifth IEEE International Conference on Soft Computing As Transdisciplinary Science and Technology*, (397-403), October, 26-30, 2008, Cergy-Pontoise / Paris. ACM.

Chalup, S. K., Henderson, N., Ostwald, M. J. & Wiklendt, L. (2009a). A computational approach to fractal analysis of a cityscape's skyline. *Architectural Science Review*, *52(2)*, 126-134.

Chalup, S. K., Hong, K. & Ostwald, M. J. (2009b). Simulating pareidolia of faces for architectural image analysis. *International Journal of Computational Intelligence Research*, In press.

Chalup, S. K. & Ostwald, M. J. (2009). Anthropocentric biocybernetic computing for analysing the architectural design of house façades and cityscapes. *Design Principles and Practices: An International Journal*, *3(5)*, 65-80.

Chang, C. C. & Lin, C. J. (2001). LIBSVM: *a library for support vector machines*. Software available at www.csie.ntu.edu.tw/_cjlin/libsvm (date of access 02.02.2009).

Choset, H., Lynch, K. M., Hutchinson, S., Kantor, G., Burgard, W., Kavraki, L. E. & Thrun, S. (2005). Principles of robot motion: *Theory, algorithms, and implementations*. Cambridge, MA, The MIT Press.

Cooper, J. C. (2000). The potential of chaos and fractal analysis in urban design. Unpublished PhD thesis, *Joint Centre for Urban Design*. Oxford, Oxford Brookes University.

Cooper, J. C. (2003). Fractal assessment of street-level skylines - A possible means of assessing and comparing character. *Urban Morphology*, 7, 73-82.

Cox, T. F. & Cox, M. A. A. (2001). Multidimensional scaling (2nd edition). *Boca Raton*, Chapman & Hall/CRC.

Damasio, A. R. (1994). Descartes' error: *Emotion, reason, and the human brain*. New York, G. P. Putnam's Sons.

Dautenhahn, K. (2007). Methodology and themes of human-robot interaction: A growing research field. *International Journal of Advanced Robotic Systems*, 4(1), 103-108.

De Valois, R. L. & De Valois, K. K. (1988). *Spatial vision*. New York, Oxford University Press.

D., Dietrich, G., Fodor, G. Zucker, & D. Bruckner, (Eds.) (2009). Simulating the mind – A technical neuropsychoanalytical approach. *Wien*, Springer-Verlag.

Doya, K. (2002). Metalearning and neuromodulation. *Neural Networks*, 15(46), 495-506.

Eco, U. (1980). Function and sign: The semiotics of architecture. In: G., Broadbent, R. Bunt, & C. Jencks, (Eds.) Signs, *symbols and architecture*, (56-65). New York, Wiley.

G. A. Edgar, (Ed.) (1993). Classics on fractals. *Reading*, MA, Addison-Wesley.

Ekman, P., Friesen, W. V. & Hager, J. C. (2002). *Facial action coding system*, the manual. Salt Lake City, UT, A Human Face.

Ekman, P. & Matsumoto, D. (1993). Combined Japanese and Caucasian facial expressions of emotion (JACFEE) and Japanese and Caucasian neutral faces (JACNeuF) datasets. Available from: www.mettonline.com/products.aspx?categoryid=3 (date of access: 02.02.2009)

Engel, S. A. (2005). Adaptation of oriented and unoriented color-selective neurons in human visual areas. *Neuron*, 45(4), 613-623.

Engell, A. D. & Haxby, J. V. (2007). Facial expression and gaze-direction in human superior temporal sulcus. *Neuropsychologia*, 45(14), 323-341.

Farah, M. J. & Aguirre, G. K. (1999). Imaging visual recognition: PET and fMRI studies of the functional anatomy of human visual recognition. *Trends in Cognitive Sciences*, 3(5), 179-186.

Finkbeiner, M. & Palermo, R. (2009). The role of spatial attention in nonconscious processing: A comparison of face and non-face stimuli. *Psychological Science*, 20(1), 42-51.

Forsyth, D. A. & Ponce, J. (2003). Computer vision: A modern approach. *Upper Saddle River*, NJ, Pearson Education, Inc.

George, N., Driver, J. & Dolan, R. J. (2001). Seen gaze-direction modulates fusiform activity and its coupling with other brain areas during face processing. *NeuroImage*, 13(6), 1102-1112.

Gonzalez, R. C. & Woods, R. E. (2002). Digital image processing (2nd edition). *Upper Saddle River*, NJ, Prentice-Hall, Inc.

Grill-Spector, K. & Malach, R. (2004). The human visual cortex. *Annual Reviews of Neuroscience*, *27*, 649-677.

Hagerhall, C. M., Purcell, T. & Taylor, R. P. (2004). Fractal dimension of landscape silhouette as a predictor of landscape preference. *The Journal of Environmental Psychology*, *24*, 247-255.

Hagerhall, C. M., Laike, T., Taylor, R. P., Küller, M., Küller, R. & Martin, T. P. (2008). Investigations of human EEG response to viewing fractal patterns. *Perception*, *37(10)*, 1488-1494.

Heath, T., Smith, S. G. & Lim, B. (2000). Tall buildings and the urban skyline: The effect of visual complexity on preferences. *Environment and Behavior*, *32(4)*, 541-556.

Hillier, B. & Hanson, J. (1984). *The social logic of space*. Cambridge, Cambridge University Press.

Hough, P. V. C. (1962). *Methods and means for recognizing complex patterns*. 3, 069,654 (patent).

Hubel, D. H. & Wiesel, T. N. (1962). Receptive fields, binocular interaction, and functional architecture in the cat's visual cortex. *Journal of Physiology*, *160(1)*, 106-154.

Hubel, D. H. & Wiesel, T. N. (1968). Receptive fields and functional architecture of monkey striate cortex. *Journal of Physiology*, *195(1)*, 215-243.

Illingworth, J. & Kittler, J. (1988). A survey of the Hough transform. Computer Vision, Graphics, *Image Process*, *44(1)*, 87-116.

Ito, K. (1987). *Encyclopedic Dictionary of Mathematics by the Mathematical Society of Japan (2nd edition)*. Cambridge, MA, The MIT Press.

I. M. James, (Ed.) (1999). *History of topology*. Amsterdam, North Holland.

Jencks, C. (1980). The architectural sign. In: G., Broadbent, R. Bunt, & C. Jencks, (Eds.) Signs, *Symbols and Architecture*, (83-85). New York, Wiley.

Johnson, M. H. (2005). Subcortical face processing. *Nature Reviews Neuroscience*, *6(10)*, 766-774.

Johnston, J. (2008). The allure of machinic life: Cybernetics, *artificial life*, and the new AI. Cambridge, MA, The MIT Press.

Kanade, T., Cohn, J. F. & Tian, Y. (2000). Comprehensive database for facial expression analysis. *In Proceedings of the Fourth IEEE International Conference on Automatic Face and Gesture Recognition*, (FG'00) (46-53). Grenoble, France.

Kanizsa, G. (1979). Organization in vision: Essays on gestalt perception. New York, Praeger.

Kanwisher, N., McDermott, J. & Chun, M. M. (1997). The fusiform face area: A module in human extrastriate cortex specialized for face perception. *The Journal of Neuroscience*, *17(11)*, 4302-4311.

Keller, J. M., Crownover, R. M. & Chen, R. Y. (1987). Characteristics of natural scenes related to the fractal dimension. *IEEE Transactions on Pattern Analysis and Machine Intelligence*, *9(5)*, 621-627.

Koffka, K. (1935). *Principles of gestalt psychology*. New York, Harcourt Brace.

Kourtzi, Z., Tolias, A. S., Altmann, C. F., Augath, M. & Logothetis, N. K. (2003). Integration of local features into global shapes - Monkey and human fMRI studies. *Neuron*, *37(2)*, 333-346.

Krampen, M. (1979). *Meaning in the urban environment*. London, Pion.

Larsson, J., Landy, M. S. & Heeger, D. J. (2005). Orientation-selective adaptation to first- and second-order patterns in human visual cortex. *Journal of Neurophysiology, 95(2)*, 862-881.

Lee, J. A. & Verleysen, M. (2007). Nonlinear dimensionality reduction. New York, Springer.

Li, J., Du, Q. & Sun, C. (2009). An improved box-counting method for image fractal dimension estimation. *Pattern Recognition, 42(11)*, 2460-2469.

Lin, H. T., Lin, C. J. & Weng, R. C. (2003). A note on Platt's probabilistic outputs for support vector machines. Technical report, *Department of Computer Science and Information Engineering*, National Taiwan University, Taipei, Taiwan.

Luger, G. F. (2009). Artificial intelligence: Structures and strategies for complex problem solving (6th edition). Boston, MA, Addison-Wesley Pearson Education.

Mandelbrot, B. B. (1983). *The fractal geometry of nature*. New York, W H Freeman.

Masland, R. H. (2004). Direction selectivity in retinal ganglion cells. In L. M. Chalupa, & J. S. Werner, (Eds.), *The visual neurosciences*, (451-462). Cambridge, MA, The MIT Press.

Mitchell, T. M. (1997). *Machine learning*. New York, McGraw-Hill.

Norberg-Schulz, C. (1980). Genius loci: *Towards a phenomenology of architecture*. New York, Rizzoli.

Oku, T. (1990). On visual complexity on the urban skyline. Journal of Architecture, *Planning and Environmental Engineering*. Transactions of AIJ, 412, 61-71.

Ostwald, M. J. (2001). Fractal architecture: Late twentieth century connections between architecture and fractal geometry. Nexus Network Journal, *Architecture and Mathematics, 3(1)*, 73-84.

Ostwald, M. J. & Tucker, C. (2007). Reconsidering Bovill's method for determining the fractal geometry of architecture. In J., Coulson, D., Schwede, & R. Tucker, (Eds.), ANZAScA 2007: Towards solutions for a liveable future: progress, practice, performance, people. Proceedings of the 41st Annual Conference of the Architectural Science Association (182-190), November 2007. Deakin University, Geelong, ANZAScA.

Ostwald, M. J., Vaughan, J. & Tucker, C. (2008). Characteristic visual complexity: Fractal dimensions in the architecture of Frank Lloyd Wright and Le Corbusier. In K. Williams, (Ed.), Nexus VII: *Architecture and Mathematics*, (217-232). Turin, K. W. Books and Birkhäuser.

Ostwald, M. J., Vaughan, J. & Chalup, S. K. (2009). A computational investigation into the fractal dimensions of the architecture of Kazuyo Sejima. Design Principles and Practices: *An International Journal, 3(1)*, 231-244.

Palensky, B. & Barnard, E. (2009). A brief overview of artificial intelligence focusing on computational models of emotions. In D., Dietrich, G., Fodor, G. Zucker, & D. Bruckner, (Eds.), Simulating the mind - A technical neuropsychoanalytical approach (76-98). *Wien*, Springer-Verlag.

Palmer, S. & Rock, I. (1994). Rethinking perceptual organization: The role of uniform connectedness. *Psychonomic Bulletin and Review, 1(1)*, 29-55.

Panksepp, J. (1998). Affective neuroscience, *the foundations of human and animal emotions*. New York, Oxford University Press.

Passino, K. M. (2005). *Biomimicry for optimization*, control, and automation. London, Springer-Verlag.

Pelphrey, K. A., Singerman, J. D., Allison, T. & McCarthy, G. (2003). Brain activation evoked by perception of gaze shifts: The influence of context. *Neuropsychologia, 41(2),* 156-170.

Picard, R. W. (1997). *Affective computing*. Cambridge, MA, The MIT Press.

Platt, J. C. (2000). Probabilistic outputs for support vector machines and comparison to regularized likelihood methods. In A. J., Smola, P., Bartlett, B. Schölkopf, & D. Schuurmans, (Eds.), *Advances in large margin classifiers*. (61-74) Cambridge, MA, The MIT Press.

Rayner, K. & Pollatsek, A. (1992). Eye movements and scene perception. *Canadian Journal of Psychology, 46(3),* 342-376.

Russell, S. & Norvig, P. (2003). Artificial intelligence: A modern approach (2nd edition). *Upper Saddle River*, NJ, Pearson Education, Inc.

Schölkopf, B., Smola, A. J., Williamson, R. C. & Bartlett, P. L. (2000). New support vector algorithms. *Neural Computation, 12(5),* 1207-1245.

Schölkopf, B., Platt, J. C., Shawe-Taylor, J., Smola, A. J. & Williamson, R. C. (2001). Estimating the support of a high-dimensional distribution. *Neural Computation, 13(7),* 1443-1471.

Schölkopf, B. & Smola, A. J. (2002). Learning with kernels: *Support vector machines, regularization, optimization, and beyond*. Cambridge, MA, The MIT Press.

Shapiro, L. G. (2006). Accuracy of the straight line Hough Transform: The non-voting approach. *Computer Vision and Image Understanding, 103,* 1-21.

Shapiro, L. G. & Stockman, G. C. (2001). Computer vision. *Upper Saddle River*, NJ, Prentice Hall.

B. Siciliano, & O. Khatib, (Eds.) (2008). *Springer handbook of robotics*. Berlin, Springer-Verlag.

Siegwart, R. & Nourbakhsh, I. R. (2004). *Introduction to autonomous mobile robots*. Cambridge, MA, The MIT Press.

Soille, P. & Rivest, J. F. (1996). On the validity of fractal dimension measurements in image analysis. *Journal of Visual Communication and Image Representation, 7(3),* 217-229.

Spillmann, L. & Ehrenstein, W. H. (2004). Gestalt factors in the visual neurosciences. In L. M. Chalupa, & J. S. Werner, (Eds.), *The visual neurosciences*, (1573-1589). Cambridge, MA, The MIT Press.

Spivac, M. (1979). *A comprehensive introduction to differential geometry*, (2nd edition). Houston, TX, Publish or Perish, Inc.

Stamps, A. E. (2000). *Psychology and the aesthetics of the built environment*. Berlin, Springer.

Stamps, A. E. (2002). Fractals, skylines, nature and beauty. *Landscape and Urban Planning, 60(3),* 163-184.

Steagdman, J. P. (1983). *Architectural morphology*. London, Pion.

Stiny, G. (1975). Pictorial and formal aspects of shape and shape grammars. *Basel*, Birkhäuser.

Taylor, R. P., Spehar, B., Wise, J. A., Clifford, C. W. G., Newell, B. R., Hagerhall, C. M., Purcell, T. & Martin, T. P. (2005). Perceptual and physiological responses to the visual complexity of fractal patterns. Journal of Nonlinear Dynamics, *Psychology, and Life Sciences, 9(1),* 89-114.

Taylor, R. P. (2006). Reduction of physiological stress using fractal art and architecture. *Leonardo*, *39(3)*, 245-251.

Tenenbaum, J. B., de Silva, V. & Langford, J. C. (2000). A global geometric framework for nonlinear dimensionality reduction. *Science*, *290(5500)*, 2319-2323.

Thrun, S., Burgard, W. & Fox, D. (2005). *Probabilistic robotics*. Cambridge, MA, The MIT Press.

Vuilleumier, P. (2007). Neural representation of faces in human visual cortex: The roles of attention, emotion, and viewpoint. In N., Osaka, I. Rentschler, & I. Biederman, (Eds.), *Object recognition, attention, and action*, (109-128). Tokyo, Springer.

Vuilleumier, P., Armony, J. L., Driver, J. & Dolan, R. J. (2003). Distinct spatial frequency sensitivities for processing faces and emotional expressions. *Nature Neuroscience*, *6(6)*, 624-631.

Wang, Y. & Guan, L. (2008). Recognizing human emotional state from audiovisual signals. *IEEE Transactions on Multimedia*, *10(4)*, 659-668.

Wertheimer, M. (1923). Untersuchungen zur Lehre von der Gestalt II. *Psychologische Forschung*, *4*, 301-350.

Westheimer, G. (1999). Gestalt theory reconfigured: Max Wertheimer's anticipation of recent developments in visual neuroscience. *Perception*, *28(1)*, 5-15.

Wiener, N. (1948). Cybernetics or control and communication in the animal and the machine. Paris, *Librarie Hermann et Cie*.

Wise, J. A. & Rosenberg, E. (1986). The effects of interior treatments on performance stress in three types of mental tasks. Technical Report, *Space Human Factors Office*, NASA-ARC, Sunnyvale, CA.

Zeng, Z., Pantic, M., Roisman, G. I. & Huang, T. S. (2008). A survey of affect recognition methods: Audio, visual, and spontaneous expressions. *IEEE Transactions on Pattern Analysis and Machine Intelligence*, *31(1)*, 39-58

In: Built Environment: Design, Management and Applications   ISBN: 978-1-60876-915-5
Editor: Paul S. Geller, pp. 147-165   © 2010 Nova Science Publishers, Inc.

Chapter 6

CREATING SUSTAINABLE HABITATS FOR THE URBAN POOR: REDESIGNING SLUMS INTO CONDOMINIUM HIGH RISES IN COLOMBO

Ranjith Dayaratne

Asst. Professor of Architecture, University of Bahrain, Bahrain.

ABSTRACT

Many cities of the developing countries have for a long time been burdened with the sprawling squatter settlements constituted of insanitary spaces in ad-hoc structures, overcrowded and inappropriate to the contemporary standards of living. Occupied by the poor who have been trapped in the vicious circle of poverty, lack of access to education, unemployment and underemployment and therefore continued poverty, squatter settlements pose an insurmountable challenge to the city authorities and governments involved in the design and management of the built-environments in cities. While approaches to breaking this cycle has shifted from housing to poverty alleviation often dependant on government subsidies, the need for sustainable development demands that these settlements be transformed in a manner that they contribute positively to social progress, economic growth and environmental improvements by themselves.

This chapter presents a case study of an approach to transforming a number of old squatter settlement in Colombo, by which the squatters have been persuaded and facilitated to voluntarily move from slums and shanties to modern, multi-storey condominium apartments built replacing the squalid housing estates. The program based on an exchange of the real estates occupied by the squatters for apartments with modern conveniences has released the valuable urban land for progressive developments while enabling the poor to get out of the poverty trap and live in decent housing. It discusses the *Sustainable Township Program* (STP), the government-owned development company that spearheaded the program; *Real Estate Exchange Limited* (REEL) and how residents of *Vanathamulla;* one of the most dense and derelict squatter settlements have been moved to the condominium high rise; *Sahaspura.* The chapter highlights the problems and potentials of this approach as a model for creating sustainable habitats in place of squatter settlements in cities.

Keywords: Sustainable housing, Colombo, squatter settlements.

INTRODUCTION: THE SLUMS AND SHANTIES OF COLOMBO

Colombo has been a city of Sri Lanka's rich, the elite and the preferred residential destination of its bourgeois society in as much as it's poor. Since Colombo became the capital during the British colonial occupation of Ceylon which introduced the tea trade, slums and shanties and under-serviced neighbourhoods have mushroomed both within the centre and the periphery of the city. Since Independence in 1948, city planners have been grappling with the wicked problem of the provision of sustainable habitations for the poor with no great success. In fact, statistics show that 51% of the population in Colombo continue to live in slums and shanties despite half a century of planned interventions (Colombo Municipality,1999). This means that more than 350,000 live in squalid conditions without appropriate housing in a capital city that is aspiring to become a global hub in south Asia.

Colombo's slums and shanties and under-serviced neighbourhoods have a remarkable history. Its beginnings can be traced back to the making of Colombo as a port city by the British for the export of tea, produced in the central hill plantations (Perera, 1998). The British built the railways terminating in Colombo and built all other infrastructure such as the warehouses around its port. The demand for labour increased tremendously prompting rural urban migration while no plans were made to provide any accommodation for the migrant labour. The end result was that the city residents cashed in on the demand for cheap accommodation by building small, poorly-serviced huts in large gardens, popularly referred to as 'watte', while those who could not afford even such facilities squatted in government lands available aplenty at that time. The 'watte's became over-crowded, lacked services, and received no maintenance creating the slums while the squatters of government lands created the shanties out of buildable waste and garbage available in the city. Understandably, many of the places within these settlements were spatially poor because needs of individuals, families and communities constantly clashed there while the debris-generated architecture was craggy, environmentally hazardous and culturally demeaning.

Cheap labour was a necessity for the industry to flourish and with meagre wages and low affordability, no better housing other than the slums and shanties could provide such accommodation. With the numbers of slum and shanty dwellers swelling which enlarged the electorate, the politicians feared to make any move to change the *status quo*, except to promise improved services to the shanties and upgrading of the slums.

Figure 1. Slums and Shanties in Colombo.

At the same time, the presence of the poor was necessary for the city to grow, because it was such poor who did the menial jobs such as cleaning the gardens or toilets, or attending to small repair works in houses. They also provided cheap labour for other industries, shops and restaurants. While a whole system of activities emerged to sustain that community itself, the informal sector became an integral part of the urban social system, its spatial conglomerate having been defined also by the presence and absence of the slums and shanties.

The planners saw them as a problem and contrived to relocate, while the politicians saw them as potential voters who could be the powers behind and therefore wanted to keep in their electorates. The traders and industries wanted them to be closer for the convenience and efficiency of productivity, although most city residents despised them for bringing down the character and quality of spaces and places of the city. Thus the slums and shanties grew at an alarming rate occupying ever increasing tracks of public land often in prime locations of the city.

The growth of these settlements in Colombo however is not an unusual happening. As is well-known, the emergence of slums and shanties in many Third world cities occurred towards the middle of the 20th century following similar patterns. Pursuant to industrialisation, people who migrated from the rural areas to the cities in search of jobs in many countries such as Nicaragua, Mexico, India and Sudan, either squatted in government lands in the absence of suitable and affordable accommodation or occupied poor quarters of the cities (See Landaeta,1994). Colombo inherited its own share of the problem from similar circumstances, although rural urban migration subsequently diminished and so did the slums and shanties.

Housing Solutions for the Low-Income

The earliest attempts of the Ministry of Housing established in 1952 to resolve this problem was to offer alternative accommodation and relocate the slums while upgrading the shanties. Shanties were often in state lands, on railway reservations and vacant government lands. Slums were often in private properties. Thus the government had a greater control over the shanties and eviction from such land was legally possible. However, housing policies were driven largely by the notion that the government must 'provide'; lands and houses built upon them, the model having evolved from government quarters introduced by the British. The process engaged land surveyors to allocate land, architects and engineers to design and contractors to build. 'Low-cost' emerged as the principle approach through which sub-standard housing was constructed in the peripheral, disadvantaged sites of Colombo to which the squatters were re-located often without their expressed consent. In the process, single-story, semi-detached houses, raw houses, and walk-up apartments were designed and built in peripheral locations of the city. However, such efforts were continuously constrained by the fact that people were considerably poor and had very limited levels of affordability even to pay for such low-cost housing. Moreover, they were often inadequate in terms of size, had compromised on arrangements and were constructed of poor quality. Above all, people preferred to live close to the work places and therefore rented them and moved to shanties again where despite the poverty of built space and design, life was colourful and offered self-help and a great sense of community.

Around 1970s, major changes of government took place which moved for social reform, distribution of resources and equity among the haves and have-nots. As a strategy, 'ceilings' were imposed on many things from food to clothing by the government and a *Ceiling on Housing and Property Law* (1973) nationalized land in excess of 50 acres and houses in excess of one property per person. The intention was equitable distribution of houses assuming that this will enable the poor to acquire housing. However, housing construction for rent came to a standstill and previous property owners themselves sometimes became homeless, because the law enabled those who were on rent over a long period of time to claim the properties in preference to the owners. Given the favour received in the law, the low-income tenants took advantage of the situation and claimed ownership to the properties often involving long drawn-out court cases in which house owners and tenants were locked in, paralyzing the entire housing market that provided even the sub standard slums. Shanties proliferated instead and the housing problem worsened.

1977s saw a major shift in policy with the change of government that moved to open-economy and the engagement of the private sector in development. Major structural changes took place with the setting up of the *Urban Development Authority* (UDA,1977) and *National Housing Development Authority* (NHDA, 1979) which were empowered to formulate the clearance of slum and shanty areas and to undertake their development. In addition, both NHDA and UDA had the powers of compulsory acquisitions under the Land Acquisition Act of July 1977. Continuing the provision of low-cost housing approach aggressively, the NHDA was entrusted with the task of constructing some 36,000 houses and flats during the period of 1978 and 1983, mostly in Colombo.

The *Ministry of Housing* launched the program known as the *Hundred Thousand Housing Program* (HTHP) to construct 100,000 houses in the entire island within five years (1978-1983) of which 36,000 were earmarked for Colombo. It articulated three strategies to achieve its objectives; direct construction, aided self-help, and offering housing loans. Undeniably, the HTHP was ambitious given the fact that the previous government had constructed only 4,700 units over the period of seven years from 1970-1977. However, HTHP suffered immensely from an absence of any new vision or strategies. Its delivery system was the same and led to the new housing projects to cost overruns and delays. Many housing schemes intended for the low-income thus became unaffordable to them. *Rukmalgama* and *Raddolugama* housing schemes for example, eventually served the middle income groups and provided no major increase in the housing stock available to the low-income. They consumed heavy subsidies and the government resources were limited in terms of finance, man-power and even materials. Most significantly, the ministry officials failed to see beyond the conventional approach of 'provision'; defined by type plan houses which led to uniform, stereo-typed settlements lacking in the qualities necessary for creating sustainable habitats and healthy communities.

From Provision to Support

Moreover, statistics showed that the actual increase of the housing stock far exceeded the numerical achievements of the state; people were building by themselves in numbers far more than the state. Housing shortage was on the increase despite the fact that both the government and the people were involved in building houses (Weerapana,1986). Ganesan (1986) reports that the cost of provision became inhibitive with the unit cost of *Aided Self-Help* housing going up by 250% while that of direct construction went up by 450% within the five year period. Public sector investment on housing had been 12% in 1979 and 1980 and as per the World Bank reports, the public investment in urban housing alone amounted to 375 million in 1979 and exceeded 1000 million in 1980 (Gunatilake, 1987).

At the same time, attitudes and arguments were also shifting in the international scene, (Turner, 1976; Habraken, 1972; Alexander, 1985) towards bottom-up approaches to housing. The economic and social pressures mounted and these led to a re-examination of the provision approach and its ability to deal with the low-income housing. By 1983, housing approaches shifted from 'provision' to 'supports', with the launch of the second phase of the HTHP in 1984, popularly known as the Million Housing Program (MHP). Sirivardana (1987) however argues that the insights for this change was derived primarily from internal Sri Lankan experiences which demonstrated that housing can be a catalyst for accelerated development and creation of wealth at the grassroots levels.

MHP reached almost every one of the 25,000 or so rural villages as much as almost all urban centres of Sri Lanka. Coupled also with a program known as '*Gam Udawa*' meaning 'awakening of villages', low income people felt that the state was interested in solving one of the burning issues of everyday life. *Gam Udawa* however was both a development program as well as a political program dealing with housing and infrastructure while keeping the people engaged in the process. However, MHP was not conceived by housing professionals and planners out of any sophisticated theoretical analysis. Nevertheless, it was conceptually radically different and practical and eventually a sophisticated and convincing theoretical position was discovered, articulated and celebrated (Sirivardana,1987;1988). Guided by a highly politically committed leadership, people-based support approach sought to enable the peoples' process of shelter (Weerapana,1986).

In this context, the urban housing sub program of the MHP took a radically different attitude towards the slums and shanties of Colombo. It recognised the slums and shanties as legitimate urban settlements and persuaded the professionals and others who were sceptical of the program to support the people to transform themselves from slum dwellers to socially accepted citizens by transforming the settlements to valued neighbourhoods. State took a 'supportive' role, standing by the side of the squatters and challenging the authorities to change rules and regulations to accommodate them. According to Jagoda (2009), the state had adopted a number of different roles with regard to such housing before; as regulator, housing promoter, developer, financier and manager, yet the role it adopted as the supporter and enabler through MHP was perhaps the most promising and revolutionary of such roles. MHP spearheaded a number of significant shifts as follows.

- It argued against the view of 'slums and shanties' as a 'problem' requiring solutions to be given by the government, and promoted a view of them as a housing situation.
- Housing was recognised as a process, and it posited that the housing of the poor encountered enormous 'obstacles' in the process.
- The role of the government was re-defined as an 'enabler' or a supporter which involved removing such obstacles and helping the process to progress in an efficient manner.

- It promoted a participatory approach where communities, 'barefoot' architects and housing professionals collectively decided the causes of actions most suitable for each community,
- It sought to situate the issues in the localities and therefore to discover specific solutions to different situations.

MHP discovered that in all situations, absence of land ownership was one of the major obstacles that hindered the transformation of shanties to well-built structures and settlements. Indeed, the shanties were occupied by squatters in state land, and none were in possession of any ownership rights despite having lived there for decades. At the same time, building regulations prevalent in the municipality areas demanded larger properties for planning approval while the squatters occupied much smaller plots which could not be enlarged. Taking up the role of the facilitator and enabler, the state transferred the land ownership rights and passed special regulations to enable people to build approved buildings in such smaller plots. Services could be provided with approved plans which were generally produced to formalise designs prepared by the people with the help of the housing professionals and were immediately approved under *Community Development Guidelines*, and *Community Building Regulations'*. The process now known as the *Community Action Planning* challenged the conventional planning practices and 'regularised' the ad-hoc settlements with infrastructure and roads (Dayaratne,1992).

Indeed, the MHPs urban housing sub programme used a number of strategies to deal with the slum and shanty problem within the support paradigm which involved land regularisation, settlement upgrading, new settlements and service provision with varying success. Many smaller shanties were transformed into well-kept neighbourhoods with house designs having been modified with the help of the housing professionals or 'draughtsmen' employed by the people while government loans enabled the purchase of good materials. Roads and services were upgraded and community committees were formed to help maintain and manage the settlements. The outcome was noticeable in the houses that were built in place of the slums, employing modern materials following contemporary ideas of built-forms and expressions. Building fronts in particular were decoratively improved with arched-windows, over-hanging gable roofs, and popular embelishemnets on walls. Undeniably, people had invested in the properties, given the transfer of land ownership and the quality of the environment brought about through improved services, infrastructure and regularisation. However, while smaller settlements such as *Navakelanipura* and *Bandaranaikepura* benefited from these practices (Dayaratne, 1992), some of the larger *'wattes'* struggled to organise the community and benefit from the new developments, because of complex social and political issues that existed there. Nevertheless, it is undeniable that many slums and shanties had been significantly transformed.

In spite of the progress achieved through the MHP, many low-income settlements remained in the prime locations of the city occupying vast tracts of valuable land upon which small, single-storey houses provided meagre accommodation for the low-income families. From the point of view of supports, the approach was successful in mobilising communities to house themselves although it did not eradicate the slums and shanties. Given the fact that more people would come and erect new shanties in wherever land was available, improvements to some of the shanties did not take away the presence of the slums and shanties from the city. Moreover, the state had 'given up' large tracts of valuable state land in which people squatted in order to achieve those changes. These settlements remained breeding grounds of squalor around creating insanitary public spaces, occupied by poor in crowded surroundings. Canals around which they lived were polluted and canal banks littered, while drainage often broke down and drains harboured polluted water. Indeed, land regularization and slum upgrading alone did not seem to transform them to cherish-able neighbourhoods.

Figure 2. Houses upgraded by supporting and enabling: *Bandaranayakapura.*

Towards the 1980s with the focus on sustainability emerging, the state looked at these issues from an environmental point of view arguing that slum and shanty settlements although upgraded also required 'environmental cleaning'. Thus the *Colombo Metropolitan Environmental Improvement Program* was launched to deal with the environmental issue of settlements which then evolved into what is known as the *Clean Settlement Program*. By the 1990s however, supporting and enabling strategies and the outcome of subsequent approaches began to be questioned, given the fact that the improvements to shanties were modest and the run-down character of the areas did not change appreciably. Despite the fact that the MHP won the prestigious Habitat Award (BSHF, UK) and was hailed as the way forward for housing problems of the Third World (Hamdi,1991), at the turn of the century, Colombo was scattered with ad-hoc structures on prime state land where squatters still lived in environmentally poor areas. It was soon re-discovered that even the environmental issues of solid waste management or environmental improvement in the shanty settlements had other significant issues underlying, which neither the provision nor the support approaches had addressed. Foremost among them was the issue of land that had been kept out of the market economy which had evolved and engulfed the rest of the city of Colombo except those occupied by the low-income communities.

The Sustainable Township Program

Thus it was clear that despite decades of interventions ranging from provision to supports, neither the government nor the Colombo Municipality was able to find sustainable, holistic solutions to the issues of housing the slums and shanty dwellers in Colombo. Colossal sums of funds had been spent and numerous concepts and strategies within both provision and support paradigms had been developed, applied and abandoned. In fact, despite some successes, most programs had perpetuated poverty, land fragmentation and ad-hoc development of common amenities while creating numerous social and environmental problems in the city. Notably, the ideas of treating housing as a welfare system, and the provision of special preferences and conditions had not yielded appreciable positive results. For example, many re-location programmes of the low income communities resulted in their eventual return because once the new houses were given, they were seen as a commodity in the market and the recipients would capitalise on the financial gains possible, sell them despite all the restrictions and return to the shanties. Settlement regularisations were not sustainable because many irregular land sub divisions and encroachments took place subsequently converting them yet again to slums and shanties. Ability of many of the housing approaches to produce meaningful sustainable urban settlements remained debatable, while funding was scarce. According to the central bank reports (2000), a sum of Rs. 50 billion would have been needed if Colombo were to undertake the task of upgrading all slums and shanties and their built-environments. If there were 50,000 slums, this is a staggering 500,000 Rs. per slum unit which the government would not have been able to afford.

Given this situation, a presidential taskforce was appointed in 1998 to re-examine housing and urban development strategies and to draw up a macro policy framework and a program of activities for short and long term development. The *Sustainable Township Program* (STP) grew out of these policy recommendations which drew attention to the plight of thousands of low income families trapped in the poverty cycle occupying a large portion of valuable land in the city while it lacked the much needed land for development. It was conceived that land and their economic values were the key to unlock the huge potentials that lied in engaging the market economies to bring people out of the poverty trap and spearhead a vibrant urban development practice in Colombo (Hettiarachchi,1999; Gamage, 2007).

STP was thus aimed at the dual tasks of providing appropriate shelter by liberating the land occupied by the poor by engaging the market economy to steer the costs and benefits of low-income housing, prime land and real estate development. It was emphasised that the housing should be 'socially acceptable, economically viable, technically feasible and environmentally friendly' and that housing and urban development should ensure a sustainable development beneficial to the *Colombo Municipality Council* in particular and the nation as a whole (Central Bank of Sri Lanka annual report, 2000).

As can be seen, the approach to STP is fundamentally different from any of the previous housing programs and moved away from some facets of enabling and supporting to 'let the market in'. It adopted a number of innovative concepts that had not been hitherto employed in the state-intervened housing sector together with the positive ideas of support housing.

These were,

- Self-financing instead of state funding.
- Voluntary relocation instead of forced eviction.
- Market-based instead of provider or welfare approach.
- Full ownership of the apartment instead of conditional tenure.
- Offered to all households living in a particular locality rather than a selected few.
- Re-housing instead of on-site upgrading.
- Enabling the urban poor to move in to the mainstream housing market.
- Bottom-up instead of top-down.

(Ekanayake, 2001:16)

According to Wickrema (2005) and Jagoda, (2009), the *Sustainable Townships Programme* envisaged to move more than 50,000 out of the total of 66,000 households living in low- income settlements to small but modern towns comprised of high-rise condominium housing, without burdening the people or the government. It was expected to free approximately 1,000 acres of prime land occupied by low-income communities on a voluntary basis and 60% of the cleared land was expected to be utilized for urban development, with the rest to be allocated for environmental and public purposes. The income so gained were to be available to the government through a system of land-based securities which could be used for constructing well-designed and serviced townships in selected locations.

Real Estate Exchange Ltd

STP required the co-operation of a number of agencies to work together; the *Colombo Municipality Council* (CMC), the *Urban Development Authority* (UDA), *National Housing Development Authority* (NHDA), *Sri Lanka Land Reclamation and Development Co* (SLLRD) and the private developers. To co-ordinate the activities of these agencies, a state-owned public company was formed known as the *Real Estate Exchange Ltd* (REEL), which was entrusted with the task of working out the modalities of identifying, persuading, negotiating and moving the slum dwellers to condominium apartments. The entire program was to work on market forces involving no welfare except initially the capital required to launch the program were provided by the state to be later recovered. REEL although was state-owned, acted independently and enjoyed the benefits of a private company where the decisions were made by a Board of Directors, nominated by the share-holding agencies; namely the STP, UDA, NHDA, SLLRDC and CMC. These shareholders held Rs. 5 million worth of equity among them but only some were main shareholders. NHDA and STP held 50,000 shares each, the SLRDC and UDA held 20,000 shares each while the chairman of REEL, UDA and SLRDCC held one share each (Wickrema, 2005).

REEL outlined its main aim as "to attract direct foreign investment to Colombo, the leading commercial centre of Sri Lanka" and argued that "it is essential to provide inner city space, infrastructure facilities as well as cheap labour. Its position was that the era of keeping economic policy related to lands in Colombo under government authority has ended and that in a transparent economy, land must be released for the use of investment promotion" (Deheragoda, 2007a). The chairman of REEL claimed that the process would eliminate land fragmentation and "do the opposite by assembling land to enable urban regeneration".

The procedure involved educating the slum dwellers on the benefits of the scheme which included a valid deed for a flat in a modern high-rise condominium in exchange for the piece of land on which they squatted for which they did not posses any ownership. Some lands were identified as designated lands and the families living in the designated households were barred from selling the properties without the consent of the STP. The volunteer-designated household families were given a voucher or an *Entitlement Certificate* that represented the value of the land. This entitled them to choose an apartment located in any of the proposed developments which were in the same or very similar locations. STP estimated that each family occupied on average 2 perches of land (1 perch= 272 Sq.ft), which at market prices were valued at 800,000 Rs (Rs. 400,000 per perch) and thus if a family would release this land, they would be entitled to an apartment of the same value in the high-rise condominium. Therefore, apartments ranging from 300-600 Sq.ft were envisaged to have different floor plans and different levels of facilities. These, it was expected, while being self-financing could lure the slum dwellers to move to the condominium high rises which will have ripple effects in the community.

Real Estate Trustee Company (RETCL)

While REEL was to coordinate the public sector agencies involved in the STP, the private sector investment agencies interested in financing the urban re-development initiatives were organised into a company called the *Real Estate Trustee Company* Ltd (RETCL). This group comprised of banks and finance companies was envisaged to promote and persuade developers to participate in developing the liberated lands. The developers were expected to compete for possible developments whereby the construction industry would also be activated creating ripple effects across the city development.

The RETCL was expected to set up *Real Estate Investment Trusts* (REIT) for each development site but REEL also had a share in this activity because it was offering the lands for development. Profits were to be shared as 15% for REEL, 25% for the developer and 60% investors. In fact, REEL was a joint investor in the development process and held the key to unlock the land while facilitating the slum and shanty dwellers to move in to the condominium high rises and thereby activate all sectors of urban development.

At the beginning, 495 settlements of the 1,506 unserviced settlements were designated for the program and seven locations were identified for the first phase. Among them, *Wanathamulla* was chosen as the model development together with the following seven locations (Hettiarachchi, 1999).

- *Bakerywatte* in *Kirula* where there were 38 housing units on 307 perches owned by the UDA
- *42 watte* on *Navam Mawatha* with 44 units on 124 perches owned by the Colombo Municipal Council (CMC)
- *Cooper Hill* in *Kollupitiya* with 17 units on 144 perches owned by the UDA
- *54 watte* in *Wanathamulla* with 207 units on 352 perches owned by the CMC
- *62 watte* in *Wanathamulla* with 13 units on 25 perches owned by the CMC
- *Samagi* on *Elvitigala Mawatha* with 92 units on 435 perches owned by the UDA
- *Stationwatte* in *Kollupitiya* with 18 units on 42 perches owned by the UDA.

Sahaspura: Exchanging Slums for Condominium Flats

Wanathamulla was one of the largest slum settlements in the city occupying 134 acres of state land. The first phase of STP targeted 3,500 families from this settlement of whom about 500-600 families were to be provided with well-designed, well-serviced new apartments. The new township to be constructed by the turn of the century was aptly named 'Sahaspura' meaning the 'millenium city' and was perceived as being comprised of high-rise condominium residencies located on a 2.3 acres of land. Although the buildings were to be high-rise, it was conceived as a model urban place the likes of which were to soon replace all slum and shanty areas of Colombo. To enable it to become such a vibrant urban place, a railway station was planned closer to the township, together with a commercial area for shops and boutiques, recreational facilities, parking spaces and a children's playground. More over, common amenities and facilities such as a number of schools, clinics and health centers were also developed in the vicinity. In other words, *Sahaspura* was not another model development. It was the beginning of a grand plan to transform Colombo and particularly its derelict settlements, with a new vision for the new millennium.

Figure 3. Sahaspura Condominium Apartments.

Once the buildings were constructed, groupings of 300 to 500 potential owner-families were guided to form a management company by buying one share for Rs. 25,000 each. The company so formed was expected to buy electricity and water in bulk and then distribute the facilities to individual families. The management company was also expected to help maintain the building with a fixed deposit in the banks from which a monthly interest could be drawn to carry out day to day maintenance work.

The Design of the Township

In keeping with the policy of the REEL which was to act more like a property developer rather than a government agency, architects were commissioned to design the township that will suit the former shanty dwellers, but will comprise a modern high-rise condominium housing complex. Many architects and builders were excited about the new program and therefore submitted a number of proposals, making serious efforts to contribute to develop the new idea of designing condominium high-rises for low-income communities. Three proposals were short-listed and presented and exhibited to the designated households who had the opportunity to vote for the final selection. 80% of the potential residents numbering 600 voted for the winning design submitted by *Rohan Fernando and Pillar Asset Architects*. The consortium which was a joint venture between Sri Lanka and Singapore has had previous experience in designing condominium housing although not in such circumstances. Construction was supervised by the *National Building and Research Organisation* (NBRO) which was a government agency also involved in consultancy and construction.

The winning design of the condominium was a 14 storied building complex with 52 apartments in upper floors providing for about 500 families while the ground floor was allocated for common spaces, commercial activities and public utility spaces. The thirteen upper floors were predominantly residential although they also had smaller commercial and public spaces and facilities. Conceptually, the design had made a significant attempt to understand the specificity of the condominium having to cater to a previously socially cohesive community. *Rohan Fernando*; the architect claims that the underlying theme came from the predominance of the streets in the squatter settlements and was replicated in the condominium design. Fernando has said that he was influenced by *Unite D'habitation De Mersaille* designed by *Le Corbusier* and the *Toa Payo* New towns of Singapore and therefore created vast, free horizontal streets in every floor that was expected to regenerate the social life that existed in the slums and shanties (Wickrema, 2005).

In the Condominium, the elevators opened into these 'streets in the air' which were wide and had half level walls behind which the apartments reflecting the houses were located. This main street branched in to smaller streets that were *cul-de-sacs* ending in the fire escapes, but were smaller and more intimate than the main street. However, streets being wide afforded many of the social activities of the residents; children could play there, elders could walk and neighbours could sit and chat. In fact, all the flats had smaller verandas overlooking the street where families could sit outside home. Some however had small vegetable plots created in these spaces or had enclosed them for privacy.

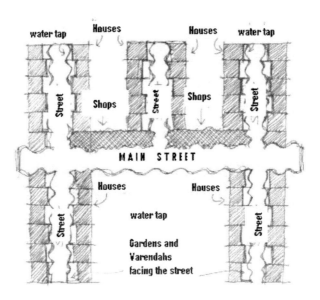

Figure 4. Conceptual sketch of the floor arrangement of the high rise.

The concept of a street has been a dominant spatial stategy that linked the condominium to the settlement it replaced and also differentiated it from the many luxury condominiums mushrooming in the city. Most people contributed to the transformation of the streets in the condominium. While they added greenery and carried out social activities there, many also dried clothes along the street which was intrinsic to all streets in low-income settlements. In fact, the main street of the condominium had spaces earmarked for commercial activities in the central locations which were quickly converted to convenient stores, saloons, clinics and small enterprises that served the community, thus recreating the ubiquitous Sri Lankan community street in the high-rise settlement.

Fernando's plan also employed the idea of clustering as a means of recreating the sense of place and sense of community in this vertical settlement. As it is well-known, low-income settlements are closely-knit with smaller areas of recognition and conceptualised boundaries that tied small social groups together. The central point of such groups had often also been the 'garden tap' that provided water to be shared. Fernando uses both ideas in the ordering of space whereby smaller clusters have been spatially defined so that the new groups of families may meet at the tap as they used to and come together again even with the new residents coming sometimes from different locations. While the clustering has been a good spatial devise to achieve some degree of togetherness and smaller community groups within such a large group as 500 families occupying the settlement, the tap has been a total misconception of the expectation and aspiration of the families. As it turned out, the tap signified the 'slum' which the residents now wanted to distance themselves from and therefore had been abandoned and transformed into places for drying clothes.

Apartment Design

The designers had prepared 10 different options to the potential residents and in the absence of a number of plots being developed simultaneously, this offered them greater flexibility. These options ranged from apartments with different floor areas ranging from 300 Sq. ft- 600 Sq. ft. with a variety of finishes and options.

Potential residents have been allowed to work out their entitlements to these apartments, on the basis of pricing used as a guide. For this purpose, the properties 'owned' in the low-income settlements were valued and priced in accordance with the facilities available there and physical characteristics of the properties. These involved,

- Floor area occupied.
- Type of the structure.
- Type and material of roof.
- Types of materials of construction of walls.
- Floor finishes.
- Doors and windows; materials and sizes.
- Age of structure.
- Existing services - water, electricity, telephone.
- Ownership - lease holder, conditional deed, 20-30 year rental.

Moreover, the householders were also allowed to contribute and top up their entitlements if they had the means to do so. For example, if a householder's entitlement worked out to be adequate only to buy a 400 Sq.ft apartment but wanted to purchase a 500 Sq. ft apartment, the additional 100 Sq. ft could be paid for with the personal funds available to them. Interestingly, REEL offered these at a staggering 70 % less than the costs and many had opted to and had the means to buy more than what they were entitled through the *Entitlement*

Certificates. In fact, these suggest the desires for upward mobility and the resourcefulness of low-income communities in Colombo which helped the project to succeed.

Difficulties and Problems

STP worked hard to improvise methods for the new system of condominium living to work, which was alien to slum dwellers. Although they did share the resources and spaces in the slums, creating organised systems for a modern condominium proved difficult. For example, Fernando points out that REEL introduced a practice to collect payments for maintenance long before people were allowed to move into the new apartments, to make sure that the building was maintained by the residents themselves. The ideas was to create a fixed deposit in a bank and the monthly interest to be used for the maintenance activities.

Despite these, the program has now been halted given the lack of interest of the participating state agencies and the loss of interest from the investors resulting from the difficulties REEL encountered in marketing land at the market prices. As Wickrema (2005) pointed out, the program had already faced major difficulties even by 2005. Wickrema outlined them as follows.

- Change of government leading to,
  - the withdrawal of political backing
  - institutional instability
- Rise in Construction cost resulting from various changes such as
  - increase of electricity tariff
  - increase in material costs and labour
  - poor construction management
- Inadequate coordination at National levels between agencies such as,
  - Board of Investment and UDA.
  - Colombo Municipality and UDA
  - lack of foresightedness in devising plans for management of the property such as,
    not requiring the maintenance contribution in full before the residents moved in inability to muster full support of the residents for the management plan.

Residents' Satisfaction

Completed in 2001, a number of researchers (Hettige et.al, 2004; Wickrema,2005) have found that most residents of *Sahaspura* expressed delight of being able to move into the new apartments and have claimed that they were in much better conditions than the other slum dwellers. The design of the complex seems to have had a contribution to this sense of comfort, where the architects had created both intimate as well as public spaces meaningfully, almost reflecting those that existed in the slum settlements while improving upon the personal spaces and individual dwellings. Most significantly, they have valued the ownership of the houses, physical quality of materials with which they had been constructed, good maintainable finishes, workmanship as well as clear order and arrangement. The houses have had constant supply of modern amenities such as water and electricity which meant that the toilets and kitchens could be kept well, while modern electrical appliances eased the daily chores. As Deheragoda (2007b) points out, the slum dwellers who moved to *Sahaspura* had made a huge leap from cluttered, chaotic and derelict life spaces to refined and progressive ones that had not been imaginable to be achieved by themselves.

Critiques of the STP however point out that STP policies are driven purely by commercial principles and cannot deal with those lands with little economic value (Ekanayake, 2001). For example, many shanty dwellers live on canal banks, railway reservations etc. which cannot be brought into the land market in order to relocate them. Such lands are environmentally relevant and such shanty dwellers also require fair solutions. Similarly, STP does not seem to be able to deal with smaller plots of low-income families living in scattered locations. However, these are not significant criticisms of the STP itself. Rather, it suggests that STP is not a *panacea* for housing the poor in cities living in all circumstances. Obviously other place-specific approaches are necessary (Dayaratne,1992).

One of the other criticisms often levelled against *Sahaspura* is that the collective community management of the property did not really work. Fernando proposing a new system for the management of maintenance of the property says that "From the Sahaspura experience we have learned that the maintenance of these common areas is highly problematic" (Fernando, 2009:10).

A more fundamental criticism comes from those who see the city as a struggle between the marginalised-poor against the state-sponsored rich. They argue that there is an inbuilt-interest in 'liberating' land from the low-income communities to be handed over to the property developers for producing places for the rich, and that the motivating factor is not that of improving the lives of the poor but the promotion of the city as a 'global economic hub' for which land should be taken from the poor. There is no denial in the STP that this is one of the objectives, but as opposed to the social welfare approach of the support paradigm that keeps the low-income people where they are, STP absorbs them into the mainstream housing market; a social ladder through which the low-income communities may come out of the poverty trap and its concomitant agonies.

Most significant questions come however from the social and cultural implications of shifting. It is often argued that slum and shanty communities are highly-knit social groups who have developed intricate social networking and strategies for survival in the context of their low-income culture. There is much invisible structures of social and cultural value in the slums and shanties despite being located in derelict environmental and physical conditions. Life is perceived to be colourful, culturally-rich and socially-enhancing although the visible built-structures that situate them are craggy and derelict, aesthetically unarticulated and visually unpleasant (Wanasinghe et.al, 1999a;1999b). In relocating vertically and in rigidly-framed spatial structures, the opportunities for re-establishing and reliving those social networks and colourful lives are highly limited creating strains upon the very practice of habitation of these communities. It is argued that the policies should also address those issues if the shifting from the slums to the condominiums are to be successful. Undeniably, architect Rohan Fernando's approach to 'creating streets in the air' in *Sahaspura* have pointed towards a positive direction although much more versatile spatial practices could be employed. Gamage et.al. (2007) however point out that the research justified this radical shift of housing policy in terms of better social lives and voluntary relocation in the face of urban land constraints, provided that their former livelihood does not depend on the residential location.

CONCLUSION

In the long and arduous series of solutions Sri Lanka has offered to creating sustainable habitats for the urban poor, it had shifted from provision to support in the 1980s with the conviction that supporting and enabling will radically change the urban landscape where the poor will house themselves without the government having to build houses (Hamdi,1991). However, it was proved to be a partial solution incapable of addressing the complexity of land, houses and communities within the context of continued progress and development in a sustainable manner. Everyday economics having been set aside for a long time for the pursuance of welfare policies, low-income housing remained outside the mainstream real estate development and practice.

What Sustainable Township Program has done is to engage the land values as determined by the everyday market forces to become a driving force of the housing process whereby the enormous potential that resided in the derelict yet valuable lands could be released. The 'real estate exchange' approach has brought in the market forces to play a significant role in funding while the state moved slowly away from the welfare-driven approaches to low-income communities. This shift is significant in that it at once moves the low-income dwellers from the position of a recipient of hand-outs, to one of proprietor and property owner whose stakes in the housing market enables the retrospective movements into the larger housing market. Until this time, despite the land and house ownerships had been passed over at concessionary prices, the recipients of such properties had been barred from entering the housing market and have been kept away from the mainstream residential sector.

As anticipated, this strategic practice has enabled the release of some valuable urban land for development while facilitating the poor to get out of the poverty trap. Even though some residents have sold their apartments and have moved out, they have not returned back to the slums as it used to be the practice. There are indications that other people who had not signed up to the program had shown a great interest in the possibilities the program offers, although the program has later on received lukewarm support from the state. Nevertheless, *Sahaspura* experience shows that high-rise apartments could indeed become a potential solution to low-income housing.

To make this a successful, reliable and sustainable practice however, a greater commitment is needed from the state, while allowing the market forces to work fully, independently and as it plays out in the market. At the moment however, it has not been the case in Colombo, and severe disruptions to the anticipated outcomes have occurred as a result of many incompatible government actions. This has stalled the process and thereby once again, the entire effort of enticing the low-income families to move from the slum and shanties to condominiums has been halted, together with the release of land that the program entailed. This is because, REEL has not been able to market the land so released at the actual market values because the Board of Investment of Sri Lanka has initiated a practice of making land available to the developers at a concessionary price in order to lure foreign investors for development. As a result, the expected benefits of REEL have not been realised and the process has been so paralysed, without anticipated funds.

Moreover, there have been other obstacles at *Sahaspura* itself that have not been well taken care of. For example, there have been numerous issues of the management of the condominium which was expected to be undertaken by the residents through a company formed by themselves. It has been observed that the system put in place has not worked properly, because, the members of the management group were both incapable and did not have either the skills or the time necessary to sustain an effective management system. In the absence of a well functioning management system, Fernando proposes to form a *Community Based Management Committee* (CBMC) where the community is considered as the main driving body of the maintenance structure. These are necessary mechanisms that need to be adopted in future projects after being well-articulated and tested under given circumstances.

Despite these problems however, It is undeniable that *Sahaspura* has transformed the lives of the low-income communities who aspired to own their own dwellings, to have such properties with separate toilets that had uninterrupted water, and continuous water supply and electricity services. As Van Horena et.al. (2009) quoting Prema kumara, (undated) reported, "low-income settlements in Colombo do not have access to most basic municipal services. About 56% of households rely on common water taps and, on average, there are 40–100 households per tap. About 67% of households in low-income settlements either share or do not have access to toilets. In the case of garbage collection, 66% of low-income communities do not have access to municipal waste collection services and throw garbage into nearby canals, drains, or reservation lands. In most low-income settlements (about 70%), paved roads and improved storm and wastewater drains have not been constructed". These, indeed are the agonies of the low-income people in Colombo and the forces that drive their settlements to become slums and shanties.

For the residents of *Sahaspura*, all of these have changed for better and the community who moved into the condominium high-rises with the streets-in-the-air does not see a major problem in adjusting to vertical living when compared with the derelict squatter settlements on the ground. *Sahaspura* has enabled the people to move away from the 'common water tap'; a signification of poverty, squalor and social under-class. Absence of it is now seen as a great relief that has helped social well-being, individual identity and even potential social coherence.

Thus the STP program has clearly demonstrated that the housing for the poor do not have to remain as welfare programs whereby they will be kept in their deteriorating settlements, while the rest of the city grows upwards. As Deheragoda (2007b) argues "… it is possible to address the shelter issue of the urban poor by adopting pro-market shelter strategies provided the necessary policy environment, political steadfastness and will and the institutional set-up is ensured by the authorities". *Sahaspura* experience has demonstrated that the principles of this practice are sound and can be employed to deal with low-income housing of squatter settlements, where the land is valued and can be exchanged to create progressive vertical communities.

REFERENCES

Alexander, C. (1985). The production of houses. USA: *Oxford University Press.*

Central Bank of Sri Lanka, *Annual Report,* 1998 & 2000

Colombo Municipality. (1999). Annual report. *CMC.* Colombo.

Dayaratne, R. (1992). Supporting people's placemaking: Theory and practice. *PhD Thesis submitted to the University of Newcastle upon Tyne*, U.K.

Deheragoda, C. (2007a). *Living high: Social dimension*. Paper presented at Living high: Sri Lankan context, organised by the Institute of Town Planners Sri Lanka and Condominium Management Authority, BMICH. Colombo.

Deheragoda, C. (2007b). Shelter issue of urban poor: Case study from Sri Lanka. Paper presented at the Meeting global challenges in research cooporation, Uppsala, Sweden. Department of Census & Statistics, *"Census of Population & Housing"*, (2002), Sri Lanka.

Ekanayake, M. (2001). Redevelopment of under served settlements in the city of Colombo, Sri Lanka. *Masters thesis submitted to the International Institute for Geo-information Science and Earth Observation*, Netherland.

Fernando, D. (2009). Sustainable Housing for Urban Poor: Design of a Preventive Maintenance & Management Mechanism for Public Urban Housing Schemes, accessed on 20.10.2009 @ http://www.lth.se/fileadmin/hdm/alumni/papers/SDD_2007_242a /Deepika_Fernando__Sri_Lanka.pdf

Gamage, T., Perera, U.,Onishi, T. & Kidikoro, T. (2007). High- Rise through Market Based Approach as a Shelter Strategy for Urban Informal Settlements? (Case study; Sustainable Township Program-Sahaspura Apartment Complex, Sri Lanka). paper presented at the International symposium on city planning,

Ganesan, (1986). *The construction industry in Sri Lanka*. ILO, World Employment Program, Research Working Paper.

Gunatilaka, V. (1987). *From provision to support*. An unpublished M.Phil thesis at the University of Newcastle upon Tyne. U.K.

Habraken, N. (1972). *Supports, an alternative to mass housing*. London: Architectural press.

Hamdi, N. (1991). *Housing without houses*. New York: Van Nostrand Reinhold.

Hettiarachchi, K. (1999). *Living on the edge*. Sunday Times Sri Lanka.

Hettige, S., Fernando, N., Mayer, M., Noe, C. & Nanayakkara, G. (2004) *Improving Livelihoods of the Urban Poor: A Study of Resettlement Schemes in Colombo*, Sri Lanka. GTZ-Sri Lanka, Colombo Municipal Council and the University of Colombo.

Jagoda, D. (2009). Sustainable Housing Development for Urban Poor in Sri Lanka: *Recommendations for the Improving Relocation Housing Projects in Colombo* @ http://www.lth.se/fileadmin/hdm/alumni/papers/ad2000/ad2000-14.pdf

Landaeta, G. (1994). Strategies for Low -income Housing A comparative study on Nicaragua, Mexico, Guatemala, Cuba, Panama, Costa Rica and El Salvador. *Dept. Of Architecture & Development Studies*, Lund University, Sweden. Lund.

Perera, N. (1998). *Society and Space: Colonialism, Nationalism, and Postcolonial Identity in Sri Lanka*. Boulder: Westview Press.

Sirivaradna, S. (1987). *Process initiatives and support. A search for fundamentals*. A revised version of the key note paper presented at the session on THE HOME at the Union of International Architects XVI congress on 'Shelter and Cities'; Building Tomorrows World, Brighton, England.

Sirivaradna, S. (1988). *Housing mainstreams: A case study in learning*. The Sri Lanka case study prepared for presentation at the Washington senior level shelter policy seminar, Washington, November 6-9 Colombo: NHDA.

Turner, J. (1976). *Housing by People – Towards Autonomy in Building Environments*, London: Marian Boyers.

Van Horen, B. & Pinnawala, S. (2009). 'Sri *Lanka' in Urbanisation and Sustainability.* Accessed on 20.10.2009 @ http://www.adb.org/Documents/Books/Urbanization-Sustainability/chapter12.pdf.

Wanasinghe Y. & Abeyratna, M. (1999a). *"Perception on Relocation: An Attitudinal Survey in the five selected under-served settlements in the city of Colombo"*, Report Submitted to the Sustainable Townships Programme Unit of the Ministry of Housing and Urban Development.

Wanasinghe Y. & Abeyratna, M. (1999b). '*Socio-Economic Survey of Under-served Settlements in the City of Colombo"*, Report Submitted to the Sustainable Townships Programme Unit of the Ministry of Housing and Urban Development.

Weerapana, D. (1986). Evolution of a support policy of shelter: The experience of Sri Lanka, *Habitat International, Vol.10. No. 3* 79-89.

Wickrema, M. (2005). Moving on up: Mainstreaming under-serviced urban communities in Colombo, Sri Lanka, a dissertation submitted to MIT for Masters in City Planning.

In: Built Environment: Design, Management and Applications   ISBN: 978-1-60876-915-5
Editor: Paul S. Geller, pp. 167-195   © 2010 Nova Science Publishers, Inc.

Chapter 7

UNCOMFORTABLE, ADJUSTABLE TRADITION: LESSONS FOR CONTEMPORARY BUILDING

Helen Wilkins
University of Sydney, Sydney, Australia.

ABSTRACT

A persistent and prevalent view exists within the architectural and ethnographic literature that the development of vernacular buildings throughout the world has been deterministically driven by a conscious desire amongst humans to create interior spaces that are both thermally 'comfortable' and thermally neutral. This was purportedly achieved through the gradual harmonisation of the buildings' construction and thermal operation with the outside climate. This view is, however, largely erroneous and is the result of an overwhelming focus on only select examples of vernacular. Studies have tended to focus on extreme and easily classifiable climates, such as hot-arid, hot-humid and very cold climates, and have overlooked or ignored other more complex types of climate, such as high altitude and temperate climates. Studies have also tended to ignore traditional buildings that by western standards of thermal comfort are thermally marginal.

A theory that attempts to explain the development of vernacular must, however, be based on a quantitative examination of a wide range of buildings from diverse climates and cultures. It must also examine thermal performance holistically, because the thermal behaviour of structures is the result of the ever-changing interaction between all their built components and thermal features. This paper outlines a quantitative study of the holistic *thermal capacity* of examples of vernacular from a wide range of climates and locations around the world, comparing the results for a range of scales of climate classification.

The paper concludes that the achievement of thermal 'satisfaction' has had a much greater influence than thermal comfort/neutrality on the types of structures that people have traditionally chosen to inhabit. Analysis of the ethnographic record indicates that people who inhabited thermally 'marginal' structures would have been thermally satisfied, if not comfortable, supporting the findings of Adaptive Comfort theory. This theory states that people appear to have a universal preference for possessing the ability (or perceived ability) to make thermal choices and to be able to enact those choices, in preference to being always thermally neutral, regardless of social and climatic differences. This has important implications for contemporary buildings and implies that a building's adjustability and *thermal capacity* is of much greater importance than thermal comfort/neutrality in achieving occupant satisfaction and, ergo, their capacity to support social and functional change.

INTRODUCTION

This paper raises issues about the relationship between society and built environments that are critical to our understanding of social sustainability. The relationship that exists between social life, which alters at will, and the built environment, which possesses huge inertia to change, are extremely complex. Yet too few studies have sought to deal with this in terms of the complexity that is inherent in the relationship, choosing instead to take a deterministic or simplistic approach to the relationship, and erroneously assuming that humans alter their built environment as spontaneously as they do their words and their actions. Yet anyone who has had to bide their time, waiting for their home renovations to begin and then, once begun, to be finalised, will understand first hand that there is inertia inherent in buildings and the consequence of this is a complex relationship between what people *want* to do with their buildings and what they *actually do* at any given moment in time.

Additionally, whilst vernacular has traditionally been approached in the architectural-historical literature from the point of view that owner-designed-and-built structures have been developed and fine-tuned over several millennia to create thermally 'comfortable' interiors (Butti & Perlin 1980; Cofaigh et al. 1996), this assumption has never been empirically tested. The term 'vernacular' refers to buildings that are ordinary, non-monumental, and not architecturally designed. Vernacular studies have generally not taken a holistic approach to the material, but have instead selectively chosen only those examples that 'support' the idea that housing has been consciously developed over time to be thermally 'comfortable'. Yet, there are more numerous examples that clearly demonstrate that this has not always been the case. Examples have largely derived from only climates that are extreme and easily classifiable, such as hot-arid, hot-humid or very cold, overlooking or ignoring other more complex types of climate, such as high altitude and temperate climates. Likewise, examples of structures that by western standards of thermal comfort are thermally marginal have also been overlooked or ignored, even though the occupants of such structures appear to have been thermally content.

Any theory that claims to be able to deterministically explain the development of vernacular or of whole classes of buildings must include a wide range of examples from diverse climates and cultures, for this is the only way to confidently assess any theory objectively. "Details that could throw doubt on your interpretation must be given, if you know them. You must do the best you can – if you know anything at all wrong, or possibly wrong – to explain it. If you make a theory, for example, and advertise it, or put it out, then you must also put down all the facts that disagree with it, as well as those that agree with it" (Feynman 1985: 341). A deterministic explanation is only as relevant as the number of entities within the set of all entities it can account for. An 'ideal' building type is only as valid as the number of actual buildings that conform to the ideal. The more buildings that deviate from the 'ideal', the less explanatory power the concept possesses.

COMFORTABLE VERNACULAR THEORY

It has traditionally been assumed that buildings and the methods by which they have been heated and/or cooled have been developed in synchronisation with their different prevailing climates and that, over time, they were developed to be attuned to the average ambient outside daily and seasonal cycle of temperatures. The ethnographic record is deemed to be the end product of this purported process. "Perhaps they did not know technically what they were doing or why, but the results were effective, comfortable and practical" (Stead 1980: 41).

The prevalent view is that buildings in cold climates were designed for heating and buildings in hot climates were designed for cooling. "Indigenous architecture is often a clever series of construction strategies to amplify the effects of a comfort source. In cold climates the source is the open flame from burning fuel. In overheated climates the source of comfort is not always such a point source, although an overhead fan is a parallel device in many locations" (Cook 1996: 289). Cook has, however, pointed out that the focus on comfort as the driving force behind climatic determinism is possibly a First World-centric cultural derivative, as more traditional societies have tended to view their structures as modifiers of the outside environment, rather than as independent environments. "One hesitates to use the terms comfort or 'environmental control' for this high performance shelter [the igloo] since our use of the word has a different meaning from the Inuit goals of environmental modification" (Cook 1996: 282).

Vernacular in Cold Climates

The view that structures in cold climates were deterministically developed with a primary need for heating was accompanied by a need for minimising air movement. Of course, without heating humans would have been unable to occupy some of the regions in which they are settled today. Winter conditions in the arctic and sub-arctic are generally so cold that human life cannot be sustained by means of passive solar heating alone and active heating, in the form of fires, has been a fundamental component of thermal systems. Fire plays a symbiotic role with structures in cold climates. The most publicised example of cold climate structures is the Polar Inuit igloo of northwest Greenland, where temperatures rise to over freezing for only one month of the year (Cook 1996: 279-280), although the Mongolian yurt has also been mentioned (Rapoport 1969: 98). The Inuit igloo is generally regarded as the 'ideal' cold climate building, being elevated and relatively warm on the inside and aerodynamic to the cold winds on the outside (Rapoport 1969: 98-99; Coch 1988: 75-85; Cook 1996: 279-282; Crouch & Johnson 2001: 90-94). Igloos, with their oil burners, their low domed roofs, their subterranean and down-wind entranceways, and their insulating internal lining of skins, have been presumed to be climatically ideal because they can create temperatures of up to 40°C higher than the outside air.

The most extreme example of structures that do not conform to the Comfortable Vernacular theory – by Eurocentric standards – are the traditional structures of the Selk'nam and Yámana cultures of Tierra del Fuego (Figure 1). These contrast strikingly with the Inuit igloo even though their occupants inhabit a similar tundra climate, where temperatures are commonly below zero (Rudloff 1981: 550). The Selk'nam structures, which were used until very recent times, consisted of no more than windbreaks and log or bark huts, permanently open to the outside (Cook 1892; Keynes 1988; McEwan *et al.* 1997). No compensation existed in the way of clothing, which was minimal and consisted of loose animal furs thrown over the shoulders. The furs were, however, used and worn in a diverse number of ways, such as blankets to lie on and as covering for only part of the body. Fires, too, which were heavily relied upon even though they had the capacity to raise a sweat on the natives (Keynes 1988: 135), were used in a diverse number of ways including accompanying the natives in their canoes (McEwan *et al.* 1997: 63).

A less extreme example is that of the winter kata of the northern Scandinavian Lapps, who inhabit a subarctic climate. These consist of tee-pee shaped layers of logs, bark, peat and turf and a central fire pit. They have been described as "cold and drafty" by comparison with the Inuit igloo (Cook 1996). The Tasmanian aborigines also, like the Fuegans, inhabited only windbreaks and bark huts and wore only loose animal furs for clothing in a regularly cold and windy climate (Gilligan 2004).

Whilst the above examples belong to mobile societies that, as a consequence of their mobile lifestyle, possess the capacity to move about the landscape as the seasons change, there are also examples of buildings that belong to sedentary societies. One well known example is that of the traditional Japanese townhouse where, in Hokkaido for example, the average winter temperatures range from –1°C to –9°C. The townhouse was traditionally constructed of lightweight plaster and timber walls with adjustable external timber shutters and internal rice-paper partitions (Engel 1964). Thickly padded clothing with woollen underwear, were worn both inside and outside the house because "there is little, if any, difference between interior and exterior temperatures" (Engel 1964: 360). Heat was provided by portable braziers, *hibachi*, which generate only local heat, although the heat was often confined to, for example, the area beneath a table around which several people could sit.

Figure 1.

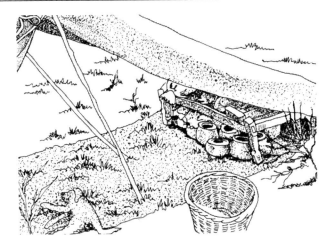

Figure 2.

Vernacular in Hot-Arid Climates

The view that structures in hot-arid climates were deterministically developed with a primary need for cooling in summer was accompanied by a secondary need for heating in winter. This type of climate has received the majority of attention, with focus almost exclusively on the 'ideal' hot-arid climate middle eastern courtyard house and the Anasazi pueblo (Pillet 1980; Stead 1980; Kriken 1983:99-103; Talib 1984: 45-65; Alp 1988; Coch 1988: 77-81; Gallo 1988; Hinrichs 1988; Cook 1991; Givoni 1991: 26-31; Facey 1997; Sayigh & Marafia 1998; Crouch & Johnson 2001: 98-103), although cursory mention has been made of the subterranean Matmata houses in southern Tunisia and the courtyard houses of other bordering regions (Rapoport 1969: 91; Butti & Perlin 1980: 2-27; Talib 1984: 45-47). These houses are presumed to be climatically 'ideal' because they provide cool spaces for use in the summer and warm spaces for use in the winter. This bipartisan usage has been widely recognised, although the full significance of the inbuilt adjustability that this gives these types of structures has not been fully understood.

In contrast to the contrary examples in cold climate mentioned above, contrary examples in hot-arid climates belong largely to nomadic societies. The most extreme examples include the climatically 'inappropriate', very minimalist structures of the Rabaris, who inhabit the very hot semi-desert areas of Kutch in western India (Rudloff 1981: 257) (Figure 2). The Rabaris are a semi-nomadic society whose structures consist of no more than cotton blankets or shoulder cloths raised on two sticks at one end and pegged to the ground at the other (Shah 1980). This simple structure is adjusted to face away from the hot northeasterly winds in summer and away from the cold westerly winds in winter.

Vernacular in Hot-Arid Climates

The view that structures in hot-humid climates were deterministically developed with a primary need for cooling throughout the year was achieved by maximising air movement and minimising solar penetration. Few thermal performance studies have, however, looked at buildings in the hot-humid tropics and the few that have have tended to focus almost exclusively on the light, raised, thatched hut that is widespread throughout Southeast Asia (Rapoport 1969: 93-95; Yuan 1987: 68-79; Coch 1988: 81-82; Givoni 1991: 21-26; Crouch & Johnson 2001: 103-106). Cursory mention has been made of the saddle-roofed New Guinean ceremonial houses and various minimalist shade structures, such as the Seminole houses and minimalist shelters of tropical America (Rapoport 1969: 95; Coch 1988: 82-83; Crouch & Johnson 2001: 103-105). The raised thatched hut is presumed to be climatically 'ideal' because the thick roof thatch minimises the amount of solar energy that can pass through to the shaded interior and the raised, open sides allow for maximum ventilation.

Contrary examples to the light, well ventilated elevated structures of Southeast Asia include the structures of the sedentary Ngelima and Nalya communities in tropical, equatorial Zaire (Figure 3). These structures are tall and narrow, wholly insulated and have only a single low opening (Denyer 1978: 59-60). The arrangement of the structures relative to each other also contrasts to that of Southeast Asia, which are largely well dispersed to maximise the ingress of the natural breezes. The Ngelima and Nalya structures are by contrast arranged close together in straight rows and without shade trees. The neighbouring Panga community inhabit identical structures, although they are arranged around open compounds, interspersed with saddleback roofed structures on raised platforms and shaded by trees (Denyer 1978: 60). In tropical, equatorial Africa there are, in fact, very few examples of elevated structures with numerous openings. The structures of the Grebo community on the border of Liberia and the Ivory Coast are characteristic of structures in the region. They are located at ground level and generally have only one opening, although they do have lightweight walls and high peaked, heavily thatched roofs (Denyer 1978).

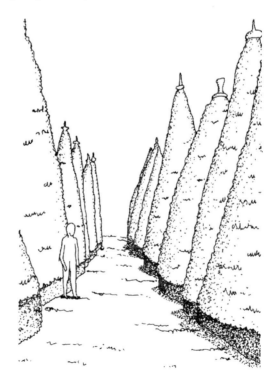

Figure 3.

Vernacular in Complex Climates

The view that structures in temperate and high altitude climates were deterministically developed with an equal need for cooling in summer and heating in winter is, however, based on the complex climates having not been as tightly classified as buildings in more severe climates. Very few thermal performance studies exist that have looked at buildings in temperate climates, although cursory mention has been made of Cape Cod cottages, Hungarian village houses and Normandy farmhouses (Rapoport 1969: 99; Cook 1996: 283-288). Likewise, very few studies have looked at high altitude climates, although cursory mention has been made of Tibetan farmhouses and palaces (Crouch & Johnson 2001: 94-98).

SUMMARY OF THE COMFORTABLE VERNACULAR THEORY

The ethnographic record therefore contains numerous examples of buildings in which, whilst their occupants may not have been always thermally 'comfortable', they were presumably not always thermally uncomfortable and, in fact, appear to have been thermally content. So what does this mean for the theory that buildings were deterministically developed over time to provide interior environments that are both thermally 'comfortable' and neutral? This question can only be answered relative to a quantitative analysis of the thermal behaviour of a wide range of ethnographic buildings from diverse climates and cultures.

Any such study must also examine the thermal behaviour of buildings holistically and not focus on only select building properties. For buildings, like other complex systems, possess emergent properties and buildings possess emergent thermal properties. The emergent properties of complex systems cannot be precisely ascertained from a knowledge of only single features, for single building features do not equate to a building's thermal performance. Buildings are *thermal machines*, whether the builders or occupants are aware of this or not, and the properties of the *thermal machine* are the product of ever-changing interactions between the vast number of physical, thermodynamic and spatial features and traits (Szokolay 1987: 21; Clarke 2001: ix). Thermal behaviour must be studied at the scale of whole interrelated assemblages of built features and traits that must include the building's primary thermal features. Buildings should also be examined in terms of their *thermal capacity*, rather than thermal 'performance'. This is because thermal performance possesses a time component. It equates to all the actual thermal states that the building has undergone from moment to moment through time and is consequently intangible at any useful level of analysis. *Thermal capacity*, on the other hand, equates to the full range of thermal states that a building is capable of experiencing or has ever undergone: the full range of thermal states that the buildings' occupants have at their disposal, whether or not they utilise the full range or not. Following is an outline of a quantitative study that examines the *thermal capacity* of a wide range of ethnographic buildings holistically, relative to their respective climate at three different scales of classificatory detail.

THE QUANTITATIVE STUDY

The quantitative study outlined here examines the generic buildings of ethnographic societies. By limiting the study to only ethnographic buildings it is possible to test for various factors that may have potentially influenced the long-term pattern of change in the *thermal capacity* of buildings without examining the vastly larger assemblage of buildings in the historical and archaeological records. This has been done elsewhere (Wilkins 2009a). The purpose of the study was to ascertain the statistical correlation that exists between the *thermal capacity* of generic ethnographic buildings and their climate, for both mobile and sedentary societies.

The set of buildings examined is based on the Standard Cross Cultural Sample, which contains 157 societies from diverse regions and climates from around the world (Cullen 1993: 215-259). The Standard Cross Cultural Sample is based on Murdock's two sets of ethnographic data – the Ethnographic Atlas (Murdock 1967) and the reduced set (Murdock and Wilson 1972). The pertinent information was the geographical bearing (from the reduced set), the cultural and social grouping (from the Atlas and the reduced set), the time period (from the reduced set), and the building features (from the Atlas). These data and their relevant climatic classifications are listed in Table 1.

Table 1. Entities (Generic Buildings) Included in Study

Simplified Climate	Broad Köppen-Trewartha Climate	Detailed Köppen-Trewartha Climate	Culture	Cultural Group	Approx. Time Period	Bearing
Tropical	*Am*	*Amaa*	Mundurucu	Equatorial	1850	7s,57w
			Saramacca	Indo-European	1928	3n,56w
		Amha	Mende	Niger-Kordofanian and Micellaneous African	1945	8n,12w
			Toda	Dravidian and Mon-Khmer/Austroasiatic	1900	11n,76e
	Ar	*Araa*	Cubeo	Equatorial	1939	1n,71w
			Mbuti	Khoisan and Mbuti	1950	1n,28e
			Nkundo	Niger-Kordofanian and Micellaneous African	1930	0,18e
			Ibo		1935	5n,7e
			Cuna	Mayan, Cariban and Macro- Chibchan	1927	9n,78w
			Cayapa		1908	1n,79w
			Carib		1932	7n,60w
			Amahuaca	Ge-Panoan and Andean	1960	10s,72w
			Kimam	Malayo-Polynesian	1960	7s,138e
			Javanese	Thai-Kadai	1954	8s,112e
			Balinese		1958	8s,115e
			Iban		1950	2n,112e
			Badjau		1963	5n,120e
			Trobrianders		1914	9s,151e
			Tikopia		1930	12s,168e
			Manus		1937	2s,147e
			Marquesans		1800	9s,140w
			Samoans		1829	14s,172w
			Marshallese		1900	6n,169e

Simplified Climate	Broad Köppen-Trewartha Climate	Detailed Köppen-Trewartha Climate	Culture	Cultural Group	Approx. Time Period	Bearing
			Trukese		1947	7n,152e
			Palauans		1947	7n,134e
		Arab	Banen	Niger-Kordofanian and Micellaneous African	1935	5n,10e
			Aymara	Ge-Panoan and Andean	1940	16s,66w
			Tanala	Thai-Kadai	1925	22s,48e
			Ajie		1845	21s,166e
		Arha	Vedda	Indo-European	1860	8n,81e
			Kwoma	Malayo-Polynesian	1960	4s,143e
			Siuai		1939	7s,155e
			Black Carib	Mayan, Cariban and Macro-Chibchan	1940	16n,89w
			Semang	Dravidian and Mon-Khmer/Austroasiatic	1925	4n,101e
			Nicobarese		1870	7n,94e
			Yapese	Thai-Kadai	1910	9n,138e
			Ifugao		1910	17n,121e
			New Irelanders		1930	2s,151e
		Arhh	Gilbertese	Thai-Kadai	1890	3n,172e
	Aw	Awaa	Warrau	Mayan, Cariban and Macro-Chibchan	1935	9n,62w
			Ashanti	Niger-Kordofanian and Micellaneous African	1895	7n,2w
			Alorese	Thai-Kadai	1938	8s,125e
		Awab	Siriono	Equatorial	1942	14s,63w
			Trumai		1938	12s,54w
			Nambicuara	Ge-Panoan and Andean	1940	12s,59w
		Awha	Tiwi	Malayo-Polynesian	1929	12s,131e
			Orokaiva		1925	8s,148e
			Timbira	Ge-Panoan and Andean	1915	6s,45w
			Fon	Niger-Kordofanian and Micellaneous African	1890	7n,2e
			Azande		1905	5n,28e
			Tiv		1920	7n,9e
			Andamanese	Dravidian and Mon-Khmer/Austroasiatic	1860	12n,93e
			Rhade	Thai-Kadai	1962	13n,108e
		Awhb	Lengua	Ge-Panoan and Andean	1889	23s,59w
			Lamet	Dravidian and Mon-Khmer/Austroasiatic	1940	20n,101e
			Burmese	Tibetan	1965	22n,96e
Temperate Hot-arid	BS	BSlk	Tehuelche	Ge-Panoan and Andean	1870	46s,70w
		BSha	Shilluk	Niger-Kordofanian and Micellaneous African	1910	10n,32e
	BW	BWha	Wolof	Niger-Kordofanian and Micellaneous African	1950	14n,15e
			Otoro		1930	11n,31e
		BWhl	Egyptians	Afroasiatic	1950	25n,33e
			Papago	Uto-Aztecan & Zuni	1910	32n,112w
		BWia	Songhai	Niger-Kordofanian and Micellaneous African	1940	17n,1w

Table 1. (Continued)

Simplified Climate	Broad Köppen-Trewartha Climate	Detailed Köppen-Trewartha Climate	Culture	Cultural Group	Approx. Time Period	Bearing
Temperate	Cr	Crak	Gheg	Indo-European	1910	42n,20e
			Japanese	Japanese/Korean	1950	34n,134e
			Creek	Macro-Penutian, Kutenai, Caddoan, Iroquoian & Natchez-Muskogean	1800	33n,85w
		Cral	Aweikoma	Ge-Panoan and Andean	1932	28s,50w
		Crbl	Maori	Thai-Kadai	1820	35s,174e
		Crhl	Atayal	Thai-Kadai	1930	24n,121e
	Cs	Csak	Riffians	Afroasiatic	1926	35n,3w
			Neapolitans	Indo-European	1960	41n,13e
		Csbk	Eastern Pomo	Salishan, Hokan & Siouan	1850	39n,123w
		Csik	Kurd	Indo-European	1951	36n,44e
		Csll	Lake Yokuts	Macro-Penutian, Kutenai, Caddoan, Iroquoian & Natchez-Muskogean	1850	35n,119w
	Cw	Cwak	Lolo	Tibetan	1910	29n,103e
		Cwhl	Santal	Dravidian and Mon-Khmer/Austroasiatic	1940	23n,87e
			Vietnamese	Dravidian and Mon-Khmer/Austroasiatic	1930	20n,105e
			Garo	Tibetan	1955	26n,91e
Temperate	DC	DCao	Koreans	Japanese/Korean	1947	38n,126e
			Omaha	Salishan, Hokan & Siouan	1860	41n,96w
		DCbo	Ainu	Far North Asian	1880	43n,143e
	DO	DOlk	Irish	Indo-European	1932	53n,10w
			Mapuche	Ge-Panoan and Andean	1950	38s,73w
		DObk	Yurok	Macro-Algonkian	1850	41n,124w
			Twana	Salishan, Hokan & Siouan	1860	47n,123w
			Klamath	Macro-Penutian, Kutenai, Caddoan, Iroquoian & Natchez-Muskogean	1860	43n,122w
		DOlk	Haida	Nadene and Eskimo	1875	54n,132w
Cold	DC	DCbc	Russians	Indo-European	1955	53n,41e
			Montagnais	Macro-Algonkian	1910	48n,72w
		DClc	Gilyak	Far North Asian	1890	54n,142e
Temperate	EO	EOlo	Eyak	Nadene and Eskimo	1890	60n,145w
			Aleut		1800	54n,167w
Cold	EC	ECbc	Saulteaux	Macro-Algonkian	1930	51n,95w
		EClc	Ingalik	Nadene and Eskimo	1885	62n,159w
		ECld	Yukaghir	Far North Asian	1900	70n,145e
			Slave	Nadene and Eskimo	1940	62n,122w
Temperate	FT	FTkk	Yahgan	Ge-Panoan and Andean	1865	55s,69w
Cold		FTkc	Lapps	Uralic	1950	68n,22e
			Yurak		1894	68n,51e

Simplified Climate	Broad Köppen-Trewartha Climate	Detailed Köppen-Trewartha Climate	Culture	Cultural Group	Approx. Time Period	Bearing
		FTkd	Copper Eskimo	Nadene and Eskimo	1915	69n,110w
			Chukchee	Far North Asian	1900	66n,177e
Tropical	GAm	GAmhl	Lakher	Tibetan	1930	22n,93e
	GAr	GArab	Kapauku	Malayo-Polynesian	1955	4s,136e
		GArbb	Ganda	Niger-Kordofanian and Micellaneous African	1875	0,32e
	GAw	GAwab	Hadza	Khoisan and Mbuti	1930	3s,34e
			Masai	Niger-Kordofanian and Micellaneous African	1900	3s,36e
			Luguru		1925	7s,38e
			Amhara	Afroasiatic	1953	13n,37e
		GAwal	Lozi	Niger-Kordofanian and Micellaneous African	1900	16s,23e
		GAwbb	Mbundu	Niger-Kordofanian and Micellaneous African	1890	12s,16e
		GAwbl	Miskito	Mayan, Cariban and Macro-Chibchan	1921	15n,91w
			Bemba	Niger-Kordofanian and Micellaneous African	1897	11s,31e
			Nyakyusa		1934	9s,34e
			Thonga		1865	26s,32e
			Kikuyu		1920	1s,37e
			Quiche	Mayan, Cariban and Macro-Chibchan	1930	15n,91w
			Kaffa	Afroasiatic	1905	7n,36e
		GAwha	Konso	Afroasiatic	1935	5n,37e
		GAwhb	Gond	Dravidian and Mon-Khmer/Austroasiatic	1940	19n,82e
		GAwhl	Huichol	Uto-Aztecan & Zuni	1890	23n,105w
Hot-arid	GBW	GBWha	Aranda	Malayo-Polynesian	1896	24s,134e
		GBWhb	Fur	Afroasiatic	1880	13n,25e
			Somali		1900	9n,47e
		GBWhk	Rwala	Afroasiatic	1913	33n,38e
			Havasupai	Salishan, Hokan & Siouan	1910	36n,112w
		GBWhl	Teda	Afroasiatic	1950	21n,17e
			Tuareg		1900	23n,6e
Temperate	GBW	GBWak	Chiricahua	Nadene and Eskimo	1875	32n,109w
			Zuni	Uto-Aztecan & Zuni	1880	36n,109w
		GBWal	Nama	Khoisan and Mbuti	1860	27s,17e
	GBS	GBSao	Pawnee	Macro-Penutian, Kutenai, Caddoan, Iroquoian & Natchez-Muskogean	1867	42n,100w
		GBSab	Kung	Khoisan and Mbuti	1950	20s,21e
Cold		GBSld	Khalka	Altaic	1920	47n,96e
Temperate	GCs	GCsao	Turks	Altaic	1950	39n,34e
		GCshk	Basseri	Indo-European	1958	29n,54e
	GCwhk	GCwhk	Comanche	Uto-Aztecan & Zuni	1870	33n,100w
Cold	GDC	GDCbc	Gros Ventre	Macro-Algonkian	1880	49n,109w

Table 1. (Continued)

Simplified Climate	Broad Köppen-Trewartha Climate	Detailed Köppen-Trewartha Climate	Culture	Cultural Group	Approx. Time Period	Bearing
			Hidatsa	Salishan, Hokan & Siouan	1836	47n,101w
			Kazak	Altaic	1885	48n,70e
	GDO	GDOlc	Bellacoola	Salishan, Hokan & Siouan	1880	52n,126w
Temperate	GDC	GDCao	Wadadika	Uto-Aztecan & Zuni	1870	43n,119w
	GDO	GDObo	Kutenai	Macro-Penutian, Kutenai, Caddoan, Iroquoian & Natchez-Muskogean	1890	49n,117w
Cold	GEC	GEClc	Kaska	Nadene and Eskimo	1900	60n,131w
Temperate	HBS	HBSlo	Lepcha	Tibetan	1937	27n,89e
Tropical	HAw	HAwll	Jivaro	Equatorial	1920	3s,78w

The Methodology: Multivariate Statistical Analysis

The entities (the generic buildings) were grouped according to the climate of their type-site, which varied in classificatory detail. The climate was examined at three different levels of classificatory detail, resulting in three different data sets. The broadest climate classification possessed fewer categories but each category applied to more buildings, and the most detailed climate classification possessed more climate categories but each category applied to fewer buildings. The buildings were treated as synonymous with their arrays of unique and idiosyncratic characteristics (the building variables). Thus each of the three data sets (tests) comprised arrays of variables grouped according to climatic similarity. Each of the three data sets was then analysed using Multivariate Statistical Analysis (MVA) for climate:*thermal capacity* correlation. The MVA techniques applied were Discriminant Analysis, because it allowed the buildings to be grouped according to their climate, and Correlation Analysis, which was able to show whether or not the results were statistically significant.

The Data: Climate

The first data set (Test 1) used the climate classification that is traditionally used in Comfortable Vernacular studies and which produces a very broad order of classification. The second (Test 2) and third (Test 3) data sets both utilised the Köppen-Trewartha classification method, the only difference being that the second is a more broad classification than the third, which is a relatively very detailed order of classification.

All three tests used the same two-step analytical process. The first step established the geographical bearing of the type-site, available from Murdock (Murdock & Wilson 1972: 278-295). The second step then established the climatic classification at that bearing, using either the traditional/simplified, the broad-Köppen-Trewartha or the detailed-Köppen-Trewartha classifications.

Table 2. The Traditional/Simplified Climate Classification

Traditional/simplified Climate Classification	
Abbreviation	Climate
C	cold
H	hot-arid
Temp	temperate
Trop	tropical (hot-humid)

Test 1: The Traditional/Simplified Climate Classification. Test 1 used the traditional method for classifying climates. This is simple and broad and uses only four different categories (cold, hot-arid, temperate and tropical) (Table 2). This classification is used in the vast majority of literature that deals with the thermal performance principles of buildings (e.g. Givoni 1998). Such literature tends to discuss detailed thermal principles at a microclimatic scale of detail, yet does so within an extremely broad climate classification. The logic of doing this is questionable given that there are as many different types of climate in the world as there are types of microclimates within a building.

Test 2: The Köppen-Trewartha Broad Climate Classification. Test 2 used the Köppen-Trewartha method for classifying climates. This is extensively used for classifying climates outside of the building thermal performance literature. It was initially developed by Wladimir Köppen, a Russian climatologist, c. 1900, and further modified by him in 1918 and 1936. The methodology combines average annual temperatures, monthly temperatures, precipitation and precipitation seasonality. There are six generally accepted principal classes of climate (tropical, subtropical, temperate, subarctic, polar and dry). These classifications were developed by Köppen, but Trewartha further classified them into 16 principal climates by temperature for the warmest and coldest months and precipitation for the wettest and driest months of the year (Rudloff 1981: 81-85). These sixteen classifications are, however, often simplified into eight broad classifications (Table 3).

Table 3. The Broad Köppen-Trewartha Climate Classification

Broad Köppen-Trewartha Climate Classification	
Abbreviation	Climate
Ar	tropical rain climate
Am	tropical monsoonal rain climate
Aw, As	tropical summer and winter rain climate
BS, BW, BM, FT, FI	steppe, desert, marine desert, tundra and ice climate
Cr	subtropical rain climate
Cw, Cs	subtropical summer and winter rain climate
DO, DC	temperate oceanic and continental climate
EO, EC	subartic oceanic and continental climate

Table 4. The Detailed Köppen-Trewartha Climate Classification (to be read in conjunction with Table 5)

Detailed Köppen –Trewartha Climate Classification	
Abbreviation	Climate
Ar	tropical rain climate
Am	tropical monsoonal rain climate
Aw	tropical summer rain climate
As	tropical winter rain climate
BS	steppe climate
BW	desert climate
BM	marine desert climate
Cr	subtropical rain climate
Cw	subtropical summer rain climate
Cs	subtropical winter rain climate
DO	temperate oceanic climate
DC	temperate continental climate
EO	subartic oceanic climate
EC	subartic continental climate
FT	tundra climate
FI	ice climate

Table 5. The 'Universal' Thermal Scale

'Universal' Thermal Scale		
Temperature Range	Term	Code
35°C to …..	severely hot	i
28°C to 34°C	very hot	h
23°C to 27°C	hot	a
18°C to 22°C	warm	b
10°C to 17°C	mild	l
0°C to 9°C	cool	k
-9°C to -1°C	cold	o
-24°C to –10°C	very cold	c
-39°C to -25°C	severely cold	d
…. to -40°C	excessively cold	e

Test 3: The Köppen-Trewartha Detailed Climate Classification. Test 3 expanded the eight Köppen-Trewartha categories to include the full sixteen (Table 4) and appended further information pertaining to the average air temperature during the warmest and coldest months according to the Universal Thermal Scale (Table 5). The climates were further detailed by prefixing a G (for mountain climate) where the altitude was at least 500m, or an H (for high mountain climate) where it was at least 2500m. The reasoning behind this is that the number and distribution of weather stations is high and small differences that may exist between the type-site climate and that of the nearest weather station can in the majority of cases be easily accommodated within the granularity of the classification system. However, in very mountainous regions the difference between the climate at the nearest weather station and the

type-site can be high. Therefore, the climate classification for the weather station that equated most closely to the type-site latitude, longitude *and altitude* above sea level was adopted, rather than simply the geographically closest.

The Data: Generic Buildings

Murdock's Ethnographic Atlas consists of 412 societies from around the world for which a wide range of social, economic and technological information is listed. These are not, however, evenly distributed around the world, but tend to cluster in certain regions. There are, for example, very few examples from Europe and Northern Asia. The problems associated with the uneven distribution inherent in the Ethnographic Atlas have been recognised (e.g. Cullen 1993: 226-228). "An 'autocorrelation' problem exists to the extent that neighbouring societies tend to be similar. In a recent unpublished paper on "Design Effects in Standard Samples", Malcolm Dow shows that use of the Ethnographic Atlas as a sample may lead to underestimates of variance by orders of magnitude" (White 1985: 5). To overcome these problems Murdock and Wilson reduced the original 412 societies to a smaller subset of 197 (Murdock & Wilson 1972). This smaller subset consists of a selection of one society from each 'cultural group', some of which originally contained several individual societies. The reduced subset includes the date to which the original information pertains, as well as bibliographical sources (Murdock & Wilson 1972: 254-255). The Standard Cross Cultural Sample is an even further reduced dataset which contains 157 societies (Cullen 1993: 215-259) and it is this Sample, with several more societies removed for reasons discussed below, that forms the basis of the datasets used here (Figure 4). Whilst the final set is low on certain specific climates (notably cold climates), each climate type is sufficiently well represented to adequately test for a possible climate:*thermal capacity* correlation. That is, regardless of bias that might exist within the data, sufficient examples exist to reveal the existence or absence of a correlation.

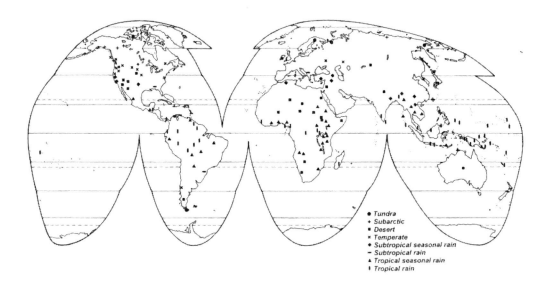

Figure 4.

The reasons for removing several further societies are, first, an absence of data about the buildings. For example, the Armenians lacked information pertaining to the building features. Secondly, examples that dated to A.D. 1750 or earlier were removed due to their non-ethnographic origins. That is, earlier examples represented an early stage in the development of the buildings within the region. For example, the Inca in the New World, c. A.D. 1530, and the Babylonians in the Old World, c. 1750 B.C. both represent an early phase in the development of buildings within the regions and are non-ethnographic. Where an alternative example from the same 'cultural group' existed this was used to replace the eliminated example. For example, the Neopolitans c. A.D. 1960 replaced the Romans c. A.D. 110. If there was no suitable replacement, however, the society was removed and not replaced. The final number of societies in the database is 145.

The Variables: Generic Building Features

The building variables used in the study consist of the various characteristics of the main settlement dwellings, plus the presence/absence of communal buildings. The presence/absence of communal buildings is included so as to be able to extend the analysis beyond just the buildings themselves and incorporate some of the *thermal capacity* available within the settlement as a whole. Buildings do not operate in thermal isolation from their surroundings and the presence of other buildings, such as communal buildings, have the capacity to extend a person's *thermal capacity*.

The variables included in the dataset are: presence of 'Large or Impressive Structures' (Column 6 from Murdock & Wilson (1972) (Table 6); 'Ground Plan of Dwelling' (Columns 80 from Murdock (1967) (Table 7); 'Floor Level' (Column 81 from Murdock (1967) (Table 8); 'Wall Material' (Columns 82 from Murdock (1967) (Table 9); 'Shape of Roof' (Columns 83 from Murdock (1967) (Table 10); and 'Roofing Material' (Columns 84 from Murdock (1967) (Table 11).

The original data was in a form that was inappropriate for MVA, which requires metric variables that are numerically graded or ranked. Murdock's variables are qualitative and not quantitative, and they were not graded. The primary data was therefore converted to graded metric values such that the numerical equivalents quantitatively defined the level of *thermal choices* and *thermal control* inherent in the variable, as per the results of a separate engineering-analysis (ref. Wilkins 2009b).

Variable 1: Large or Impressive Structures. The presence of Large or Impressive Structures in a settlement increases both the *thermal choices* and *thermal control* of the occupants of the settlement in proportion to their accessibility. A structure that is fully publicly accessible extends the *thermal capacity* of all occupants, but a structure that is only partially or selectively accessible extends the thermal environment of only those occupants who have access to it. Note that this is the only variable of those included in the dataset in which there is a direct correlation between the properties of the variable and both *thermal choices* and *thermal control*. That is, both *thermal choices* and *thermal control* increase proportionately with public accessibility to communal buildings, and decrease as a result of inaccessibility (Table 6). With each other variable, as the properties of the variable are enhanced in one direction *thermal choices* increase and *thermal control* decreases, and as the properties are enhanced in the opposite direction *thermal choices* decrease and *thermal control* increases.

Table 6. Discriminant Values for Large or Impressive Structures

Large or Impressive Structures				
Murdock		Converted Variable and Value Used in the Study		
Code	Description	Value	Description	
A	Presence of public building (assembly hall etc.)	*1*	Public building fully accessible	
T	Presence of religious or ceremonial building (church etc.)			
F	Presence of military installation (fort etc.)	*2*	Public building selectively accessible	
E	Presence of economic or industrial building (storehouse, factory etc.)			
R	Presence of impressive residence (owned by headman etc.)	*3*	Private buildings only	
O	No structures larger than usual residential dwellings			

Variable 2: Ground Plan of Dwelling. A structure that is only semi-enclosed to the outside or quadrangular in plan offers enhanced *thermal choices*, because it offers choice via the variety inherent in the external environment, but it offers minimal *thermal control*, for the same reason. Conversely, a circular structure offers very few *thermal choices,* but it does offer enhanced *thermal control,* due to the homogeneousness of the spatial layout. The more thermally homogeneous an environment, the more easily controlled it is and the more easily it can be selectively altered, compared with a variable one (Table 7).

Table 7. Discriminant Values for Ground Plan of Dwelling

Ground Plan of Dwelling				
Murdock		Converted Variable and Value Used in the Study		
Code	Description	Value	Description	
		1	Semi-enclosed	
Q	Quadrangular or partially quadrangular	*2*	Quadrangular	
R	Rectangular or square	*3*	Rectilinear	
E	Elliptical		Elliptical	
S	Semicircular		Semicircular	
P	Polygonal	*4*	Polygonal	
C	Circular	*5*	Circular	

Table 8. Discriminant Values for Floor Level

Floor Level				
Murdock		Converted Variable and Value Used in the Study		
Code	Description	Value	Description	
P	Raised substantially on piles etc.	1	Raised substantially on piles etc.	
E	Slightly elevated on a raised platform etc.	2	Slightly elevated on a raised platform etc.	
G	Floor at ground level	3	Floor at ground level	
S	Subterranean or semi-subterranean	4	Subterranean or semi-subterranean	

Table 9. Discriminant Values for Wall Material

Wall Material				
Murdock		Converted Variable and Value Used in the Study		
Code	Description	Value	Description	
		1	No walls present	
M	Mats, latticework or wattle	2	Permeable light thermal mass	
H	Hides or skins			
G	Grass, leaves or other thatch			
B	Bark			
F	Felt, cloth or other fabric			
W	Wood, logs, planks, poles, bamboo or shingles	3	Non-permeable medium thermal mass	
P	Plaster, mud and dung, or wattle and daub			
A	Adobe, clay, or dried brick	4	Non-permeable heavy thermal mass	
S	Stone, stucco, concrete, or fired brick			
		5	Subterranean	

Variable 3: Floor Level. A structure that is raised high above the ground, such that it can interface with the external environment in all directions (below as well as outwards and above) offers enhanced *thermal choices*, but minimal *thermal control*, because it interfaces more with the uncontrolled natural elements. Conversely, a subterranean structure offers very few *thermal choices*, but it does offer enhanced *thermal control*: the further below ground the space, the more homogeneous is the environment and therefore the more easy it is to selectively and controllably alter (Table 8).

Table 10. Discriminant Values for Roof Shape

Wall Material					
Murdock		Converted Variable and Value Used in the Study			
Code	Description	Value	Description		
B	Beehive shaped with pointed peak	1	Beehive or pyramidal		
H	Hipped or pyramidal				
D	Dome shaped or hemispherical	2	Dome		
S	Shed, ie. with one slope	3	Skillion, ie. with one slope		
G	Gabled, ie. with two slopes	4	Gabled, rounded and semi-hemispherical		
R	Rounded or semi-cylindrical				
E	Semi-hemispherical				
C	Conical				
F	Flat or horizontal	5	Flat		

Table 11. Discriminant Values for Roofing Material

Roofing Material					
Murdock		Converted Variable and Value Used in the Study			
Code	Description	Value	Description		
M	Mats	1	Permeable light thermal mass		
F	Felt, cloth or other fabric				
H	Hides or skin				
B	Bark				
W	Wood, logs, planks, poles, bamboo or shingles	2	Non-permeable medium thermal mass		
T	Tile or fired brick				
S	Stone or slate				
G	Grass, leaves, brush or other thatch				
P	Plaster, clay, mud and dung, or wattle and daub	3	Non-permeable heavy thermal mass		
I	Ice or snow	4	Non-permeable very heavy thermal mass		
E	Earth or turf				
		5	Subterranean		

Variable 4: Wall Material. A structure that has walls with a light thermal mass, such that they interface closely with the external environment, offers enhanced *thermal choices* but minimal *thermal control.* Conversely, a structure that has a very heavy thermal mass offers

very few *thermal choices*, but it does offer enhanced *thermal control* because the interior environment will be more homogeneous (Table 9).

Variable 5: Roof Shape. A tall, pointed, domed roof offers enhanced *thermal choices* because it accentuates the vertical temperature distribution, generating hotter air at the peak relative to the air at floor level, but it will have minimal *thermal control* for the same reason: the vertical temperature distribution has an inherent variability. Conversely, a flat roof offers very few *thermal choices*, but it does offer enhanced *thermal control*, because the flat roof minimises the vertical temperature distribution, creating an homogeneous internal environment (Table 10).

Variable 6: Roofing Materia. A structure that has a roof with a light thermal mass operates in a similar manner to its having walls with a light thermal mass. A structure with a light roof will interface closely with the external environment and offers enhanced *thermal choices*, but minimal *thermal control*. Conversely, a structure with a light roof will offer very few *thermal choices*, but it will offer enhanced *thermal control* (Table 11).

ANALYSIS

The results of the Discriminant Analysis are shown visually on the scatterplots (Figures. 5-7). Each point on the plot represents the *thermal capacity* of one generic building (the sum of its interrelated variables) correlated with its respective climate classification. At all three levels of detail of climate classification the scatterplots show a wholly random pattern of *thermal capacity*. This thus indicates that no correlation exists between the *thermal capacity* of ethnographic buildings and climate, regardless of the degree of climatic classificatory detail.

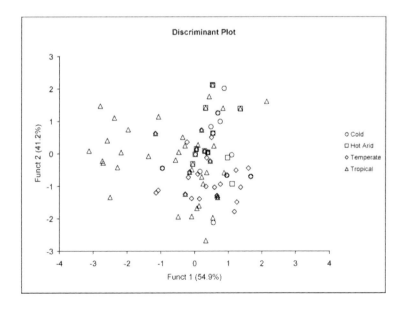

Figure 5.

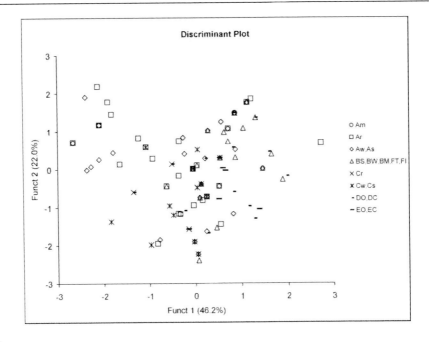

Figure 6.

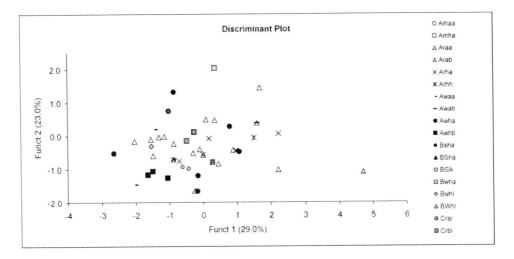

Figure 7.

This visual non-correlation is supported by the Correlation Analysis, which shows empirically that no statistical correlation exists between the *thermal capacity* of ethnographic buildings and climate at all levels of climatic classificatory detail: traditional/simple method (r = 0.247), broad Köppen-Trewartha method (r = 0.285) and detailed Köppen-Trewartha method (r = 0.348). However, the detailed Köppen-Trewartha classification gives a result that is statistically significant (p = 0.005), whereas the broad Köppen-Trewartha method and the traditional methods are progressively unreliable (p=0.062 and p = 0.177 respectively).

It can, therefore, be stated that, whilst studies examining the thermal behaviour of classes of buildings in relation to progressively more detailed climatic data produce progressively more reliable results, even the most detailed set of climatic data shows that no statistical correlation exists between the *thermal capacity* of ethnographic buildings and climate. This result thus directly and quantitatively challenges the Comfortable Vernacular theory.

ADAPTIVE COMFORT THEORY AS EXPLANATION

So what then does this non-correlation mean? Has the pattern of thermal change in buildings been wholly random over time? Are humans wholly random in their thermal responses and expectations? The answer is, regrdless of the non-correlation with climate, thermal behaviour of humans is not random at a statistical level and the pattern of thermal change in buildings has likewise not been random over time. The way in which buildings have behaved thermally since the time of the earliest built structures has progressed along a trajectory, the boundaries of which are quantifiably definable. The trajectory has followed a transitional path towards maximum possible *thermal choices* and *thermal control*, even though achieving this becomes increasingly difficult as the 'point of maximum' thermal choices and control is neared (Wilkins 2009a). Throughout history, those classes of buildings that have fallen outside of the boundaries of the trajectory have fallen out of general use, not to reappear. One such example was the prehistorical, almost universal transition from brushwood huts to pithouses to above-ground rectilinear buildings. This transition constituted an example of architectural convergence, having occurred in numerous culturally unrelated regions around the world (Wilkins 2009b). Today very few societies inhabit brushwood huts and pithouses as their predominant class of building.

The explanation for the existence of this long term trajectory or trend lies with what has become known as Adaptive Comfort theory (ACT). Within the ACT model quantitative trends in human thermal perception and adjustment have been observed and recorded. Whilst, again, not all human thermal behaviour falls within the boundaries of the trend, the boundaries are quantifiably definable. ACT thus offers a viable and convincing explanation for the non-correlation between the *thermal capacity* of buildings and climate, because it offers a non-deterministic, quantifiably definable explanation for the immense variability that exists within human thermal behaviour and their built structures. If humans had been deterministically directing the course of change in buildings and built environments over the past millennia, why then does history abound with anomalous, non-'ideal' buildings? Any theory that purports to be able to explain the nature of the change must account for *all* change and *all* anomalies. Comfortable Vernacular theory cannot account for examples of change in buildings which resulted in supposedly less thermally comfortable buildings, such as the pithouse-to-pueblo transition, or for the presence of thermally anomalous buildings, such as the structures of Tierra del Fuego. Adaptive Comfort Theory can do this.

The current interest in human thermal 'comfort' and adjustment began in 1936 when focus shifted away from examining physiological thermal responses in humans and moved towards examining their subjective responses (Bedford 1936). However, it was not until the results of the early studies were compiled by Humphreys in the mid-1970s (Humphreys 1975; Humphreys 1978) that the theory was accepted within the mainstream architectural profession, where it became known as Adaptive Comfort theory. Adaptive Comfort theoreticians, such as Humphreys, Auliciems, de Dear, Brager and others, as well as the American Society for Heating, Refrigerating and Air-Conditioning Engineers (ASHRAE), have since undertaken numerous studies of large numbers of building occupants from sedentary societies (Humphreys 1975; Humphreys 1978; Auliciems 1983; Humphreys 1995; de Dear 1998a). The global database (Humphreys 1975; Humphreys 1978; de Dear 1994; de Dear & Brager 1998) now contains over 21,000 measurements from numerous regions around the world. Although the underlying causal mechanisms driving the common behaviours in human thermal adjustments indoors has been questioned (de Dear 1994: 111-113; Brager & de Dear 1998: 88-92), the central tenets of the theory have remained unchanged.

The central tenets state that, first, occupant 'satisfaction' is higher when there is a perceived level of thermal control, even if the occupant confesses to a level of thermal discomfort, in preference to being always thermally neutral. This is regardless of social and climatic differences. Secondly, humans can experience physiological thermal stress without showing signs of psychological strain, in the form of mental dissatisfaction or distress. Thirdly, the average temperatures inside buildings at which the (sedentary and habitual) occupants are thermally 'satisfied' is quantifiable, falls within a 4°C band of the neutral temperature that is dependant on the average monthly outdoor temperature, and appears to track the average outdoors monthly temperatures. Again, this is regardless of climatic and/or social differences. This thus indicates that an epiphenomenal relationship exists between humans and their selective thermal environments (Humphreys 1975; Humphreys 1978; Auliciems 1989; de Dear 1994; Nicol et al. 1994; Williamson et al. 1995; Baker & Standeven 1996; Nicol & Raja 1996; Humphreys 1997; Brager & de Dear 1998; de Dear & Brager 1998; McCartney et al. 1998; de Dear & Brager 2001; Nicol & Humphreys 2001). Significantly, the international standards for thermal environmental conditions inside buildings, which are set by ASHRAE and which have to date been limited to only mechanically-conditioned buildings, were amended in 2004 to include buildings in which the occupants can manually control their thermal environments (ASHRAE 2004). Past and continuing ASHRAE funded research (Plant Engineering 2005) has shown that where occupants can open and close windows their thermal responses will "depend in part on the outdoor climate and may differ from thermal responses in buildings with centralized HVAC systems primarily because of the different thermal experiences, changes in clothing, availability of control and shifts in occupational expectations" (ASHRAE 2004: 9-10; Olesen & Brager 2004). In recent years there has also been a growing number of studies that have investigated the thermal responses of people whilst out of doors and the incorporation of the results of these studies into the ACT model adds further weight to its validity: access to transient and outdoors environments extend a person's capacity to make selective thermal adjustments.

Thus it can be stated that people have a dynamic relationship with their immediate thermal environment and humans make use of both ambient and transitional thermal spaces. It is now generally agreed within the ACT model that people are active participants in person-environment feedback loops, adjusting naturally to changing thermal conditions (Brager & de Dear 1998). If thermal discomfort is experienced, people may react in ways to alleviate the discomfort, as long as they have the ability to do so, making adjustments to either their person or to their surroundings. They may also consciously choose to be 'uncomfortable' if that satisfies other (functional) needs. If people are able to make selective thermal adjustments they will tolerate greater degrees of discomfort without experiencing accompanying psychological strain. If, however, they do not possess this capacity they will experience psychological strain, even at thermally neutral temperatures. Thermal 'satisfaction' is thus seen as being synonymous with a lack of psychological strain, rather than a lack of thermal stress, although this distinction has not always been clear (Auliciems 1983: 71). People gain thermal 'satisfaction' from being able to choose between different thermal states and/or to alter their thermal environment at will. Possessing the capacity to make a compromise between contradictory thermal factors appears to be of much greater importance than the possession of thermal neutrality/comfort in influencing the types of thermal environments people choose to inhabit.

CONCLUSION

When buildings are statistically analysed at the scale of generic buildings, with features and traits belonging to whole classes of ethnographic buildings, no correlation appears to exist between the *thermal capacity* of buildings and their climate, although it should be noted that statistically significant results can only be achieved with greater, rather than lesser climatic information. This therefore challenges the Comfortable Vernacular theory. Humans have and do inhabit a wide range of classes of structures irrespective of the type of climate in which they live/d. "Mental well-being needs varied perception, as many researchers have pointed out... the implication is that the change of stimulation, rather than the absolute level involved is the more important consideration" (Gerlach 1974: 15).

Thermal environments that induce thermal strain in people should not, therefore, be automatically treated as deleterious but, rather, as environments in which people can make satisfactory thermal compromises between a wide range of contradictory thermal states. "Traditional buildings provide a variety of thermal environments between rooms and within a single room. The bedroom of an English house was colder than the living room, but the bed is warm and snug. A coal-burning fire gives a focus of radiant heat, which provides a variety of environments depending on how close you are to it, and so on. In the middle east, houses owned by the wealthy would have many rooms each with its own particular thermal characteristics from which the inhabitants could choose the most appropriate at different times of day and year" (Nicol & Raja 1993: 3).

So, if buildings have developed along a trajectory towards increasing selective thermal adjustability, how do we account for those buildings and classes of buildings that have fallen (and do fall) outside of the trend and which did not/do not offer their occupants the capacity to implement an instantaneous thermal change? Two factors need to be considered. First, it is far easier to effect an instantaneous change via more ephemeral ways of adjusting thermally, such as via one's clothing and activity, than it is to make a change via adjusting the building itself, particularly at a structural level.

Secondly, inertia exists inherently in the built environment along a scale of minimal to maximal. Smaller scale changes to a building such as, for example, opening and closing a window or even installing a new window, is an easier way to implement a satisfactory thermal change than, for example, implementing a larger structural change to a building. And structural changes to smaller, light weight, non-load-bearing buildings are more easily implemented than they are to larger, more massive, load-bearing buildings, which thus possess more inertia. In this regard it is significant that the number and diversity of thermally altering built components in buildings have generally increased in number and complexity over time (Wilkins 2009a). Windows, for example, which add significantly to a building's selective thermal adjustability, post-dated the construction of buildings by at least several thousand years: to date the earliest excavated hut dates to the Palaeolithic woven brushwood hut at Ohalo II, c. 17,000 B.C. (Nadel & Werker 1999) and the earliest archaeological evidence for windows in the same region dates to the late Neolithic, c. 6500 B.C. (Nissen et. al 1987: 93-94). Yet since that time external windows and doors have become ever more prevalent and mechanistically complex (Brand 1997: 144-145).

At the scale of whole buildings, the Japanese urban house is an ethnological class of building that is mechanistically complex and thermally highly adjustable, but one which has persisted in the face of climatic 'inappropriateness'. "The traditional Japanese house is essentially a tropical house, being lightweight in construction with highly permeable external walls... at first inspection these houses seem inappropriate to this harsh [cold] climate. However, a more thoughtful analysis reveals several interesting insights. Firstly, the internal spatial volume can be modified seasonally to be subdivided into smaller spaces in winter to limit airflow and reduce draughts and heat losses. Secondly, thermal comfort in winter was achieved by employing small, portable braziers, *hibachi* or *kotatsu*, [that could be] placed under tables and carried from room to room and were quite efficient in directly heating the individual by radiation" (Forwood 1995: 176).

Understanding the implications of the issues raised here is vital if we hope to build structures that are thermally, and therefore socially, satisfactory in the future. For not only is a building's longevity intrinsically linked to its capacity for thermal selective adjustability, but a built environment's capacity for selective thermal (and structural) adjustability is therefore one of the key factors in facilitating society's capacity to change freely. The thermal environments provided by buildings are the context within which social life operates. Adjustable built environments generate diverse thermal environments that increase the range of social options that the building milieu can accommodate, compared with less adjustable built environments, because they are more readily able to accommodate changing social options, circumstances and climate, all of which are unpredictable and unknowable in the long term. That is, a highly adjustable built environment is more likely to be occupied for a longer period of time prior to undergoing a system-level alteration because it can accommodate more internal social change and more external environmental and climatic change. Conversely, an inflexible built environment is more likely to become redundant sooner because it is more likely to sooner encounter change that it cannot accommodate without major system-level alteration. Social, climatic and environmental change is a reality, whether or not human-induced global warming is a reality or not, and if we are to create built environments that are capable of supporting our social needs, they must be able to accommodate future change in all its unpredictable and unknowable forms, in the short term and in the long term.

Of great importance then is the general international shift towards, and reliance on, mechanical heating, ventilation and air-conditioning systems (HVAC). HVAC offers supremely enhanced selective thermal adjustability, which presumably accounts for its success, particularly in tropical regions where otherwise the potential for such is seriously capped. But ultimately a reliance on HVAC reduces the flexibility demanded of the building itself. Thus the societies that rely on such systems become proportionately dependent on climatic and environmental stasis. If HVAC were to become unviable in the future, for what ever reasons, the buildings that utilise it may not be able to provide an equivalent level of selective thermal adjustability without it and will therefore cease to be able to accommodate the functions that they are relied on to accommodate. This is the essence of social sustainability and there are historical reasons for why we need to take this seriously. The past abounds with examples of societies and cultures that have faced similar situations, wherein their built environments became insufficiently selectively adjustable to accommodate changing times and circumstances, and which accordingly either died out or faded away (Wilkins 2005, 2009a).

REFERENCES

Alp, A. (1988). Vernacular Climate Control in Desert Architecture. *Energy and Buildings for Temperate Climates - A Mediterranean Regional Approach*, Proceedings of the Sixth PLEA Conference, Porto, Portugal, Pergamon Press.

ASHRAE. (2004). Standard 55-2004 *Thermal Environmental Conditions for Human Occupancy*. Atlanta, American Society of Heating, Refrigerating and Air-Conditioning Engineers, Inc. www.ashrae.org.

Auliciems, A. (1983). Psycho-Physiological Criteria for Global Thermal Zones of Building Design. *Biometeorology: Supplement to International Journal of Biometeorology 1982, 26 8(2)*, 69-86.

Auliciems, A. (1989). Thermal Comfort, in Ruck, N. (Eds.) *Building Design and Human Performance*. New York, Van Nostrand Reinhold, 3-28.

Bedford, T. (1936). Report *no.76, The Warmth Factor in Comfort at Work*. Report Industrial Health Research Board. London, H.M.S.O.

Brager, G. & de Dear, R. (1998). Thermal Adaptation in the Built Environment: A Literature Review. *Energy and Buildings, 27*, 83-96.

Brager, G. & de Dear, R. (2001). Climate, Comfort and Natural Ventilation: A New Adaptive Comfort Standard for ASHRAE Standard 55. *Moving Thermal Standards into the 21st Century*, Cumberland Lodge, Windsor, UK, Oxford Brookes University.

Brand, S. (1997). *How Buildings Learn: What Happens After They're Built* (revised). London, Phoenix Illustrated.

Butti, K. & Perlin, J. (1980). *A Golden Thread: 2500 Years of Solar Architecture and Technology. Palo Alto, Cheshire Books*.

Clarke, J. A. (2001). *Energy Simulation in Building Design*. Oxford, Butterworth-Heinemann.

Coch, H. (1988). Bioclimatism in Vernacular Architecture, in Gallo, C., Sala, M. & Sayigh, A. (Eds.) *Architecture: Comfort and Energy*. Amsterdam, Elsevier, 67-88.

Cofaigh, E. O., Olley, J. A. & Lewis, J. O. (1996). *The Climatic Dwelling: An Introduction to Climate-Responsive Residential Architecture*. London, James & James (Science Publishers).

Cook, J. Cpt. (1892). The *Voyages of Discovery of Captain James Cook*. London, Ward, Lock, Bowden & Co.

Cook, J. (1996). Architecture Indigenous to Extreme Climates. *Energy and Buildings, 23 (3)*, 277-291.

Crouch, D. & Johnson J. (2001*). Traditions in Architecture - Africa, America, Asia and Oceania*. New York, Oxford University Press.

Cullen, B. (1993). *The Cultural Virus*. Unpublished PhD Thesis, Department of Archaeology, University of Sydney.

de Dear, R. (1994). Outdoor Climatic Influences on Indoor Thermal Comfort Requirements, in Oseland, N. & Humphreys, M. (Eds.) *Thermal Comfort: Past, Present and Future*. Garston, UK, B.R.E.: 106-132.

de Dear, R. (1998). A Global Database of Thermal Comfort Field Experiments. *ASHRAE Transactions, 104(1B)*, 1141-1153.

de Dear, R. & Brager, G. (1998). Developing an Adaptive Model of Thermal Comfort and Preference. *ASHRAE Transactions, 104(1)*, 1-18.

de Dear, R. & Brager, G. (2001). The Adaptive Model of Thermal Comfort and Energy Conservation in the Built Environment. *International Journal of Biometeorology 45*, 100-108.

Denyer, S. (1978*). African Traditional Architecture: An Historical and Geographical Perspective*. London, Heinemann.

Engel, H. (1964). *The Japanese House: A Tradition for Contemporary Architecture*. Rutland, Vermont, Charles E. Tuttle Co.

Facey, W. (1997). *Back to Earth: Adobe Building in Saudi Arabia*, Al-Turath.

Feynman, R. P. (1985). "Surely You're Joking Mr. Feynman!": *Adventures of a Curious Character as Told to Ralph Leighton*. London, Vintage.

Fletcher, R. J. (1995). *The Limits of Settlement Growth*. Cambridge, University Press.

Forwood, B. (1995). Redefining the Concept of Thermal Comfort for Naturally Ventilated Buildings. *Ecolological Perspectives and Teaching Architectural Science*, 29th Conference of the Australian and New Zealand Architectural Science Association, Canberra.

Gallo, C. (1988). The Utilisation of Microclimate Elements, in Gallo, C., Sala, M. & Sayigh, A. (Eds.) *Architecture: Comfort and Energy*. Amsterdam, Elsevier: 89-114.

Gerlach, K. (1974). Environmental Design to Counter Occupational Boredom. *Journal of Architectural Research, 3(3)*, 15-19.

Gilligan, I. (2004). Another Tasmanian Paradox. Unpublished *M. Phil.* Thesis, Department of Archaeology, University of Sydney.

Givoni, B. (1998). *Climate Considerations in Building and Urban Design*. New York, Van Nostrand Reinhold.

Givoni, B. (1991). Urban Design for Hot Humid and Hot Dry Regions. *Architecture and Urban Space*, Proceedings of the Ninth International PLEA Conference, Seville, Spain, Kluwer Academic Publishers.

Hinrichs, C. (1988). The Courtyard Housing Form as a Traditional Dwelling in the Mediterranean Region. *Energy and Buildings for Temperate Climates* - A Mediterranean Regional Approach, Proceedings of the Sixth PLEA Conference, Porto, Portugal, Pergamon Press.

Humphreys, M. (1975). Field Studies of Thermal Comfort Compared and Applied. Garston, *Building Research Establishment*.

Humphreys, M. (1978). Outdoor Temperatures and Comfort Indoors. *Building Research and Practice, 6(2)*, 92-105.

Humphreys, M. (1995). Thermal Comfort Temperatures and the Habits of Hobbits: Comfort Temperatures and Climate, in Nicol, F., Humphreys, M., Sykes, O. & Roaf, S. (Eds.) *Standards for Thermal Comfort: Indoor Air Temperature Standards for the 21st Century*. London, Chapman & Hall: 3-13.

Humphreys, M. (1997). An Adaptive Approach to Thermal Comfort Criteria, in Clements-Croome, D. (Eds.) *Naturally Ventilated Buildings: Buildings for the Senses, the Economy and Society*. London, E & FN Spon: 129-137.

Keynes, R. D. (Eds.) (1988). *Charles Darwin's Beagle Diary*. Cambridge, University Press.

Kriken, J. (1983). Town Planning and Cultural and Climatic Responsiveness in the Middle East, in Golany, G. (Eds.) *Design for Arid Regions*. New York, Van Nostrand Reinhold Co., 97-120.

McCartney, K. J., Nicol, J. F. & Stevens, S. (1998). *Comfort in Office Buildings: Results from Field Studies and Presentation of the Revised Adaptive Control Algorithm*. CIBSE National Conference, Bournemouth.

McEwan, M., Borrero, L. & Prieto, A. (Eds.) (1997). *Patagonia: Natural History, Prehistory and Ethnography at the Uttermost End of the Earth*. Princeton, N. J., Princeton University Press.

Murdock, G. P. (1967). *Ethnographic Atlas*. Pittsburgh, University of Pittsburgh Press.

Murdock, G. P. & Wilson, S. F. (1972). Settlement Patterns and Community Organization: Cross-Cultural Codes 3. *Ethnology XI*, 254-295.

Nadel, D. & Werker, E. (1999). The Oldest Ever Brush Hut Plant Remains from Ohalo II, Jordan Valley, Israel (19,000 BP). *Antiquity 73,* 755-764.

Nissen, H. J., Muheisen, M. & Gebel, H. G. (1987). Report on the First Two Seasons of Excavations at Basta. *Annual of the Department of Antiquities, Jordan, 31,* 79-119.

Nicol, J. F. & Humphreys, M. A. (2001). Adaptive Thermal Comfort and Sustainable Thermal Standards for Buildings. *Moving Thermal Standards into the 21st Century*, Cumberland Lodge, Windsor, UK, Oxford Brookes University.

Nicol, F. & Raja, I. (1993). Thermal Comfort - *A Handbook for Field Studies Towards an Adaptive Model*, University of East London.

Olesen, B. W. & Brager, G. S. (2004). A Better Way to Predict Comfort. *ASHRAE Journal 46(8),* 20-27.

Pillet, M. L. R. (1980). *Pueblo House Design as a Response to the Arid-Zone Climate*, in Golany, G. (Eds.) Housing in Arid Lands: Design and Planning. London, The Architectural Press, 85-90.

Rapoport, A. (1969). House, Form and Culture. *Englewood Cliffs*, N. J., Prentice-Hall Inc.

Rudloff, W. (1981). *World Climates: with Tables of Climatic Data and Practical Suggestions*. Stuttgart, Wissenschaftliche Verlagsgesellschaft mbH.

Sayigh, A. & Marafia, A. (1998). Vernacular and Contemporary Buildings in Qatar, in Gallo, C., Sala, M. & Sayigh, A. (Eds.) *Architecture: Comfort and Energy*. Amsterdam, Elsevier, 25-37.

Shah, M. (1980). Nomadic Movements and Settlements of the Rabaris, in Murotani, B. (Eds.) *Indigenous Settlements in Southwest Asia*. Tokyo, Process Architecture Publishing Co. Ltd., 49-66.

Stead, P. (1980). Lessons in Tradition and Vernacular Architecture in Arid Zones, in Golany, G. (Eds.) *Housing in Arid Lands: Design and Planning*. London, The Architectural Press: 33-44.

Szokolay, S. V. (1987). *Thermal Design of Buildings*. Canberra, RAIA Education Division

Talib, K. (1984*). Shelter in Saudi Arabia*. London, Academy Editions.

White, D. R. (Eds.) (1985). Murdock and White's Standard Cross-Cultural Sample Codes 1. *World Cultures Journal, 1(1). Electronic journal.*

Wilkins, H. (2005). From Massive to Flimsy: The Declining Structural Fabric at Mohenjo-daro, in Franke-Vogt, U. & Weisshaar, H. J. (Eds.) *South Asian Archaeology 2003: Proceedings of the Seventeenth International Conference of the European Association of South Asian Archaeologists*: 137-146. Aachen, Linden Soft.

Wilkins, H. (2009a). *The Evolution of the Built Environment: Complexity, Human Agency and Thermal Performance*. Oxford, Archaeopress: British Archaeological Reports International Series.

Wilkins, H. (2009b). Transitional Change in Proto-Buildings: A Quantitative Study of Thermal Behaviour and its Relationship with Social Functionality. *Journal of Archaeological Science, 36*, 150-156.

Yuan, L. J. (1987). The Malay House: Rediscovering Malaysia's Indigenous Shelter System. *Pulau Pinang*, Phoenix Press.

In: Built Environment: Design, Management and Applications   ISBN: 978-1-60876-915-5
Editor: Paul S. Geller, pp. 197-214   © 2010 Nova Science Publishers, Inc.

Chapter 8

A SECONDARY MORTGAGE MARKET IN GHANA: ANTIDOTE OR POISON[1]?

Franklin Obeng-Odoom

Department of Political Economy, University of Sydney, Australia

ABSTRACT

There is considerable support for the development of the secondary mortgage market in Ghana. A secondary mortgage market, proponents argue, would be a panacea to the housing problems in Ghana. However, the studies on which a secondary mortgage market has been recommended have focused only on the process and structural pre-requisites needed for establishing such a market. As such, we do not know what the outcome of a secondary mortgage market would be in the Ghanaian context. This study takes up the challenge by analyzing the possible outcome of a secondary mortgage market in Ghana. It finds that a secondary mortgage market in Ghana would not resolve the housing problems of the poor and would condemn a majority of people to homelessness. It presents alternative approaches likely to offer a more lasting solution for the housing crisis in Ghana.

Keywords: Housing, Affordable, Ghana, Mortgage, Poverty

INTRODUCTION

Addressing the first ever West African affordable housing conference in Accra, Ghana, the executive director of UN-Habitat, Mrs. Ana Tibaijuku, dismissed any attempt by government to provide its citizens with affordable housing as an 'illusion'. Rather, she called for the institution of a secondary mortgage market to solve the housing crisis in Africa. In this direction, she called for pro-poor mortgage schemes to alleviate Africa's housing problems. In her own words, 'it was America's best practices in housing finance which attracted UN-

---

[1] This paper benefits from helpful comments by Prof. Frank J.B. Stilwell. The standard disclaimer applies.

Habitat to seek their expertise in finding workable solutions for Africa's housing crisis.' The four-day conference, a replica of a similar one in East Africa in 2005, brought together representatives of the government of Ghana, the United States and the UN-Habitat. Also in attendance were professionals from banks and other sectors of finance (UN-Habitat, 2006).

Many other housing experts, both local and international, have recommended a secondary mortgage market as a panacea to the housing problems in Ghana. One such advocate is Peter Asiedu, a Ghanaian real estate expert who has wondered why Ghana has failed to adopt this simple panacea for its housing problem. To him, Ghana should learn from the 'lofty' USA model. He draws striking contrasts between USA and Ghana in the following words:

> Walk into any banking institution in the USA and the first thing that greets you when you enter is a big banner proclaiming the easy availability of their HOME EQUITY LOANS. Indeed, right from the airport into town when you come to any part of the USA, on billboards, radio, TV and newspapers, the advertisements and commercials are all touting the easy availability of home equity loans and the need for the American people to take advantage of the so-called "American Dream" [of] home ownership.
>
> On the other side of the coin, come to good old mother Ghana and drive from Kotoka International Airport into town and pay a few visits to banking institutions in Accra. What do you find? Nothing about home equity loans! Indeed, on the few trips I have made home within the last few years, I have always wondered why (Asiedu, 2004).

It is not only housing professionals like Asiedu (2004) who hold this view. Academics at Ghana's premier department for the study of land economy, often trained in the UK at the prestigious Cambridge University, have also recommended a secondary mortgage market as the cure for Ghana's housing crisis. For example, Frank Gyamfi-Yeboah, a lecturer in land economy, and Nicholas Boamah note that a secondary mortgage market should be created in Ghana to ensure 'a sustainable flow of funds to the housing sector thus making it possible for many Ghanaians to be able to afford their own homes. This will, of course, be a major step in our bid to create a property owning democratic society' (Gyamfi-Yeboah and Boamah, 2003a). Elsewhere, the same authors make a passionate plea for government to take their advice for the creation of a secondary mortgage market in Ghana seriously. In their own words:

> It is our hope that government will give the attention that this sector deserves by facilitating the establishment of the REIT structure in Ghana. This, obviously, should not be a difficult task as experiences of other countries, particularly the US, can provide a useful guidance (Gyamfi-Yeboah and Boamah, 2003b).

In the face of such enthusiastic calls for the development of a secondary mortgage market in Ghana, it is puzzling that all the studies (e.g. Asare and Whitehead, 2006; Attakora-Amaniampong, 2006) on this subject have focussed on the *process and pre-requisite factors* needed for establishing such a market in Ghana, we do not know what empirical effect a secondary mortgage market would actually have on the housing problem. This paper takes up the challenge by analysing the possible *outcome* of a secondary mortgage market in Ghana. In order to do so, it analyses the current housing market in Ghana and gives a panoramic view of the various housing policies pursued so far to solve the housing problem. These analyses pave the way for a more systematic study of the possible effects of a secondary mortgage market in Ghana.

The Housing Market in Ghana

According to the 2000 population and housing census in Ghana, there are 2,181,975 houses in all the 10 regions[2] of Ghana. The average number of people per household is 8.7. The Ashanti region (15.08%), Greater Accra region (13.19%) and the Eastern region (12.99%) lead in the distribution of houses in Ghana. The three northern regions trail behind: Northern region (8.15%), Upper west region (4.05%) and Upper east region (3.70%). So, there are about the same number of houses in the Ashanti region alone (15.08%) as there are in all the three northern regions combined (15.89%). In between these two extremes, the distribution of houses in the four other regions in Ghana is as follows: Volta region (12.12%), Western region (11.91%), Central region (10.23%) and Brong Ahafo region (9.91%) (Ghana Statistical Service, 2000).

Housing is a major problem in Ghana. From 1970 to 2000, the housing deficit[3] has increased by over 207 per cent, and though 110,000 to 140,000 units are needed annually to correct this deficit, currently only 30,000 units are produced per year (Daily Guide, 2009).

The problem is more acute in urban areas. In only a year's time, 2010, there will be more people living in urban areas than in rural areas in Ghana, urban housing constitutes only 34.1 per cent of the total housing stock (BOG, 2007).

As one would expect in a market economy, the lack of housing supply relative to the growing demand has led to high housing prices relative to incomes in Ghana. In Accra, two bedroom houses are priced in the range of $30,000 to $60,000 depending *inter alia* on location, type of building finishes and type of house (detached, semi detached etc). Land values command prices in the range of $13,000 to $25,000 depending on location. First class areas like Cantonments, Airport Residential Area and Labone command the highest values (Obeng-Odoom, forthcoming (a)). Though the socio-economic profile of the people has improved over the years, the standard of living is still quite low in Ghana. For instance as of 2009, the GNI per capita in Ghana was $590 (World Bank, 2009) and 28.5 per cent of the population live under the poverty line (UNDP, 2007).

Table 1. Population and Housing in Ghana 1970 – 2000

YEAR	POPULATION SIZE	HOUSING DEMAND	HOUSING SUPPLY	HOUSING DEFICIT
1970	8,559,313	1,678,296	941,639	736,657
1984	12,296,081	2,410,996	1,226,360	1,184,636
2000	18,912,079	3,708,250	2,181,975	1,526,275

Source: Ghana Statistical Service (Population and Housing Census 1970 -2000)

---

[2] Note that the population in each of these regions varies.
[3] Housing deficit is calculated by subtracting the existing houses from houses required. I do not assume a particular preferred household size here.

For these reasons, the house price/income ratio in Ghana is high compared to the average in Africa (12.5) (UN-Habitat, 1998). Out of 26 cities for which the UN-Habitat had data in 2003, only 5[4] (Abidjan, Tanta, Monrovia, Maputo and Jinja) had a higher house price/income ratio than Accra (14.0), Ghana's capital city (UN-Habitat, 2003).

The average ratio for African cities (12.5) is more than twice that of cities in Highly Industrialised Countries (5.8) (UN-Habitat, 1998). This high house price/income ratio in Africa is partly attributed to bureaucratic building permit procedures and colonial building laws which depict houses constructed with local materials as sub-standard, and so bias the taste of people towards purchasing imported building materials (Njoh, 2000). Ghana depicts the polar extreme of this dependence on colonial planning laws and regulations. According to Konadu-Agyemang (2001, p.29), 'these laws [in Ghana] were implemented by bureaucrats, who mystified and bound them in red tape, excluded citizen participation of any kind in the planning process, and imposed on the citizens whatever they have to offer. This continued through to independence in 1957, when the Ghanaian elite took over the administration, and have not only maintained the status quo, but have created a litany of other legislation most of which are based on British models.' Under the current laws, a junior high school teacher with 20 years of experience would need to save *all* his/her *gross* salary for 6 years in order to build a simple 3 bedroom house in the city (Konadu-Agyemang, 2001). So, in Berekum for instance, where planning laws are less rigidly enforced, the cost of housing is lower than in Kumasi and Accra, even when account is taken of the differences in the value of land (Tipple *et. al.,* 1997).

In such circumstances, rental housing may be a viable alternative to owner occupation. But, in Ghana, even rental housing is a difficult option, because landlords demand so many years (between 3-4 years) rent in advance (without giving tenants the benefit of paying less rent in future) (Obeng-Odoom, 2009). For all these reasons, it is rare for one to own a house in Ghana before one attains the age of 40 (Tipple and Kroboe, 1998).

There are dire consequences of this state of housing in Ghana. A recent study by Schroeder *et al.* (2001) found that lack of accommodation is one of the most difficult problems for Ghanaian teachers. Not finding any other way out of this housing problem, the study revealed that, the teachers have resorted to 'Having faith in God' and 'taking time to think about the problem', in order to mitigate their accommodation hardships (Schroeder *et al.*, 2001). The plight of these teachers is echoed in Sam Sarpong's article, 'Ghanaian teachers dying of frustration' (Sarpong, 2002). Students in public universities also experience the accommodation problem. To cope with the situation, the students 'perch' or squat with other students who are admitted to the university halls. So, there is overcrowding in university accommodation and rooms originally meant for 4 students now accommodate 8 students, because each of the original owners has a percher[5] (another student who squats on the bed of his/her friend's bed). In extreme cases, there could be as many as 10 students in a room meant for 4, as the 2 'extra' students who cannot 'perch' on a bed may find a sleeping place *under the two double decked beds* (meant for four students). This state of affairs prompted the

---

[4] And only 6 cities (Abidjan, Tanta, Monrovia, Maputo, Jinja and Antananarivo) had a higher house price/income ratio than Kumasi (11.6), Ghana's second largest city (UN-Habitat, 2003).

[5] The author was a 'beneficiary' of this system. Apart from his first year of undergraduate study, all the other three years saw the author 'perching'. The author salutes all his hosts during those years; but for them, undergraduate work could have been interrupted for want of accommodation.

University of Ghana's Visitation Panel of Scholars[6], chaired by Sir John Daniels, to recommend abolishing this practice with its cataclysmic effects (Joy FM, 2007). But, 'perching' is not the only effect of the housing problem; its more pervasive manifestations are slum formation and sub-standard housing conditions.

Slums have become a feature of urban life. Currently, about 70 per cent of the urban population in Ghana lives in buildings characterised by the absence of sanitation, water and secure tenure. These conditions make life miserable. Some of the slums are called 'Sodom and Gomorrah' to mimic the Biblical account of the wretched life in Sodom and Gomorrah (Obeng-Odoom, 2007; 2009). But, slums are the polar extreme of poor quality housing in Ghana. There are as many as 1,695,443 that are dwelling units but are not fit to be called houses (BOG, 2007). These are not slums; they are poor quality housing. According to one observer:

> Mother Ghana has indeed moved from where she was, after her independence, into a very promising nation. ... Ghana... is the gateway to West Africa... However, one of the major problems, confronting the development of this country, is the poor attitude of its stakeholders towards rehabilitation, and maintenance of its structures and facilities...Our Structures and Premises ...have taken the form of hutches. Numerous school buildings are like cattle ranches (Mensah, 2008).

Many housing studies (e.g. Konadu-Agyemang, 2001) attest to the fact that these poor quality houses do not arise from the use of poor building materials, but rather poor maintenance of already built houses. As noted by Graham Tipple, a household name in housing research in Ghana[7]:

> A lack of maintenance tends to create the impression of poor quality construction. Leaking water pipes and drains, badly maintained roofs, unrepaired and unpainted woodwork, and unkempt and damaged grounds and access ways, all lead to accelerated decline and the poor appearance symptomatic of housing at the end of its economic life or housing in need of more than ordinary levels of maintenance (Tipple, 1999, p.29).

These 'lukewarm' houses, neither good nor bad, exist because of poor maintenance culture, the lack of maintenance professionals and the lack of rent payments for certain types of houses and low levels of income (Obeng-Odoom, forthcoming (b)).

The frustrations people face in the Ghanaian housing market have also led to dramatic informality. According to Kasim Kasanga, a leading researcher in land/housing market in Ghana, 'If the government were to charge people for defying its legislation affecting land, almost ...the whole country would be on trial' (cited in Gough and Yankson, 2000, p. 2489). 80 per cent of lands in Ghana have no proper documentation (GNA, 2009), but even the remaining 20 per cent with documentation cannot be said to be secure as many are the subject of litigation (Abdulai, 2006). Because of these housing problems in Ghana, the government has employed a plethora of policies over the years to improve housing in the country. It is important to briefly examine some of these policies.

---

[6] The panel was tasked to review the university's work over the last 60 years

Changes in Housing Policy in Ghana

To understand the ongoing reforms in housing policy and the current push for the creation of a secondary mortgage market, it is useful to analyse housing policies in the colonial period, independence and post independence era.

The Colonial Period

Housing provision in the colonial period was mainly government-led. The government built houses for civil servants and as far back as 1939, and government was providing housing for the staff of the Railway Department who lost their houses in the 1939 earthquake. The government also actively controlled rent and offered subsidies for renting and subsequent purchase of government houses (Tipple and Korboe, 1998). To give impetus to public housing, the State Housing Corporation and the Tema Development Corporation were established in the 1950's. But, there were also some few private firms that were engaged in housing development. Some of the pioneering private firms were the City Furniture and Construction Company (West Africa) limited (CFC Ltd.) and the Pioneer Company. However, government constructed the majority of the houses (Mohammed-Abibu, 1989).

The first attempt to develop a housing policy in Ghana was in 1944 by Sir Alan Burns, the then Governor of the Gold Coast. This saw the birth of the Four-Year Plan of 1944 and another policy to cover the 1951-58 planning period by Sir Charles Noble Arden Clarke. This latter plan could not run its full length, because Ghana obtained independence in 1957 (Ministry of Works and Housing, 1986).

Independence and Post Independence Era

The first president of Ghana, Kwame Nkrumah, pursued social housing vigorously. While his 1959-1964 plan could be described as a continuation of pre-colonial housing policies, he initiated specific programs (e.g. Roof Loan Scheme) to assist rural housing. He also instituted a seven-year development plan to cover 1963/64 to ‘1970/71. While comprehensive, the plan did not live long enough to achieve its desired impacts, because of the 1966 coup, which toppled the Nkrumah government (Ministry of Works and Housing, 1986).

From 1966 to 1981, Ghana went through a chequered political history, with military insurgencies like the regime of Colonel Kutu Acheampong and elected governments like the Limann government. Among the many parties that took charge of affairs in the country were the National Liberation Council (N.L.C.), the Progress Party, the National Redemption Council, the Supreme Military Council and the People's National Party (P.N.P.). Significantly though, each of these governments, whether military or civilian, pursued a public sector-led housing development. But, during ‘the lost decade’, the 1980s, Jerry

---

[7] Graham Tipple has observed that poor maintenance of housing is also a feature of many government-built housing in developing countries, particularly Bangladesh, Egypt, Ghana, and Zimbabwe.

Rawlings, then leader of the PNDC, sacrificed the country to the IMF policy of economic reforms, locally called 'Economic Recovery Program' (ERP).

The Reform Period

Ghana has sometimes been called 'the star pupil of the IMF' because of the zeal with which it implemented the structural adjustment programs. In terms of housing, the reforms included state withdrawal from direct housing production and financing; active encouragement of private real estate firms through tax holidays, liberalizing the land market and reforming the public housing institutions (Arku, 2009, pp.263 – 264). As a result, budgetary funding for the housing sector dropped. For example, in the 1950s, about 10 – 12 per cent of government revenue went into housing, but as of the 1990s, government was investing only 1 – 2 per cent of its revenue in housing-related investments. The corollary of this is that the State Housing Corporation and the Tema Development Corporation now have to operate on commercial basis. Government has also withdrawn its support to quasi-housing institutions like the Social Security and National Insurance Trust (SSNIT) whose public housing development has fallen by some 50 per cent over time (Arku, 2009, p.264). To entice the private sector to fill in the void, which has been created by the withdrawal of the public sector from housing development, private real estate companies have been offered many incentives. Some of these include the reduction of Corporate Tax from 55 to 45 per cent; a 5 year tax holiday, Stamp Duty exemptions and tax benefits for companies that invest part of their profit in real estate development.

The private sector has responded positively to these incentives. Between 1995 and 2005, a period of 10 years, 81 housing estates were developed in Ghana by firms from Europe, Asia, America and elsewhere in Africa (Grant, 2009). The 'intervention' of the private sector has led to the construction of 10, 954 houses since it congregated under GREDA in 1988 (BOG, 2007). However, the private sector targets the middle and high income groups while still using the mantra of 'affordable housing' (Obeng-Odoom, 2009). Richard Grant has noted that the liberalisation of the housing market has led to the development of 'gated communities' especially in Accra, which are the havens for the affluent in society. These estates include the Lagoon View Estates in Sakumono and Royal Palms in Spintex and Trasacco Valley in East Legon extension. In Accra and Tema, Richard Grant estimates that there are about 3, 644 units in these communities, representing an investment of about US$434.8 million in 2005 values. Such a sum, according to Grant, 'would cover all of the costs for building more than 17, 300 middle-class houses at Accra's average construction prices in 2005 and a considerably greater number of lower income houses (Grant, 2009, p.51). With low living standards for the majority of the population in Ghana, how are most people able to pay for expensive housing in Ghana? While there may be many reasons for this 'paradox', remittances account for a large part of the 'high class' housing development. Many such houses have been financed through a process of 'long distance building' by Ghanaians in countries like UK, USA and Australia or they are caretakers for such long distance builders (see for e.g. Diko and Tipple, 1992; Grant, 2009). So what happens to the poor with no relatives abroad? Civil society groups like the Habitat for Humanity Ghana attempt to assist them poor through low income; self help projects (Obeng-Odoom, 2008).

Because its housing policies have unsatisfactorily addressed the housing problems in Ghana and have even become contradictory, the government of Ghana has desperately been seeking new solutions to the housing problem. As usual, the Breton Woods Institution, the World Bank, claims to have that magic cure. According to the Bank, a 'well functioning housing sector' typified by private financing of the housing sector, guarantees that:

> Everyone is housed, with a separate unit for every household. Housing does not take up an undue portion of household income. House prices are not subject to undue variability. Living space is adequate. Structures are safe and provide adequate protection from the elements, fire, and natural disasters. Services and amenities are available and reliable. Location provides good access to employment. Tenure is secure and protected by due process of law. Households may freely choose among different housing options and tenures (owning versus renting). Finance is available to smooth expense over time and allow households to save and invest. Adequate information is available to ensure efficient choice. Housing consumers are able to participate in policy decisions that affect their housing and neighbourhoods (World Bank, 1993, pp. 15-16).

Many housing scholars in Ghana (e.g. Attakora-Amaniampong, 2006; Gyamfi and Boamah, 2003a; Gyamfi-Yeboah and Boamah, 2003b; Asiedu 2004) have accepted this prescription. According to Asare and Whitehead (2006):

> The development of a formal housing finance system remains a core requirement within the Ghanaian economy. The rationale behind the use of the system *per se* is not defective. The issue lies in how to implement the process and the lack of the foundational elements to ensure the smooth operation of the property mortgage market (Asare and Whitehead, 2006, p.24).

Such statements suggest that a secondary mortgage market would deliver security of homeownership and correct the housing deficit in Ghana. To assess this claim, careful political economic analysis is needed.

Would a Secondary Mortgage Market Provide Security of Homeownership in Ghana?

Mortgages may come in about four shades[8] (in terms of risk levels): agency, jumbo, alt-A, and subprime. Agency loans are the most secure because they relate to borrowers who are considered 'less risky' or 'prime' and usually conform to government underwriting standards. Jumbo loans are also secure because the borrowers here are also 'prime' though they have a principal balance larger than government standards. On the other hand, Alt-A loans relate to borrowers not conforming to government limits due to high leverage or lack of documentation for income. The last type of mortgage, subprime mortgage, is the polar extreme of agency mortgages. Subprime mortgages are advanced to people with a probability of a high default rate (O'Hara, 2009, p.10).

Currently, the Home Finance Company (HFC) offers prime mortgages. According to Arku, HFC mortgages cater exclusively for 'property ... situated in economically viable areas, mainly in Accra, Tema, Kumasi and Takoradi, and loans are only offered to those with

---

[8] Other forms of classification are discussed by Apraku-Kyei (2009, pp. 10-14).

regular employment who can afford to repay within 10 years' (Arku, 2009, p.267). Also, HFC gives loans for the purchase of only houses built by registered private developers, even though the majority of builders in Ghana are informal private individuals (Tipple and Korboe, 1998). Even this restricted lending appears to be aimed at Ghanaians abroad who can borrow up to $80,000 with 20 per cent down payment (Grant, 2009). As such, mortgages offered by HFC (vis-à-vis other loans) are limited and have even been decreasing. So in 2006, for example, mortgage loans as a proportion of total loans advanced were 38.2 per cent but, by 2007, such loans had reduced to 25.6 per cent. Thus, in just a year, HFC mortgages contracted by about 13 per cent (HFC, 2007). HFC has not been able to extend its loan facility to the majority of Ghanaians, because 80 per cent of Ghanaians live under $2 a day and the average per capita income is $590 (Gary, 2009). These considerations evidently suggest that any mortgage scheme that can attend to the needs of the majority of Ghanaians, as proponents argue, would be the subprime mortgage, as it is in poor communities in countries with better developed secondary mortgage markets like USA (Sassen, 2008).

Subprime mortgages have some positive effects. One such positive role is that they extend finance to an otherwise unqualified person. The feeling that one is about to own a house provides psychological excitement, enhances self-esteem and fosters greater community acceptance. At the neighbourhood level, that feeling may encourage him/her to take better care of the home and greater part in local activism, which may go a long way to provide healthier neighbourhoods (Girardi and Willen, 2008, pp.4-6). However, subprime mortgages also have some shortcomings, particularly so when their viability is hinged on the regularity of premiums paid by the mortgagor.

In the event of default of mortgage payments, a mortgagee has several rights against a mortgagor. One key and often used weapon against the defaulting mortgagor is foreclosure. One study by Geradi and Willen (2008) found that borrowers who purchase their homes with subprime mortgages have an almost 20 per cent chance of losing them to foreclosure. Another study by Immergluck and Smith found that for every 100 additional subprime mortgages from 1996 to 2001, there were an additional 9 foreclosure starts in 2002. In a comprehensive study of over six million mortgages from 1998 to 2004, Schloemer *et al.* (2006) find that 1 out of every 5 mortgagors with subprime mortgages end up in foreclosure, hence only few people with subprime mortgage are able to 'graduate' to become *homeowners*. All these studies were conducted in the USA, for which reason they may not be relevant to the Ghanaian case. So, we should draw lessons from the Habitat for Humanity Housing Scheme in Ghana (HFHG), which approximates a subprime mortgage scheme.

The scheme has been in Ghana since the 1980s and operates in 7 out of 10 regions. It produces affordable housing in poor communities where the beneficiaries help in reducing the cost of construction by offering their labour in the construction process and repay their loans by contributing about one and a half bags of cement every month in lieu of cash payments (Adarkwa and Oppong, 2007). Even here, the rate of default is high. One recent study of the scheme by Obeng-Odoom (2008), found that 3 out of every 5 beneficiaries default on the loan repayment. So, while it is difficult to generalise from the HFHG scheme, it is well established that the poor have a high rate of default (because being risky, they tend to be charged higher interest rates, they tend not to have stable income and employment) and there is no reason to expect high repayment of mortgage in a low income country like Ghana. But, unlike the HFHG scheme whose nature - the centrality of welfare and the nature of the loan agreement –

does not lead to automatic repossession upon default (Adarkwa and Oppong, 2007; Obeng-Odoom, 2008), defaulting on a subprime mortgage leads to repossession/foreclosure.

Such foreclosures lead to the loss of accumulated equity and impede one's access to further credit. Other non-monetary effects exist. These could be psychological hardships of losing a home and its associated effects like divorce and health problems. Again, foreclosure may set in motion a vicious cycle in which neighbourhood property values fall and the foreclosed properties are left unmaintained, vacant and hence become a haven for criminal activities (Girardi and Willen, 2008, pp.6-8) and set in motion a vicious cycle of poverty (Squires, 2008, p.5). In all these, the banks that lend the money (mortgagees) and the investors that buy the mortgage instruments would be the least losers. As noted by Saskia Sassen (2008):

> The house itself functions as collateral only for those who own the instrument which, in a fast moving market of buying and selling, may just last for two hours. Thus, when an investor has sold the instrument, what happens to the house itself is completely irrelevant; it does not even matter as collateral. Most investors can escape the negative consequences of home mortgage default because they buy these mortgages in order to sell them; there will be many winners and only a few losers. But no homeowner can escape the consequences of not paying her own mortgage. Thus investors can relate in a positive way to even the so-called subprime mortgages (poor-quality instruments), which is bad for homeowners. We see here yet another sharp inequality in the current condition.

In short, there is no reason to suggest that introducing a secondary mortgage market would guarantee homeownership in Ghana because the socio-economic profile of the country and the logic of the subprime mortgage would combine to protect the mortgagee and assault the poor mortgagor who may be rendered homeless upon foreclosure. Providing housing, not for its *use value* but its *exchange value*; not for *human need*, but for *profit* through the medium of the *market*, cannot guarantee homeownership for Ghanaians (see also Jafee *et al.,* 2009; Marcuse, 2008).

A secondary mortgage market in Ghana may reduce the housing deficit to the extent that it may lead to the construction of more houses, however, because of the likelihood of default in a poor country and the cost of default – repossession and foreclosure – Ghana would be ushered into the paradox in advanced capitalist societies of more houses and more homelessness. We therefore need to re-diagnose the housing problem, in order to suggest an alternative solution.

Re-Diagnosing the Housing Problem in Ghana: Towards a Solution

The housing problem in Ghana (and in many other developing countries) has been depicted as a function of credit. So, the solution, the argument goes, lies in the provision of credit. Proceeding on this (mis) interpretation, there has been a host of suggestions as to how this credit should be provided. The Peruvian economist, Harnando De Soto, for example has suggested that the surest way to obtain this credit is through the formalisation and regularisation of indigenous land tenure (de Soto, 2000).

However, as we have seen, providing credit to the poor, even when such credit is couched in terms like 'pro-poor' or 'affordable' as in the case of a *subprime mortgage,* would not solve the housing problem. It would simply keep the poor in bondage and perpetuate the master-servant relationship between the rich mortgagee and the poor mortgagor. In such a lopsided relationship, the poor would be forced to do any work, in any condition, to help them pay their mortgage. As noted more than a century ago by German philosopher and social scientist Frederick Engels,

> ….the workers must shoulder heavy mortgage debts in order to obtain even these houses and thus they become completely the slaves of their employers; they are bound to their houses, they cannot go away, and they are compelled to put up with whatever working conditions are offered them….(Engels, 1872)

So, the housing problem is not a function of credit; it is a function of poverty. This is why in Ghana, the housing problem is not a problem for the rich as they live in decent accommodation, while the poor live in decrepit housing in slums or are entirely homeless and live on the streets and in lorry parks, as recently uncovered by Richard Grant (Grant, 2009). Therefore, it is important to analyse how this poverty comes about and how we can possibly reduce or uproot it.

The neoliberal approach is to focus on the individual poor by way of empowering him/her through education and better healthcare. The logic underpinning this approach is that the poor caused their own poverty. So, enhancing the human capital of the poor should enable them to solve their poverty through the labour market. According to the World Bank, *individuals* should be 'empowered', given 'opportunity' and 'secured' (pp. 6-7) by making *'markets work better for poor people'* (p.1) (World Bank, 2001) – and, *eureka,* poverty is out of the door! The lure of this thesis is, perhaps, its simplicity - or as some prefer to call it, its 'practicality'.

The Ghanaian government has bought into this reasoning and has therefore implemented so called 'pro-poor' policies aimed at relieving the severity of poverty on individuals, like the *School Feeding Program and* the *Livelihood Empowerment Action Plan.* But, as Watchel has argued, often the poor have no control of their poverty and, while human capital development is adequate, it is inadequate to even improve one's position in the labour market. According to Watchel, the following determine one's status in the labour market:

1. 'Individual characteristics over which the individual has no control e.g. age, race, sex, family class status and region of socialisation;
2. Individual characteristics over which the individual exercises a degree of control – education, health, region of employment and personal motivation;
3. Characteristics of the industry in which the individual is employed e.g. technology, unionisation, relation of industry to government;
4. Characteristics of the local labour market – structure of the labour demand, unemployment rate and rate of growth' (Watchel, 1971, p.5).

Except point 2, which can be called 'human capital', all other features lie outside the individual's scope. Even with point 2, the UNDP has frequently argued that the human capital of an individual is tied to the education and social standing of his/her parents, for example (UNDP, 2006).

Even when one is able to find a place in the labour market, one may be only ushered into a new form of exploitation, social injustice and poverty. Such is the case because the very organisation of work in capitalist society, wherein workers sell their labour power to the owners of capital who exploit them either through the mechanisation of work or by forcing them to work long hours in order to produce surplus value, is the source of yet further inequality and poverty. As noted by Engels:

> The pivot on which the exploitation of the worker turns is the sale of labour power to the capitalist and the use which the capitalist makes of this transaction in that he compels the worker to produce far more than the paid value of the labour power amounts to. It is this transaction between capitalist and worker which produces all the surplus value which is afterwards divided in the form of ground rent, commercial profit, interest on capital, taxes, etc, among the various sub-species of capitalists and their servants....(Engels, 1872).

So, every policy to address the housing question - or any social problem for that matter - should also aim at abolishing such a system that thrives on inequality. Regrettably, a secondary mortgage market would leave this system intact. Already, Ghana has become highly polarised since adopting the neoliberal approach to poverty reduction in the 1980s. Hence, even though in 1987, the richest 20 per cent of the population already controlled as high as about 40 per cent of income; by 1999, the share of income of the richest 20 per cent had risen to about 50 per cent (see various UNDP reports). As expected, Ghana's Gini co-efficient of income inequality has worsened (1987 (0.38), 1998 (0.42) and 2007 (0.55)) around the same time (Vanderpuye-Orgle, 2002; UNDP, 2007b; Earth Trends, 2003). This rising inequality has prompted the Ghana Statistical Service to observe that 'If Ghana had experienced no change in inequality during the last seven years, the actual decline in poverty of 10.4 [per cent] would have been 13.8 [per cent].... the decline in poverty would have even better if it had not been offset by increasing inequality, particularly since 1998/99' (GSS, 2007, p.17).

This income inequality has adverse effects for economic growth, happiness, poverty reduction and internal security (Birdsall et al., 1994; Stilwell and Jordan, 2007, p.4; Ayelazuno, 2007). Recent econometric study of global data from 1977 to 2004 by Augustine Fosu shows that 'for all three poverty measures – headcount, gap and squared gap – the impact of GDP growth on poverty reduction is a decreasing function of initial inequality. [Also], higher rates of increases in inequality tend to exacerbate poverty' (Fosu, 2009, p.738).

A secondary mortgage market not only leaves inequality intact, it also perpetuates it by further privileging the creditors, the banks and the rich over and above the peasants, the poor and the workers. As such, a secondary mortgage market in Ghana would only worsen and intensify the housing problem. So, the only way to uproot the root causes of the housing problem is to end poverty. To lubricate the process, a welfare state could pursue certain policies to reduce the cost. I will suggest a few of such policies.

Reducing the Cost of Housing

In order to reduce the cost of housing, policies must be designed to bring down the cost of physical development *per se* and the cost of the site. Regarding the former, the cost of physical improvement, direct punitive measures for using imported building materials, beggar-thy-neighbour style, while possibly useful in the short term, are not recommended as they may set in motion higher costs in the long run. But it may be useful to actively promote the development and use of local and cheaper building materials while abolishing colonial planning laws and regulations. This means that a willing government may *tie* the tax holidays that the private sector currently enjoys to building *real* affordable homes that *use local materials*. So, only firms that build affordable homes - defined by local governments as (1) using local materials and (2) affordable to their residents- should qualify for the breaks.

Another way by which housing cost can be reduced is by minimising the cost of sending remittances to Ghana for building purposes. Any policy to reduce the cost of remittances may have wide impact because in Accra alone, about 50 per cent of all new houses are built or bought through remittances (Grant, 2009).

Apart from reducing the cost of the house *per se, the value of land* may also be reduced or removed completely. The latter may be possible in communities where land is still communally owned, in which case land is virtually free for members of the community. In the *Habitat for Humanity Housing Scheme* earlier described, the community, via the (landowning) stool, 'donates' land to the NGO, Habitat for Humanity Ghana, which then charges the beneficiaries of the scheme for only the cost of housing *per se*. For this reason[9], the prices of the houses developed tend to be about 12 times lower than similar houses in the market. In order to curtail profiteering by members of the community by way of selling their cheaply acquired houses, they are barred by the scheme from selling these houses (Obeng-Odoom, 2008). The scope of such schemes could be widened, by making them self-financing. Here, we can envision a system in which 'visitors' to this community may be allowed to buy some of these houses for *a period of time* (leasehold interests) after which the houses are transferred back to the community. Also, businesses may be actively attracted to such a community and also transfer their assets back to the community on a *Build Own Operate and Transfer* (BOOT) basis. In order to maintain infrastructure in the community, the members of the community who essentially are *usufructs* could contribute a fee depending on the size of their houses. The 'rent' paid by the visitors would be held by the stool on behalf of the community and could be used to further develop affordable housing for other members of the community (Turnbull, 2007; 2009).

But what about communities/settlements where such communal land arrangements do not exist and the majority of land is in private ownership? An oft-adopted approach is the *site and services scheme,* in which the state services land in order to take away the burden of providing such facilities by the individual. An objection to this system is that it creates a free rider problem in which those with nearby sites obtain a rise in their land values for no work done, thus creating higher land values and inequality in society. An analysis of inequalities associated with land, drawing on the insights of David Ricardo can also be extended to consider the distinctive policy remedy proposed by Henry George (Stilwell, 2006, pp.89-91; Stilwell and Jordan, 2004). George identified this problem of rich landowners benefitting from their land as a result of public expenditure and advocated that all land be taxed with, the resulting revenues being reinvested for the benefit of the poor. Unfortunately in Ghana, the

---

[9] In addition to the fact that local materials are used and labour is virtually free because of the sweat equity concept (see Obeng-Odoom, 2008).

existing property rate falls on buildings not on land *per se*. As stated in section 96 (7) of the Local Government Act, Act 462:

> … the premises rateable under this section shall be premises comprising buildings or structures or similar development.[Further, subsection 9 of section 96 provides, inter alia, that] … the rateable value of premises shall be the replacement cost of buildings, structures and other development comprised in the premises …

This legislation not only legitimises land speculation but also encourages it. Property rates in Ghana should fall on the value of the land itself, not the improved capital value. This new formula would be a fairly simple, equitable and effective as a way of ending land speculation in urban areas, because unlike capital gains and property rates which tax productive behaviour, building (property rates) and adding value to the building over time (capital gains) - the 'new property rate' would punish speculation. By dissuading speculative behaviour, the land tax would serve as a way of redistributing wealth from the rich whose properties benefit from sites and services scheme to the poor. An objection to land tax is that it may make people who are asset rich but income poor - like retirees - worse off because they do not actually earn income from the land except at the point of sale. For all such cases, the tax could be deferred (perhaps with interest) until the time of sale – a point argued in the Australian case by Stilwell and Jordan (2004, p.18).

Extra funds from such tax may be put to affordable housing in three main ways. First, the state can develop affordable housing to *rent out* to the low income people in society since not all of them can *own outright*. Second, the state can develop *new* housing *for sale* at affordable prices (see Gilbert, 2003 for full discussion). Third, the state can *rehabilitate and renovate* existing housing that may not be suitable as dwellings and let or sell them at affordable prices. Such uses of the public purse may be better than (1) using the money to subsidise the purchase of homes supplied by the private sector or (2) trying to control/subsidise rent of houses supplied by the private sector. In the case of the former, subsidising the purchase of homes, the policy is self-defeating because in the end, it increases demand for housing and may push the house prices up. Rent control, on the other hand, tends to restrict housing supply, whereas subsidies for renting private housing may eventually benefit private owners and worsen the distribution of wealth (Stilwell, 1992). As such, generally, 'measures to increase finance for public funding are more efficacious because they directly increase the supply of housing and reduce the overall inflationary effect of excess demand' (Stilwell, 1992, p.48).

CONCLUSION

Ownership of a home *is* different from ownership of a mortgage. The latter has severe disabilities because it enslaves the mortgagor and threatens him/her with homelessness in the event of default in mortgage payments.

Subprime mortgages are even worse. From our analysis of the Habitat for Humanity Housing Scheme, we expect that more than 7 out of 10 beneficiaries of a secondary mortgage market would default in Ghana. One notorious remedy to the mortgagee is foreclosure. So, subprime mortgages protect the mortgagee and expose the mortgagor to the hopelessness of homelessness.

A policy to reduce the cost of housing both in terms of physical development and site value may help in making housing accessible to many. Mechanisms should also be put in place to facilitate remittances to Ghana that may be channelled into the housing sector. So should government impose land tax to cream off unearned profit to further invest in housing development. To be sustainable, these approaches must go hand-in-hand with the use of local building materials.

These policies may be regarded as cosmetic, because they tackle only the symptoms, not the root cause, of the housing problem - poverty. Yet, no consensus[10] exists on how to tackle poverty. It is now accepted, even by liberals, that inequality is anathema for poverty reduction. How this inequality arises, however, is the source of bitter controversy. Nonetheless, it is not difficult to see that a system of production in which the triumph of one class necessarily means the exploitation of the other is destined to generate inequality. It is hard to escape the conclusion that solving poverty and *hence the housing question* requires fundamental change to the capitalist system of production. Extension of public ownership and social housing necessarily has more prominence in such a political program. Creating more forms of finance for private sector housing debt has much more problematic consequences. Recommending the latter at a time when the global economy is in turmoil because of the destabilising effect of subprime mortgage loans seems particularly inappropriate.

REFERENCES

Abdulai, T. (2006). 'Is land title registration the answer to insecure and uncertain property rights in sub-Saharan Africa?' *RICS Research paper series, Volume 6, Number 6*, 1-27.

Apraku-Kyei, J. (2009). 'The crisis in the United States mortgage market – causes and solutions', MSc thesis submitted to the Dept of Real Estate and Construction Management, Div of Building and Real Estate Economics, Master of Science Thesis *no. 464, KTH*, Sweden.

Adarkwa, K. & Oppong, R. A. (2007). 'Poverty reduction through the creation of a liveable housing environment. A case study of Habitat for Humanity International Housing Units in rural Ghana', *Property Management, vol. 25, no.1*, 7-26.

Arku, G. (2009). 'Housing policy changes in Ghana in the 1990s', *Housing Studies, vol.24, no.2*, 261-272.

Asare, E. (2006). 'Formal Mortgage markets in Ghana: nature and implications', *RICS Research Paper Series, 6(13).*

Asiedu P. (2004). '*Why Is Ghana Missing Out On Home Equity Loans?*', modernghana.com, http://www.modernghana.com/news/113784/1/why-is-ghana-missing-out-on-home-equity-loans.html (accessed 24-4-09).

Attakora-Amaniampong, E. (2006). '*Residential Development and Borrowing in Ghana: A Challenge for Banks and Private estate developers*', MSc Thesis no. 334 submitted to

---

[10] The Millennium Development Goals do not represent a consensus on *how* to end poverty. It is a consensus on *what goals we should achieve in our fight against poverty.*

Division of Building and Real Estate Economics School of Architecture and the Built Environment Royal Institute of Technology, Stockholm.

Augustin, F. (2009). 'Inequality and the Impact of Growth on Poverty: Comparative Evidence for Sub- Saharan Africa', *Journal of Development Studies*, vol. 45, no. 5, 726-745

Ayelazuno, J. (2007). Democracy and Conflict Management in Africa: Is Ghana a Model or a Paradox? *African Journal of International Affairs*, vol. 10, Nos. 1 & 2, 13-36.

Bank of Ghana (BoG). (2007). *The housing market in Ghana*, The Research Department, Bank of Ghana, Accra.

Birdsall, N., Ross, D. & Sabot, R. (1994). '*Inequality and Growth Reconsidered*', Paper for the Annual Meetings of the American Economics Association, Boston, January 3-4, 1994.

Daily Guide. (2009). '*Boost for public sector housing*', http://news.myjoyonline.com/ business/200904/29283.aspmyjoyonline.com (accessed on 30-4-09).

De Soto, H. (2000). *The Mystery of Capital: Why Capitalism Triumphs in the West and fails everywhere else*, Bantam Press, London.

Diko, J. & A. G. Tipple. (1992). 'Migrants build at home: long distance housing development by Ghanaians in London', *Cities*, 9, 4 November, 288-94

Earth, Trends. (2003). '*Country Profile – Ghana*', http://earthtrends.wri.org.text.economics-business.country-profile-72.html, (accessed 25-9-08).

Engels, F. (1872). *The Housing Question*, Co-operative Publishing Society of Foreign Workers, Moscow.

Gary, I. (2009). *Ghana's big test: Oil's challenge to democratic development*, Isodec Ghana and OXFAM America.

Gerardi, K. & Willen, P. (2008). '*Subprime Mortgages, Foreclosures, and Urban Neighbourhoods*' *Public Policy Discussion Papers, No. 08-6*, Reserve Bank of Boston.

Ghana Statistical Service (GSS). (2007). *Pattern and Trends of Poverty in Ghana*, 1991-2006, Ghana Statistical Service, Accra.

Ghana Statistical Services. (1970). *Population and Housing Censuses 2000*, Ghana.

Ghana Statistical Services. (1984). *Population and Housing Censuses 2000*, Ghana.

Gilbert, A. (2003). Rental housing: *An essential option for the poor in developing countries*. Nairobi: United Nations Human Settlements Programme (UN-HABITAT).

GNA. (2009). '80% of lands without proper documentation' ghanaweb.com http://www. ghanaweb.com/GhanaHomePage/NewsArchive/artikel.php?ID=163975 (accessed 24-06-09).

Gough, K. & Yankson, P. (2003). 'Land markets in African cities: the case of peri-urban Accra' *Urban Studies, vol. 37, no. 13*, 2485-2500.

Grant, R. (2009). *Globalizing City, the urban and economic transformation of Accra*, Ghana, Syracuse University Press, New York.

Gyamfi-Yeboah, F. & Boamah, N. (2003a). 'Toward a Sustainable Mortgage Market in Ghana' modernghana.com, http://www.modernghana.com/news/112054/1/toward-a-sustainable-mortgage-market-in-ghana.html (accessed on 24-04-09).

Gyamfi-Yeboah, F. & Boamah, N. (2003b). 'Attracting Capital to the Real Estate Sector' modernghana.com, http://www.modernghana.com/news/112213/1/attracting-capital-to-the-real-estate-sector.html (accessed 26-04-09).

Harvey, D. (1973). Social Justice in the City, Edward Arnold, London.

HFC. (2007). Annual reports and consolidated financial statements for the year ended 31 December 2007, HFC Bank Ltd., Accra.

Jafee, D., Lynch, A., Richardson, M. & Nieuwerburgh, S. (2009). *Mortgage orientation and securitization in the financial crisis*, Chapter 1: Executive summary, New York University Salomon Center and Wiley Periodicals, Inc.

Konadu-Agyemang, K. (2001). 'A survey of housing conditions and characteristics in Accra, an African city' *Habitat International, vol. 25*, 15- 34.

Mensah, O. R. (2008). *'A change in attitude: Culture of maintenance in Ghana'* modernghana.com, http://www.modernghana.com/news/176588/1/a-change-in-attitude-culture-of-maintenance-in-gha.html (accessed 20-04-09).

Ministry of Works & Housing. (1986). *'National Housing Policy and Action Plan 1987-1990'* Ministry of Works and Housing, Accra.

Mohammed-Abibu, A. (1989). *'The Role of the Real Estate Development Companies in Housing Development in Ghana'*. Unpublished thesis submitted to Department of Land Economy, KNUST.

Njoh, A. (2000). 'Some development implications of housing and spatial policies in sub-Saharan African countries with emphasis on Cameroon' *International Planning Studies, vol. 5, no.1*, 25-44.

O'Hara, P. (2009). 'The Global Securitized Subprime Market Crisis' *Review of Radical Political Economics*, doi: 10.1177/0486613409336179.

Obeng-Odoom, F. (2007). 'Enhancing urban productivity in Africa'. *Opticon 1826 Journal, vol.2, no. 1*. 1-9.

Obeng-Odoom, F. (2008). 'Has the Habitat for Humanity Housing Scheme achieved its goals? A Ghanaian case study', *Journal of Housing and the Built Environment, vol. 24, no.1*, 67-84.

Obeng-Odoom, F. (2009). 'The Future of Our Cities', *Cities, vol. 26, no.1*, 49-53.

Obeng-Odoom, F. *forthcoming (b)*, 'Housing policy changes in Ghana in the 1990s – A commentary', *Housing Studies*.

Obeng-Odoom, F. *forthcoming (a)*, 'The real estate agency market in Ghana: A suitable case for regulation?' *Regional Studies*.

Pearce, A. & Stilwell, F. (2009). 'Green collar jobs': employment impacts of climate change policies" *Journal of Australian Political Economy, No. 62*, 120-138.

Sarpong, S. (2002). 'Ghanaian teachers dying out of frustration' *NewsfromAfrica*, http://www.newsfromafrica.org/newsfromafrica/articles/art_848.html (accessed 28-04-09).

Sassen, S. (2008). 'Mortgage capital and its particularities: A new frontier for global finance' *Journal of International Affairs*, (October).

Schloemer, E. Li, W. Ernst K & Keest, K. (2006). 'Losing Ground: Foreclosures in the Subprime Market and Their Cost to Homeowners' paper prepared for the *Center for Responsible Lending*.

Schroeder, M. Dakota, S. & Apekey, K. (2001). 'Stress and Coping among Ghanaian School Teachers', *FE PsychologIA, Vol. 9, No1*. 2001, 89-98.

Squires, G. (2008). 'Do subprime loans create subprime cities? Surging inequality and the rise in predatory lending' *EPI Briefing Paper*, Washington DC.

Stilwell, F. (2006). *Political Economy: the contest of economic ideas*, Oxford University press, Melbourne.

Stilwell, F. (2008). 'Sustaining what?', *Overland, No. 193*, November.

Stilwell, F. & Kirrily, J. (2007). *Who Gets What?: Analysing Economic Inequality in Australia*, Cambridge University Press, Melbourne.

Stilwell, Frank. (1992). *Reshaping Australia; Urban Problems and Policies*, Pluto Press, Sydney.

Tipple, G. (1999). 'Transforming Government-Built Housing: Lessons from Developing Countries' *Journal of Urban Technology, Vol. 6, No. 3*, 17-35.

Tipple, G., korboe, D. & Garrod, G. (1997). 'Income and Wealth in house ownership studies in urban Ghana' *Housing Studies, vol.12, no.1*, 111-126.

Tipple, G. & Korboe, D. (1998). 'Housing policy in Ghana: Towards a supply-oriented future' *Habitat International, vol.22, no.3*, 245-257.

Turnbull, S. (2007). 'Affordable housing policy: Not identifiable with orthodox economic analysis', working paper archived on the Social Science Research Network, http://ssrn.com/abstract=10279201 (accessed on 28-05-09).

Turnbull, S. (2009). 'Affordable housing policy: Not identifiable with orthodox economic analysis' *ICFAI Journal of Urban Policy, vol.4, no.1*, 21-43.

UNDP. (2006). *Human Development Report*, Palgrave Macmillan, New York.

UNDP. (2007). *Ghana Human Development Report 2007*, Combert Impressions, Accra.

UNDP. (2007b). *Human Development Report*, Palgrave Macmillan, New York.

UN-Habitat. (2006). 'Mrs. Tibaijuka calls for pro-poor mortgage schemes in Africa' *UNHabitat*, http://www.unhabitat.org/content.asp?cid=4144&catid=198&typeid=6&subMenuId=0 (accessed 24-4-09).

UN-Habitat. (2003). *The Challenge of Slums*, Earthscan, London and Sterling.

UN-Habitat. (1998). 'Urban Indicators 98' *UN-Habitat*, www.urbanobservatory.org (accessed 1-07-09).

Vanderpuye-Orgle, J. (2002). *'Spatial Inequality and Polarisation in Ghana, 1987-1999'* Paper presented at ISSER/Cornell Universities Conference, 'Ghana at the Half Century' Accra, July 18-20.

Watchel, H. (1971). 'Looking at poverty from a radical perspective', *Review of Radical Political Economics, vol. 3., no. 1*, DOI: 10.1177/048661347100300301, 1 – 19.

World Bank. (1993). *Housing, Enabling Markets to work, File no. 11820*, World Bank, Washington DC.

World Bank. (2009). *World Development Report 2009: Reshaping Economic Geography*, The World Bank, Washington DC.

World Bank. (2001). *World Development Report 2000/1: Attacking Poverty*, Oxford University Press, Oxford.

In: Built Environment: Design, Management and Applications   ISBN: 978-1-60876-915-5
Editor: Paul S. Geller, pp. 215-237   © 2010 Nova Science Publishers, Inc.

Chapter 9

FAR, FAR AWAY – TRIP DISTANCES AND MODE CHOICE IN THE CONTEXT OF RESIDENTIAL SELF-SELECTION AND THE BUILT ENVIRONMENT

*Joachim Scheiner**

Technische Universität Dortmund, Faculty of Spatial Planning,
Department of Transport Planning, 44227 Dortmund, Germany

ABSTRACT

Access to activity opportunities depends, among other factors, on land-use mix and density at an individual's place of residence. Travel mode choice varies considerably depending on distances to facilities as well as on realised trip distances. This chapter presents findings from an empirical study conducted in the region of Cologne. Standardised data collected from 2,000 residents are used for an in-depth analysis of the interrelation between trip distances and mode choice. Based on structural equation modelling, life situation (sociodemographics), car availability, location preferences and the built environment are taken into account. In addition, lifestyles receive consideration in the study of leisure trips. The findings show the impact life situation, car availability and the built environment have both on trip distances and mode choice, as well as the interrelation between the latter two. What is more, significant effects of lifestyles and residential self-selection on travel behaviour are found. The chapter closes with discussion of consequences for urban planning, making a case for more balanced, mixed-use patterns in urban development.

Keywords: trip distance, travel mode choice, residential location choice, residential self-selection, structural equation modelling

---

* Corresponding author: E-mail: joachim.scheiner@tu-dortmund.de phone ++49-231-755-4822, fax ++49-231-755-2269

1. INTRODUCTION

Recent years have seen considerable scholarly interest in the effects the built environment has on travel behaviour. Transport researchers have highlighted the importance of a number of key attributes of the built environment in this context: density, land-use, distance to the nearest centre, and the connectivity of transport networks (see overviews in Boarnet and Crane, 2001; Stead and Marshall, 2001; Cervero, 2006). The spatial determinants of travel have been summarised using keywords such as the 'three D's' – density, diversity, design (Cervero, 2002).

The built environment has a particularly strong impact on two elements of travel behaviour: trip distances and travel mode choice. A dense, mixed-use urban structure allows residents to make short trips due to the proximity of housing to other activities. As a consequence, a comparatively large proportion of the trips may be undertaken by non-motorised transport modes (NMT, i.e. by bicycle or on foot). The impact of the built environment on mode choice is thus in this case indirect: mode choice depends on trip distance, and trip distance depends on distance to facilities (i.e. on the least required distance). These interrelations have been found in numerous empirical studies (see for overviews and empirical studies Cervero, 2002; Schwanen et al., 2004; Guo and Chen, 2007 and other papers in the same issue).

However, there is also a direct interrelation between the built environment and travel mode choice. As well as NMT, public transport (PT) accounts for a relatively high share of trips in dense, mixed-use structures for two reasons. Firstly, high population density and/or activity density often goes along with restrictions for the private car, such as lack of parking space, high traffic density, and low travel speed, and therefore reduces the comparative disadvantage of PT against the private vehicle (in some cases even producing the opposite effect). Secondly, the high demand density encourages an attractive PT system.

The interrelations between the built environment and trip distances as well as between the built environment and travel mode choice have been investigated in detail in numerous studies. However, the same is not true for the interrelations between trip distances and mode choice. This chapter aims to shed light on this issue by simultaneously studying trip distances and mode choice for maintenance trips, leisure trips, and all trips a person makes put together. Life situation variables, the built environment, accessibility preferences and car availability are taken into account as determinants. Lifestyles are considered additionally when leisure trips are studied. The following section provides a brief literature overview, followed by some hypotheses. The data and methodology employed are introduced subsequently, followed by the results. The chapter concludes with discussion of consequences for urban planning and research.

2. TRIP DISTANCES, MODE CHOICE AND THE BUILT ENVIRONMENT: THE STATE OF THE RESEARCH

2.1 Travel Mode Choice and Distance to the Nearest Facility

In contrast to realised trip distances, minimum necessary distances to nearest facilities such as shopping centres, leisure opportunities or schools can be directly affected by spatial planning concepts. Available research fairly consistently suggests that the provision of activity facilities in the neighbourhood may result in more walking and/or bicycle use and thus indeed affects mode choice (see references cited above). However, there is no common understanding of the details of these interrelations, e.g. with respect to distance thresholds.

One reason for this may be that many studies work with rather generalised distance categories. For instance, the German state North Rhine-Westphalia implemented a programme intended to promote rail-oriented housing schemes. This programme was largely based on the observation that car use among those living in a catchment area of one kilometre around a railway station was slightly lower than outside this radius (ILS, 1999). However, the possible existence of a decline in car use within the 1-km-radius has not been explored. Such a decline is highly likely. E.g., based on German data Scheiner (2010a) found that 94 percent of trips shorter than 200 m are undertaken on foot, whereas in the distance band 800-1,000 m walking holds a share of no more than 38 percent.

Holz-Rau (1991) examined shopping trips in a Berlin neighbourhood on a micro-spatial basis. His results show that car use for shopping increases rapidly among motorised households from a distance of 325 m or more to the nearest grocery store. For distances exceeding 670 m motorised households hardly use any transport modes other than the car. Individuals without access to a car tend to switch to the bicycle when the distance exceeds 325 m.

Holz-Rau et al. (1999) found considerably longer, but less frequent shopping trips in poorly served (mono-functional) residential neighbourhoods, as compared to mixed-use neighbourhoods with good shopping facilities. The proportion of car trips among all shopping trips was markedly higher in mono-functional neighbourhoods, although in absolute terms the differences were less pronounced due to the lower trip frequency. Owing to the longer distances in the mono-functional neighbourhoods, the car is in any case used much more for shopping.

For the U.S., Handy & Clifton (2001) found that proximity to shopping facilities did not seem to reduce car travel. Yet, in another study distance to the nearest facility turned out to be an important factor influencing the frequency of shopping trips on foot (Cao et al., 2006).

The same study, however, shows that the subjective importance individuals assign to the accessibility of shopping facilities increases the likelihood of walking as well, suggesting that people sort themselves into the environments they prefer. For shopping trips the effect of self-selection is stronger than for strolling (Cao et al., 2006). Scheiner and Holz-Rau (2007) carried out a similar study on the basis of structural equation modelling. The results confirm the important effect of residential self-selection on travel mode choice, even when objective attributes of the neighbourhood are being controlled for. These papers contribute to the recent debate about residential self-selection. They suggest that the conditions of the built environment, which are reflected in distances to certain destinations, should not be regarded as a fixed pre-condition of life (see also Bohte et al., 2009, Cao et al., 2009).

To sum up, results from existing studies are not entirely consistent although there is general agreement that proximity to facilities is associated with more NMT and less motorised travel (particularly, less car travel). It should also be noted that not all trip purposes are equally suited for use in such studies because for some types of activity there is a lack of clarity about the extent of the appropriateness of opportunities for certain requirements. It is usual to assume that a shopping centre or a grocery store suits the needs of all. However, the same assumption may not be valid for specialised retail branches, and certainly cannot be made with relation to the appropriateness of the nearest workplace or of a particular location for a stroll.

2.2 Travel Mode Choice and Realised Distance

Transport planning commonly assumes that walking is the fastest travel mode for distances under one kilometre, the bicycle for distances between one and six kilometres, and the car for distances over six kilometres; access and egress time being taken into consideration (Zumkeller and Nakott, 1988).

The effect of realised trip distances on mode choice is, if at all, mostly examined in rather coarsely meshed distance categories. The lowest category typically includes trips of up to about one to two kilometres, or one mile (Kloas and Kunert, 1993; Bahrenberg, 1997; Schlossberg et al., 2006; DfT, 2006). Working with these relatively broad categories may mask considerable variation within the categories and (in longitudinal studies) important shifts over time. Exceptions include Vågane (2007) who provides differentiated analyses for Norway, and Scheiner (2010a) who does the same for Germany.

Distance thresholds vary between regions and countries (Badland and Schofield, 2005, p. 186). This may have cultural, economic, topographic or climatic reasons. In many highly motorised countries the car is the preferred mode of transport even for short trips, while in developing countries very long distances are covered on foot. Among the more developed countries there are considerable differences as well. In the UK 76 percent of trips up to one mile (1.6 km) are undertaken on foot (DfT, 2006, p. 15-16). In Germany the equivalent figure is only 60 percent for trips up to one kilometre in length (calculated from DIW/INFAS, 2003), and in Norway it is 53 percent (Vågane, 2007). Methodological differences may also play a role here.

There are also marked international differences in bicycle use. While the proportion of short trips undertaken on foot is lower in Germany than in the UK, this is largely compensated by the bicycle. Generally, the bicycle is widely used in the Netherlands and in Denmark, in Germany less so, in the UK, France and the US even less so (EU, 2000; Giuliano and Dargay, 2006; Gatersleben and Appleton, 2007). Among other factors this has to do with the bicycle being seen as a serious transport mode. There are also clear inter-urban differences within individual countries which seem to suggest the existence of 'bicycle cultures', but may also indicate different socio-demographic compositions. For instance, in the German university towns of Freiburg, Munster and Erlangen the bicycle plays a much more significant role in daily travel than in most other German cities (BMVBW, 2002).

However, I am not aware of any comparisons that take trip length categories into consideration.

To the best of my knowledge, there are hardly any spatially differentiated studies on this topic concerning trends over time. The results found by Scheiner (2006, 2010a) point towards an increasing divide in travel mode choice between cities and smaller communities: in cities, car use increases more slowly and from a lower level than in smaller communities.

The decision in favour of a certain travel mode obviously does not only depend on trip distance. Important factors affecting the propensity to walk are individual motivation, available resources (transport means, financial resources, health), the attractiveness of the route, and social roles and needs, which are reflected in employment, gender and age, among other variables. For instance, adolescents cover considerable distances to school on foot (Schlossberg et al., 2006). The socio-demographic factors may be summarised under the term life situation (Scheiner and Holz-Rau, 2007)[1]. The availability of a car plays a central role in this respect, because the speed that can be achieved by car is a precondition for long trips under a restricted time budget. Car availability consequently turns out to be an important pre-decision for an individual's travel behaviour, which has a notable impact on his or her activity spaces, trip distances and travel mode choice (Simma and Axhausen, 2001; Scheiner and Holz-Rau, 2007).

In addition, trip purpose has an important influence on the propensity to walk. This is self-evident with respect to trips with an intrinsic motivation to walk, such as strolling or hiking. The requirement of carrying (shopping) goods impacts or even determines the chance of walking. In addition, the high economic cost of travel time (job trips, business trips) may limit the acceptability of slow travel.

2.3 Hypotheses and Study Approach

Travel behaviour is always based on a number of interrelated decisions an individual has to make, including decisions about the activity, an activity location (and the distance he or she has to cover to get there), a travel mode, and a date and time of activity/travel. What is more, travel behaviour is characterised by considerable social interdependencies, e.g. when it comes to decisions concerning shared trips and activities. To put it simply, one has to decide: what, where, when, how, and with whom? This framework does not assume that transport participants actually make deliberate decisions on a day-by-day basis. Employees clearly do not decide on the location of their workplaces every day. However, they all have made a decision about their particular workplace location at a certain point in time in the past. It thus seems obvious to study various elements of travel behaviour simultaneously rather than separately.

This said, it may not sound particularly spectacular to say that travel mode choice corresponds closely to travel distances. Nevertheless, the interrelation between realised or necessary trip lengths on the one hand and travel mode choice on the other hand has not yet been conclusively determined.

---

[1] In transport studies these factors are usually denominated as socio-demographics, though this is just a formal term that does not say anything about the reasons why the underlying variables should influence travel behaviour, whereas life situation explicitly points to an individual's personal circumstances (e.g. social roles, social contact) relevant for his or her travel.

Before examining this interrelation, it is important to note that there is no clear causal relationship between trip length and travel mode choice from a theoretical point of view. It seems plausible that increasing trip lengths may cause a shift towards the car. However, the causality might also be the other way round. Given a travel time budget which is by and large stable over time, the increase in car use may allow for longer trips. The study by Ye et al. (2007) may serve to support the former direction of causality. They model the interrelation between trip chain complexity and transport mode, and conclude that models in which chain complexity affects mode choice outperform models which assume reverse causality. This is true both for work trips and non-work trips. Although trip chain complexity and trip length are certainly two different things, this result suggests that mode choice might be an outcome of what somebody has to do and where he or she has to do it.

This interpretation is further supported by the findings of Lanzendorf (2001, p. 205ff) on the sequence of decisions about activities, activity places and travel modes in leisure travel. His findings indicate that in the overwhelming majority of cases people first decide on the destination, before they decide on the travel mode. Taken together, these findings might be interpreted as an indication that people decide on a certain destination (and thus, implicitly, on a certain trip length) and the mode choice decision tends to be 'at the end of the pipeline', even if there are certainly trips for which it is the other way round, e.g. trips to the countryside without a predetermined destination. Hence, in this chapter mode choice is treated as dependent on travel distance.

Explanatory variables considered for modelling include various measures of life situation, location preferences, the built environment, and the availability of a car. For the study of leisure trips, lifestyles are taken into account as well. For work and maintenance trips lifestyles have been found to be of minor or no relevance in extensive analyses of the same data (see Scheiner, 2009b, 2010b). All variables represent standard knowledge and/or more recent developments in transport research, as derived from the literature review above. The analysis is stratified by trip purpose. The general model structure used is shown in Figure 1, and more theoretical considerations are outlined elsewhere (Scheiner and Holz-Rau, 2007, Scheiner, 2009b, 2009c).

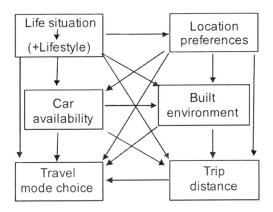

Source: own concept

Figure 8. Model structure.

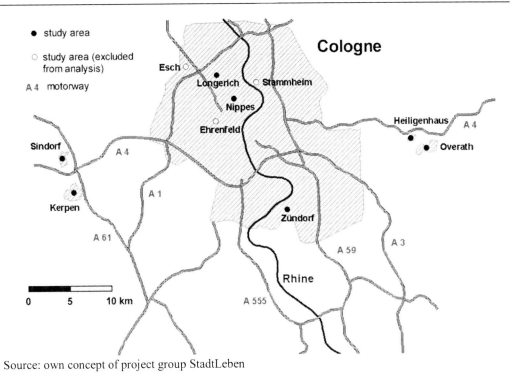

Source: own concept of project group StadtLeben

Figure 9. Location of the study areas in the region of Cologne.

3. METHODOLOGY

3.1 Data Used

The data used in this chapter were collected in a standardised, cross-sectional household survey within the framework of the project StadtLeben[2]. The survey was undertaken in ten study areas in the region of Cologne in 2002 and 2003. 2,691 inhabitants took part in extensive face-to-face interviews about their travel behaviour, housing mobility, life situation, lifestyle, location preferences and residential satisfaction. The response rate was 27 percent of those asked. This appears to be a reasonable rate, given the high respondent burden (the average interview duration was 58 minutes).

Due to reasons of project flow some of the relevant variables were recorded using different questionnaires in some of the study areas. The analysis is therefore based on seven study areas only, all of which were surveyed in 2003. Depending on the model, the resulting net samples have a value of about n=2,000. The working samples have a size of about n=1,000 due to the split of the sample (see below). The sampling procedure was based on

---

[2] "StadtLeben – Integrated approach to lifestyles, residential milieux, space and time for a sustainable concept of mobility and cities" (2001-2005). Project partners: RWTH Aachen, Institute for Urban and Transport Planning (coordination); FU Berlin, Institute of Geographical Sciences, Department of Urban Research; Ruhr-University of Bochum, Department of Cognition and Environmental Psychology; Dortmund University of Technology, Department of Transport Planning (see http://www.isb.rwth-aachen.de/stadtleben/).

random route. The sample is representative for the general population aged 16 and older in the study areas.

The five types of area are each represented by two study areas (Figure 1, study areas excluded from analysis in squared brackets): high density inner-city quarters of the 19th century ('Wilhelminian style': Nippes, [Ehrenfeld]); medium density neighbourhoods dating from the 1960s ('modern functionalism') with flats in three- or four-story row houses (Longerich, [Stammheim]); former villages located at the periphery of Cologne which since the 1950s have experienced ongoing expansion with single-family row houses or (semi-) detached single occupancy houses (Zündorf, [Esch]); small town centres in the suburban periphery of Cologne (Kerpen-Stadt, Overath-Stadt); and suburban neighbourhoods with detached single occupancy houses (Kerpen-Sindorf, Overath-Heiligenhaus). The four suburban neighbourhoods are all about 30 km away from Cologne.

As each of the two areas belonging to one type is clearly different, the areas are very varied with regard to spatial location, transport infrastructure, central place facilities and socio-demographic structure. Nonetheless it has to be noted that spatially or socially 'extreme' areas were not purposely targeted. There are no obvious high income areas, and only one distinct low income area (Stammheim; excluded from the analysis).

Heiligenhaus represents the most peripheral neighbourhood. There are no retail facilities or services worth mentioning, and PT is limited to an irregular bus service. However, one has to keep in mind that even this area is located within the outskirts of the city of Cologne. It is thus not particularly remote when seen in the context of the spatial variety of the whole of Germany.

The region of Cologne is a polycentric agglomeration with the clearly dominating centre of Cologne. The population trend is slightly positive, and the housing market is largely supply dominated. The opportunities for different population groups as defined by lifestyle or life situation to realise a specific location choice that meets their needs and wishes are thus limited. This is an important condition for the interpretation of the results.

3.2 Variables

Travel mode choice and trip distances are studied in this chapter in the context of life situation, lifestyle, location preferences, and car availability. These concepts can be specified with a low or high degree of complexity. Because of the many interdependencies, an attempt is made to keep the degree of complexity in the model components as low as possible. This is achieved by using dimensions of lifestyle, preferences and the built environment that seem relevant for mode choice and/or trip distances from a theoretical point of view, rather than including all available dimensions. The following components are used:

Life situation was measured by a set of seven observed variables, namely gender, age, number of children in the household, total household size, education level, per capita household income (with children counting as 0.8 persons) and employment. Some transformations of the ordinal-level variables education level and employment were undertaken in order to achieve metric variables. Education level was transformed into an estimated number of years in school. Employment (full-time, part-time, marginal, none) was similarly transformed into an estimated number of working hours per week.

In extensive attempts, a measurement model of life situation was developed in which household size, number of children, age and income were allowed to load on one latent variable, which was called 'family'. It should be noted that this variable refers to individuals living in a certain household type (family) rather than to households as units, as all analyses are based on individuals. Education level, employment, income and age are allowed to load on a second latent variable called 'social status'. Gender operates as an exogenous variable and rendering its binary scale unproblematic.

Lifestyles are presented in the data by four domains: leisure preferences, values and life aims, aesthetic taste, and frequency of social contacts. These were represented by a total of 34 items measured by five-point Likert-type answer scales. The scales were constructed so that they came as close to an interval scale as possible (see Rohrmann, 1978). In order to keep the models as simple as possible, only a few items are selected to represent lifestyle. What is more, the use of lifestyles is limited to the leisure trip models, as noted above. As some respondents are more inclined than others to generally agree with items, the answers were normed by subtracting a respondent's mean answer to all the items from the respective value. This results in normalised variables that take any individual tendency to generally agree or disagree into account.

Lifestyle is represented in the leisure models by the strength of out-of-home leisure preferences, a latent variable based on the items 'going to the movies/theatre/concerts' and 'attending training/education courses'. This latent variable is assumed to be related to a large variety of out-of-home leisure needs, relatively long trips and, thus, low shares of NMT, but high shares of motorised modes. In earlier attempts, the 'familial leisure preference' was used as a second lifestyle dimension. This was based on the two items 'play with children' and 'engage with my family'. However, it was found to be of minor importance for mode choice as well as trip distances. It was therefore excluded from the models for the sake of parsimony.

Individual *location preferences* were measured using subjective importance ratings of neighbourhood and location attributes, again measured by five-point Likert-type answer scales. Information was gathered as part of the survey by asking 'How important are the following features of the neighbourhood for your personal decision in favour of a certain place of residence?' The attributes were then listed, for instance 'accessibility of the city centre' or 'access to public transport'. Again, the scales are normalised in order to take any individual inclination to generally agree or disagree into account. ˙

Specifically, when examining maintenance trips, the importance of proximity to shops and services is used as an indicator of location preferences, while in the leisure models, the importance of proximity to leisure facilities is used (see overview in Table 1). The former was measured by a latent variable based on two observed variables: 'proximity to shops' and 'proximity to services'. In the latter, a single item measuring the importance of proximity to leisure facilities for adults is used. When examining the sum of all recorded trips, general measures of urbanity and/or transport service seem to be more appropriate than proximity to a certain type of facility. Hence, when examining car use and PT use, the importance of proximity to PT is used ('importance of PT'). For NMT, the quality and quantity of local activity opportunities is more important than the quality of the local PT system. Thus, the subjective importance of proximity to retail and services is used here.

The *built environment* at the place of residence is studied with regard to specific attributes of the neighbourhood that are selected in accordance with the location preferences. In the maintenance trip models, the supply of retail and services is used. Similarly, the supply of leisure opportunities is used in the leisure trip models. Both indicators are measured as the number of opportunities within a straight-line distance of 650 m around the place of residence. These indicators are calculated separately for all individuals[3]. In the models of all trips taken together, I use the quality of the local PT system when car share or PT share is studied. When the dependent variable is the share of non-motorised modes, the built environment is measured by the supply with shopping opportunities and services again.

Car availability was measured as an ordinal variable which can take on four values (see Table 2). This ordinal variable could be interpreted as metric once the distances between the four values were equal. Actual car use in the four groups suggests that *cum grano salis* this is approximately true (Table 2). None of the groups are either extremely close together or extremely far apart.

Travel behaviour was recorded by applying the frequent activities method. Activity frequency, usual travel mode, destination and travel distance were surveyed for selected activities including work, education, daily grocery shopping, weekly shopping, event shopping, personal errands at public authorities, private visits, sports, visits to restaurants or pubs, cultural events and sport events, discos and concerts, walks, and excursions.

Table 1. Measurement of location preferences and the built environment in the models

Travel mode studied	All trip purposes	Maintenance trips	Leisure trips
	Location preferences: importance of proximity to...		
Car	PT	shopping	leisure facilities
PT	PT	shopping	leisure facilities
NMT	shopping	shopping	leisure facilities
	Built environment: quality of...		
Car	PT	shopping	leisure facilities
PT	PT	shopping	leisure facilities
NMT	shopping	shopping	leisure facilities

Source: author's concept.

Table 2. Car use by car availability (means)

	car use per week		
Car availability	distance	frequency	n
no car in household	12.2	0.8	305
car in household, but not available	34.9	2.2	88
car in household, partly available	67.0	4.8	273
car in household, available at any time	104.3	6.7	1,454
All	83.2	5.4	2,120

Source: author's analysis. Data: StadtLeben.

[3] The mapping of opportunities was undertaken by the RWTH Aachen and the Ruhr Universität Bochum. Leisure opportunities include sites of informal activity, such as chance meeting points in public space. I extended this survey beyond the borders of the study areas to meet the full radius of 650 m even for respondents living close to the border of an area.

Trip distances were examined on the basis of mean values for selected activities an individual reported having made. Maintenance activities include daily grocery shopping, weekly shopping, event shopping and administrative transactions at public authorities. Leisure trips include all other activities noted above except for work and education. Job trips and education trips are not studied separately, as the data only include discrete measures of the usual travel mode for these two travel purposes, preventing the construction of metric variables.

Travel mode use was examined here with respect to relative shares of various travel modes among all trips reported by an individual in the respective activity category. Three transport modes are considered: the private car (including marginal shares of motorcycles), PT, and NMT (bicycle and walking). A related paper (Scheiner, 2009c) focuses on absolute frequencies of mode use. However, these not only depend on an individual's inclination to use a given mode, but also on his or her level of mobility, i.e. on activity frequency. In any case, comparative studies of absolute frequencies v. relative shares of mode use generally yield similar results (see Scheiner, 2009a).

It should also be noted that mode use is considered here on an individual rather than a trip level. Thus, mode choice is treated not as discrete choice, as one would normally assume, but as continuous variables. Likewise, trip distances are studied here as mean distances covered by an individual for certain purposes. This means that the interrelation between distance and mode choice is blurred to a certain extent by aggregating trips and, thus, this interrelation is likely to be underestimated.

3.3 Methodology of Structural Equation Modelling

The interrelations discussed above can be studied with structural equation modelling. This method is being increasingly used in transportation studies (Golob, 2003). Structural equation modelling can be described as a combination of factor analysis and a generalised form of regression analysis. The technique used is not described in detail here due to lack of space (see Scheiner and Holz-Rau, 2007 for more details).

There is much debate about the conditions under which the classical Maximum Likelihood (ML) approach can be regarded as superior to non-parametric procedures even when the normality assumption is violated (e.g., Hoogland and Boomsma, 1998, see also Golob, 2003 for transport applications). The available sample of about n=2,000 seems well appropriate for a robust application of the ML procedure, even if the sample is split into two halves (see below). The asymptotically distribution-free (ADF) procedure then reaches the limit of reliability, but seems still to be acceptable. Ultimately, a rather rigorous approach was applied. First, the sample was split into two halves by a random procedure. Then each model was estimated in four versions:

1. ML estimation of a theoretical model with the main sample
2. Empirical fitting of the model to the data
3. ADF estimation of the theoretical model
4. ML estimation of the theoretical model with the second sample for validation.

Version 2 only serves to verify the coefficients in the theoretical model version when fitted to the data, while my substantial interest lies in the theoretical models. Each of the four model versions was compared to the others with respect to the strength and sign of the effects. The results for each four versions turn out fairly stable and may clearly be interpreted in terms of the sign and strength of the effects. By contrast, modelling all trip purposes together without taking more than one measure of travel behaviour simultaneously into account resulted in considerable variations of effects between each of the four versions.

The interpretations are based on direct as well as total effects. Total effects of one variable on another are calculated as the sum of direct and indirect effects, the latter being mediated by intervening variables. Taking total effects into account allows for a more thorough interpretation of interrelations. An example for the calculation can be found in the text below Figure 3. The analyses were undertaken with the programme AMOS 5.0 to 7.0 (Analysis of Moment Structures).

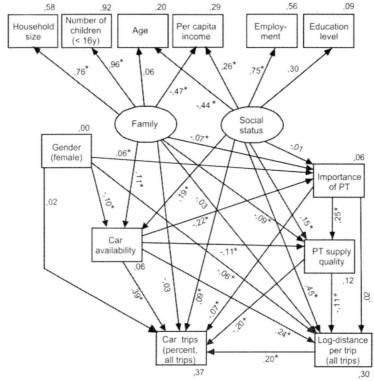

Theoretical model (version 1, ML estimation). This and the following figures show the estimated standardised path coefficients and the proportion of explained variance of the endogenous variables, the latter being indicated next to the variable boxes. Significant coefficients (p=0.05) are marked with an asterisk. The rectangles are observed variables, the ovals are latent constructs. The total effect a variable has on another variable is calculated as the sum of direct and indirect effects. For instance, the total effect of 'importance of PT' on 'car trips' equals -0.07 + 0.25*-0.20 + 0.25*-0.11*0,20+0,02*0,20 = -0.12.

Source: Author's analysis. Data: StadtLeben.

Figure 10. Model of trip distances and car share – all trip purposes.

Table 3. Goodness-of-fit and degrees of freedom for the models

model	indicator of goodness-of-fit				df	
	RMSEA		Hoelter (p=0.05)			
	decision rule					
	<0.05 good >0.1 n.a.		≥ 200 good			
	figure	best (version 2)	figure	best (version 2)	figure	best (version 2)
car share, all trips	0.127	0.035	84	678	34	26
car share, maintenance	0.119	0.044	91	505	46	33
car share, leisure	0.111	0.043	102	499	48	41
PT share, all trips	0.112	0.043	107	505	34	28
PT share, maintenance	0.118	0.046	93	469	45	34
PT share, leisure	0.133	0.047	77	477	34	25
NMT share, all trips	0.117	0.039	93	582	46	34
NMT share, maintenance	0.117	0.028	94	857	45	30
NMT share, leisure	0.112	0.036	102	616	48	39

n.a.: not acceptable The table gives values for two goodness-of-fit indicators as well as the degrees of freedom (df) for the model shown in the following figures and for the empirically fitted 'best' model version.RMSEA (Root Mean Square Error of Approximation) measures discrepancy between the model implied and the true population covariance matrix, in relation to degrees of freedom. This ratio is related to sample size. In cases of close fit, RMSEA approaches zero. Values smaller than 0.05 indicate a close fit. The Hoelter statistics specifies the required sample size (critical n) to reject the model at a given significance level. Large values indicate a close fit. Values larger than 200 can be regarded as good. The table gives decision rules for the two indices.

Source: Author's analysis. Data: Project StadtLeben.

4. RESULTS

There are a number of heuristic indicators to assess the goodness-of-fit of structural equation models. For most of these indicators there are decision rules available and they have been tested in methodological studies. The indicators are based on different principles, for instance on discrepancies between theoretical and empirical covariance matrices, or on mean differences between expected and estimated values of various parameters. Two of these indicators, along with the corresponding decision rule, are given in Table 3 for the models shown in the figures below and for the respective best model version (i.e. the ones that have been empirically fitted to the data, see above). The fit values of the theoretical models fail to meet a satisfactory level, but the values of the fitted models are satisfactory to close.

Turning our attention to the effects found in the models, social status is generally the most important element of life situation. High social status is associated with markedly longer trips and higher shares of car use *as well as* PT use (Figure 3, Figure 6). At the same time, individuals with high social status have lower NMT shares (Figure 9). These findings are true for maintenance trips and for leisure trips as well as for all trips put together. The effects of social status on travel mode choice are a result of direct effects as well as indirect effects mediated by trip distances and residential location choice. Individuals with high social status tend to locate in central areas. These residential location decisions encourage PT and NMT use and counterbalance car use to a certain extent, but without fully offsetting the positive relationship between status and car use.

As far as gender is concerned, women make slightly shorter trips than men. This mainly relates to the models that include all trip purposes (Figures 3, 6, 9), less to the maintenance models (Figures 4, 7, 10), and even less to the leisure models (Figures 5, 8, 11). Women's shorter job trips account for a large part of the significant gender difference when all trip purposes are considered. Gender does not have much of an impact on travel mode choice. Thus, gender differences seem to be more a matter of activity spaces.

The household type family is associated with longer than average maintenance trips. This finding corresponds with the high mode share of the private car among individuals living in a family. PT use (particularly for leisure trips) is slightly weaker among individuals living in family households than among others.

Lifestyle is one of the strongest impact factors for leisure trip distances (Figures 5, 8, 11), besides social status and car availability. A strong out-of-home leisure orientation is associated with markedly longer trips and, accordingly, with low shares of NMT and high shares of motorised modes. Concerning the latter, the effect of lifestyle on PT use (Figure 8) is considerably stronger than its effect on car use (Figure 5). It is perhaps surprising that a strong out-of-home leisure orientation seems to strengthen PT use more than car use. However, one has to note that individuals with this lifestyle are often young adults such as students and trainees, many of whom do not have access to a car. The positive effect of lifestyle on trip distances exists despite the tendency of individuals with strong out-of-home leisure preferences to locate in neighbourhoods with good leisure supply. In other words, although these individuals tend to live in neighbourhoods that match their preferences they tend to make long trips to fulfil their diverse leisure needs. This means that without the mediating residential location effect their leisure trips would be even longer.

The availability of a car also strongly increases trip distances and, obviously, car use, but decreases the use of PT as well as NMT. The impact the car has on PT use is considerably stronger than its impact on NMT. In the leisure sector the *direct* effect of the car on NMT is actually negligible. Even so, the overall effect of the car is negative in any case. This is mainly a result of the tendency for motorised individuals to locate in peripheral locations and to undertake longer trips.

With respect to the built environment, a well-developed supply of retail outlets and services is strongly associated with relatively short trips and lower proportions of car use among all maintenance trips (Figure 4). The low car share is both a direct and an indirect effect mediated by trip distance. Similarly, a good leisure supply structure leads to shorter leisure trips (Figures 5, 8, 11), although the effect is not as marked as for maintenance trips. What is more, a large number of leisure opportunities in the neighbourhood leads to markedly less car use (Figure 5) and a higher NMT share (Figure 11).

When all recorded activities are considered together, the quality of PT supply is used as the built environment indicator in the models of car use and PT use. PT supply corresponds with relatively short trips (Figure 3), which is likely to be an effect of the generally higher urbanity found at residential locations with a good PT system than of the PT system per se. The effect PT has on travel mode choice is even stronger than its effect on trip distances: living at a place with a good PT system leads to more PT use and less car use (Figures 3, 6).

A high subjective preference for PT affects PT use more strongly than the objective quality of the system (Figure 6). Thus, the subjective importance of PT appears to be even more important than the PT system itself. Residential self-selection according to specific location preferences is particularly pronounced in this case.

On the other hand, a high preference for proximity to shopping or leisure facilities does not lead either to short maintenance trips or short leisure trips, nor does it have significant effects on mode choice. Although such proximity preferences are clearly associated (particularly in the case of shopping preferences) with residential location choice, their total effects on travel behaviour are negligible.

Last but not least, there is a clear and strong relationship between trip distance and travel mode choice: the longer the trips, the greater the car use and the PT use (Figures 3, 6), and the weaker the use of NMT (Figure 9). Thus, the interrelation between distance and mode choice does indeed capture the two different elements outlined in section 2. Firstly, mode choice depends on realised trip distances. These may be affected at best indirectly by urban policy concepts, e.g. by impacting on the monetary or generalised costs of travelling. Secondly, the *necessary* distances to potential destinations – i.e. to opportunities – may be affected directly by land-use planning. It should be encouraging for policy makers and planners that land-use does indeed seem to have a strong independent impact on travel behaviour, irrespective of people's attitudes and social background. The data analysed in this chapter reveals the effect of the built environment in terms of the number of opportunities in the respondents' neighbourhoods. The effects found are particularly strong for maintenance trips, but are also significant for leisure trips as well as for all trips recorded when put together.

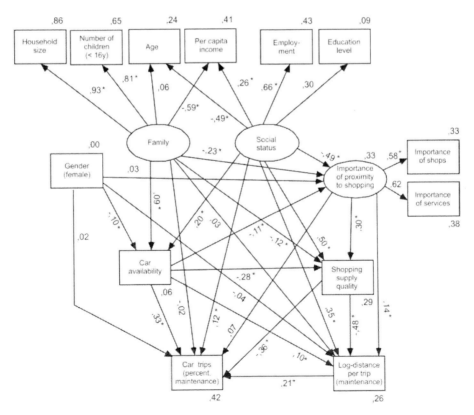

Figure 11. Model of trip distances and car share – maintenance trips.

Theoretical model (version 1, ML estimation).
Source: Author's analysis. Data: StadtLeben.

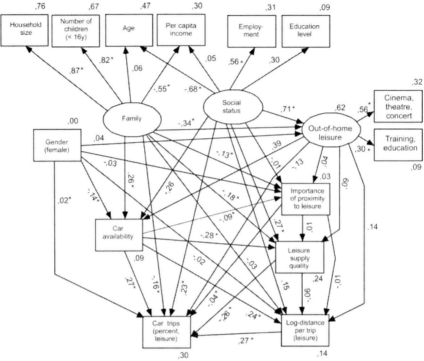

Theoretical model (version 1, ML estimation).
Source: Author's analysis. Data: StadtLeben.

Figure 12. Model of trip distances and car share – leisure trips.

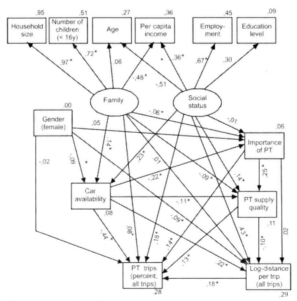

Theoretical model (version 1, ML estimation).
Source: Author's analysis. Data: StadtLeben.

Figure 13. Model of trip distances and PT share – all trip purposes.

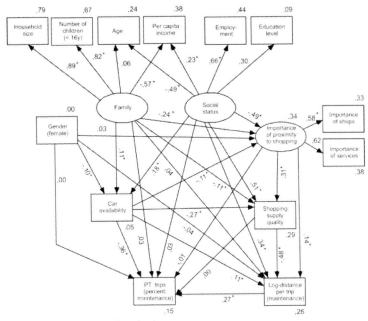

Theoretical model (version 1, ML estimation).
Source: Author's analysis. Data: StadtLeben.

Figure 14. Model of trip distances and PT share – maintenance trips.

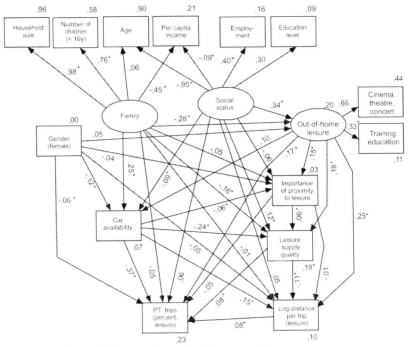

Theoretical model (version 4, ML estimation, validation sample).
Source: Author's analysis. Data: StadtLeben.

Figure 15. Model of trip distances and PT share – leisure trips.

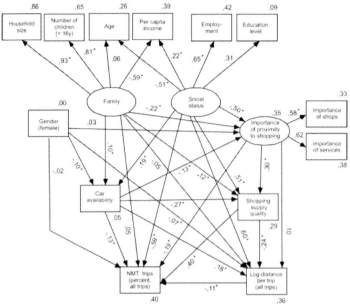

Theoretical model (version 1, ML estimation).
Source: Author's analysis. Data: StadtLeben.

Figure 16. Model of trip distances and NMT share – all trip purposes.

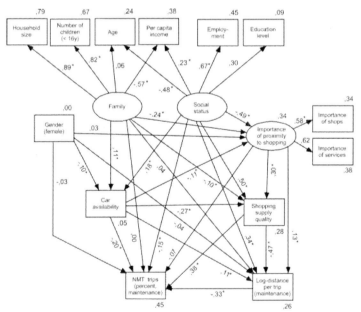

Theoretical model (version 1, ML estimation).
Source: Author's analysis. Data: StadtLeben.

Figure 17. Model of trip distances and NMT share – maintenance trips.

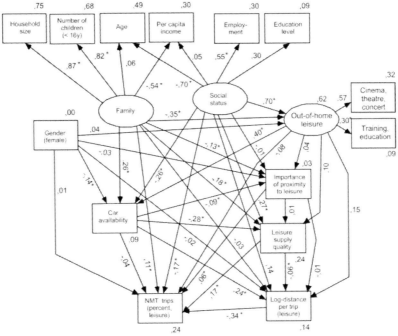

Theoretical model (version 1, ML estimation).
Source: Author's analysis. Data: StadtLeben.

Figure 18. Model of trip distances and NMT share – leisure trips.

5. CONCLUSION

The built environment matters. This is the key result of the analysis presented in this chapter. The built environment plays a substantial role in terms of people's travel distances as well as in terms of their mode choice. Social inequalities may be even more important – an observation which has been made before (Hanson and Schwab, 1995; Stead et al., 2000) –, but it seems nonetheless that the provision of a range of shopping and leisure facilities and a decent PT system has substantial potential to make people use environmentally friendly travel modes and undertake relatively short trips. It has been shown that in such urban contexts even car owners may choose to leave their car at home and walk more than elsewhere (Scheiner, 2010a).

On the other hand there are at least four caveats worthy of note.

Firstly, residential self-selection may play an important role for travel. It seems that people tend to choose a place of residence that matches their travel and accessibility preferences, and while the above findings do not support the notion of self-selection effects on trip distances with respect to shopping or leisure preferences, they suggest that PT preferences may indeed be more important for PT use than the actual PT system. Residential self-selection, however, is not only a matter of location attitudes. People also tend to sort themselves according to their life situation and lifestyle. Social status has been identified as being one of the strongest impact factors for mode choice and trip distances. In the leisure sector, lifestyle is among the most prominent impact factors for trip distances as well. These social inequalities in travel behaviour are partly indirect effects mediated by residential sorting. One has to note, however, that the existence of residential self-selection does not mean that the built environment is irrelevant. In order to give people the chance to choose between different locations, these differences have to exist. That is, residential self-selection effects are themselves evidence that the built environment matters (Naess, 2009).

Secondly, short trip distances and high shares of environmentally friendly modes can be found mainly in the inner districts of large cities, i.e. in spatial contexts that are extremely over-supplied with central place facilities. These districts do not only serve their own population with shopping and leisure opportunities, but large numbers of incoming consumers as well. This means that the lessons learned from these urban districts cannot be directly transferred to other contexts, such as small towns or suburban neighbourhoods, as the over-provision of facilities in the central locations of core cities and the under-provision in the surrounding suburbs are two sides of the same coin – they are mutually dependent.

Thirdly, this notion also implies that it does not make much sense to blame peripheral locations for high per-capita car travel volumes and at the same time praise urban locations for their low car travel volumes, as exactly the same urban locations may be blamed for their potential to attract incoming commuters, shopping and leisure consumers who are exactly those who live in the suburbs and cover long distances to the centres which attract them.

This mutual dependency, however, appears to be a dilemma only at first glance. The solution would be to aim for a spatial structure that attempts to avoid the under-supply *and* the over-supply of opportunities. For workplaces (as one particular example of opportunities) it has been shown that a spatially balanced job-housing mix is associated with less job travel (Cervero and Duncan, 2006 for the US, Holz-Rau and Kutter, 1995 for Germany).

Fourth, perhaps the most salient problem with small-scale, neighbourhood-oriented spatial distributions of opportunities is the level of specialisation required for economic viability in individualised, affluent societies. Even for retail outlets such as groceries it has been argued that neighbourhood-scale 'walk-to shops' may have limited chances of survival (Bartlett, 2005). This may be even truer for more specialised opportunities such as leisure and cultural facilities, or workplaces.

However, the question of economic viability should not lead us to conclude that there is no point in encouraging people to undertake short trips by providing relatively small-scale opportunities and a relatively balanced spatial distribution of land-uses. The chances to do so may be very limited under the current economic and societal conditions, but they do exist. And what is more, one clearly has to recognise that the current economic and societal conditions are unlikely to remain stable in the near future. Energy and, thus, transport prices are likely to further increase substantially. More than any urban policy, this may foster more balanced, mixed-use spatial patterns and a more 'traditional' development of neighbourhoods and cities. A principal task for urban policy could well be to recognise and take advantage of this opportunity to provide the conditions to facilitate more balanced, mixed-use patterns in order to be prepared for future challenges.

ACKNOWLEDGMENT

The research in this chapter was funded by the German Research Foundation (DFG) as part of the project 'Wohnstandortwahl, Raum und Verkehr im Kontext von Lebensstil und Lebenslage' (Choice of residential location, built environment and transport in the context of lifestyle and life situation, 2006-2008).

6. REFERENCES

Badland, H. & Schofield, G. (2005). Transport, urban design, and physical activity: an evidence-based update. *Transportation Research, D 10 (3)*, 177-196.

Bahrenberg, G. (1997). Zum Raumfetischismus in der jüngeren verkehrspolitischen Diskussion. In U. Eisel, & H. D. Schultz (Eds.) Geographisches Denken. *Urbs et Regio, 65* (345-371). Kassel: Universität.

Bartlett, R. (2003). Testing the 'popsicle test': realities of retail shopping in new 'traditional neighbourhood developments'. *Urban Studies, 40(8)*, 1471-1485.

BMVBW (*Federal Ministry of Transport, Building and Housing*, Ed.) (2002). National cycling plan 2002-2012: ride your bike! Berlin: BMVBW.

Boarnet, M. & Crane, R. (2001). *Travel By Design*. Oxford: Oxford University Press.

Bohte, W., Maat, K. & van Wee, B. (2009). Measuring attitudes in research on residential self-selection and travel behaviour: a review of theories and empirical research. *Transport Reviews, 29(3)*, S. 325-357.

Cao, X., Handy, S. L., Mokhtarian, P. L. (2006). The influences of the built environment and residential self-selection on pedestrian behavior in Austin, Texas. *Transportation, 33(1)*, 1-20.

Cao, X., Mokhtarian, P. & Handy, S. (2009). Examining the impacts of residential self-selection on travel behaviour: a focus on empirical findings. *Transport Reviews, 29(3)*, 359-396.

Cervero, R. (2002). Built environments and mode choice: toward a normative framework. *Transportation Research, D 7(4)*, 265-284.

Cervero, R. (2006). Alternative approaches to modeling the travel-demand impacts of smarth growth. *Journal of the American Planning Association, 72(3)*, 285-295.

Cervero, R. & Duncan, M. (2006) Which reduces travel more: Jobs-housing balance or retail-housing mix? *Journal of the American Planning Association, 72(4)*, 475-490.

DfT (Department for Transport) (Eds.) (2006). Transport statistics bulletin. *National travel survey 2005*. London: DfT.

DIW/INFAS (2003). Mobilität in Tabellen. Tabellenprogramm zur Erhebung Mobilität in Deutschland 2002. Berlin: DIW.

EU (Eds.) (2000). *Transport in figures* 2000. Luxembourg: European Commission.

Gatersleben, B. & Appleton, K. M. (2007). Contemplating cycling to work: attitudes and perceptions in different stages of change. *Transportation Research, A 41(4)*, 302-312.

Giuliano, G. & Dargay, J. (2006). Car ownership, travel and land use: a comparison of the US and Great Britain. *Transportation Research, A 40(2)*, 106-124.

Golob, T. F. (2003). Structural equation modeling for travel behavior research. *Transportation Research, B 37(1)*, 1-25.

Guo, J. Y. & Chen, C. (2007). The built environment and travel behavior: making the connection. *Transportation, 34(5)*, 529-533.

Handy, S. L. & Clifton, K. J. (2001). Local shopping as a strategy for reducing automobile travel. *Transportation, 28(4)*, 317-346.

Hanson, S. & Schwab, M. (1995). Describing Disaggregate Flows: Individual and Household Activity Patterns. In S. Hanson (Eds.) *The Geography of Urban Transportation*, (154-178). New York, London: The Guilford Press.

Holz-Rau, C. & Kutter, E. (1995). Verkehrsvermeidung. Siedlungsstrukturelle und organisatorische Konzepte. *Materialien zur Raumentwicklung, 73*. Bonn: BfLR.

Holz-Rau, C., Rau, P., Scheiner, J., Trubbach, K., Dörkes, C., Fromberg, A., Gwiasda, P. & Krüger, S. (1999). Nutzungsmischung und Stadt der kurzen Wege: Werden die Vorzüge einer baulichen Mischung im Alltag genutzt? Bonn: Bundesamt für Bauwesen und Raumordnung.

Holz-Rau, H. C. (1991). Verkehrsverhalten beim Einkauf. Internationales *Verkehrswesen, 43(7-8)*, 300-305.

Hoogland, J. J. & Boomsma, A. (1998) Robustness studies in covariance structure modeling. *Sociological Methods and Research, 26(3)*, 329-367.

ILS (Institut für Landes- und Stadtentwicklungsforschung Nordrhein-Westfalen) (1999). Baulandentwicklung an der Schiene. Dortmund: ILS.

Kloas, J. & Kunert, U. (1993). Vergleichende Auswertungen von Haushaltsbefragungen zum Personennahverkehr (KONTIV 1976, 1982, 1989). Forschungsprojekt im Auftrag des Bundesministers für Verkehr (FE-Nr. 90361/92). Berlin: DIW.

Lanzendorf, M. (2001). Freizeitmobilität. Unterwegs in Sachen sozial-ökologischer Mobilitätsforschung. *Materialien zur Fremdenverkehrsgeographie 56*. Trier: Universität.

Naess, P. (2009). Residential self-selection and appropriate control variables in land use: travel studies. *Transport Reviews, 29(3)*, 293-324.

Rohrmann, B. (1978). Empirische Studien zur Entwicklung von Antwortskalen für die sozialwissenschaftliche Forschung. *Zeitschrift für Sozialpsychologie, 9(1)*, 222-245.

Scheiner, J. (2006). Does individualisation of travel behaviour exist? Determinants and determination of travel participation and mode choice in West Germany, 1976-2002. *Die Erde, 137(4)*, 355-377.

Scheiner, J. (2009a). Sozialer Wandel, Raum und Mobilität. Empirische Untersuchungen zur Subjektivierung der Verkehrsnachfrage. Wiesbaden: VS Verlag.

Scheiner, J. (2009b). Objective and subjective socio-spatial inequalities in activity patterns. *Swiss Journal of Sociology, 35*, (accepted for publication).

Scheiner, J. (2009c). Is travel mode choice driven by subjective or objective factors? In C. Holz-Rau, & J. Scheiner (Eds.) Subject-Oriented Approaches to Transport. *Dortmunder Beiträge zur Raumplanung, Verkehr, 6 (53-69)*. Dortmund: IRPUD.

Scheiner, J. (2010a). Interrelations between travel mode choice and trip distance: trends in Germany 1976–2002. *Journal of Transport Geography, 18(1)*, 75-84.

Scheiner, J. (2010b). Social inequalities in travel behaviour: trip distances in the context of residential self-selection and lifestyles. *Journal of Transport Geography*, (accepted for publication).

Scheiner, J. & Holz-Rau, C. (2007). Travel mode choice: affected by objective or subjective determinants? *Transportation, 34(4)*, 487-511.

Schlossberg, M., Greene, J., Phillips, P. P., Johnson, B. & Parker, B. (2006). School trips: effects of urban form and distance on travel mode. *Journal of the American Planning Association, 72(3)*, 337-346.

Schwanen, T., Dijst, M. & Dieleman, F. M. (2004). Policies for urban form and their impact on travel: the Netherlands experience. *Urban Studies, 41(3)*, 579-603.

Simma, A. & Axhausen, K. W. (2001). Structures of commitment in mode use: a comparison of Switzerland, Germany and Great Britain. *Transport Policy, 8(4)*, 279-288.

Stead, D. & Marshall, S. (2001). The relationships between urban form and travel patterns. An international review and evaluation. *European Journal of Transport and Infrastructure Research, 1(2)*, 113-141.

Stead, D., Williams, J. & Titheridge, H. (2000). Land Use, Transport and People: Identifying the Connections. In K. Williams, E., Burton, & M. Jenks (Eds.) *Achieving Sustainable Urban Form* (174-186). London: E & FN Spon.

Vågane, L. (2007). *Short Car Trips in Norway: Is there a potential for modal shift?* Paper presented at the European Transport Conference, Leiden, Netherlands, 17-19 October, 2007.

Ye, X., Pendyala, R. M. & Gottardi, G. (2007). An exploration of the relationship between mode choice and complexity of trip chaining patterns. *Transportation Research, B 41(1)*, 96-113.

Zumkeller, D. & Nakott, J. (1988). Neues Leben für die Städte: Grünes Licht fürs Fahrrad. *Bild der Wissenschaft, 5*, 104-113.

In: Built Environment: Design, Management and Applications   ISBN: 978-1-60876-915-5
Editor: Paul S. Geller, pp. 239-259   © 2010 Nova Science Publishers, Inc.

Chapter 10

ON THE EMPIRICS OF HOUSING ILLEGALITY: A CASE STUDY IN HONG KONG

Yung Yau[1], Daniel Chi Wing Ho[2] and Kwong Wing Chau[2]*
[1]Department of Public and Social Administration,
City University of Hong Kong, Hong Kong
[2]Department of Real Estate and Construction, The University of Hong Kong, Hong Kong

ABSTRACT

Unauthorized building works (UBWs), particularly those attached to external walls of buildings, have posed serious threats to the safety of the community in Hong Kong. As estimated by the government, there exist around 0.75 million UBWs in some 39,000 private buildings throughout the territory. In spite of the plenteous literature on this topic, associated empirical studies have been relatively rare, and this is not constructive for the exploration of the causes of UBW proliferation. To straddle the existing research gap, this study empirically explores what types of unauthorized appendages dominate and what factors affect the degree of proliferation of unauthorized appendages in multi-storey residential buildings in Hong Kong. Given that the majority of Hong Kong's populace live in this type of housing, residents and the general public are prone to UBW hazards in their environments. Therefore, an empirical study like this one is necessary for providing valuable insights to public administrators for making more informed decisions. Through appraising 429 multi-storey residential buildings in Yau Tsim Mong, the Eastern District and Kowloon City, a profile of unauthorized appendages in these buildings were obtained. The analysis results indicated that over 98 percent of these appendages were not constructed purportedly for increasing usable floor space, but for enhancing the amenities enjoyed by the residents. Besides, building age, development scale and management characteristics have a strong bearing on the number of unauthorized appendages present in a building. These findings have far-reaching implications on the formulation of government policies regarding building safety in Hong Kong.

---

* Corresponding author: Department of Public and Social Administration, City University of Hong Kong, 83 Tat Chee Avenue, Kowloon Tong, Kowloon, Hong Kong OR E-mail: y.yau@cityu.edu.hk.

Keywords: Building safety; housing illegality; unauthorized building works; multi-storey residential buildings; Hong Kong.

INTRODUCTION

To many visitors to Hong Kong, one of the most unforgettable attractions of the city is her neon-lit cityscape at night (Macgregor & Price, 2002). As in other places like Shinjuku and Les Vegas, the urban areas of Hong Kong were brightened by the glittering neon lights and fluorescent advertisement signs in the night-time. The glowing streetscape, particularly along busy and hustle roads, is an appealing characteristic of the city. That explains why the Hong Kong Tourism Board has sold the city as a place with zillions of blazing neon signs for years (Henderson, 2001). Nevertheless, what is behind the beauty is danger because many of these neon signs are unlawful structures or so-called unauthorized building works.

The term "unauthorized building works" (UBWs) has not been explicitly defined by *the Laws of Hong Kong*. It, in principle, refers to building works carried out with no prior approval and consent obtainable from the public authority. Apart from advertisement signs, UBWs frequently mentioned can also be found in form of flower rack, drying rack, supporting frame for air-conditioner, metal cage, cockloft, illegal roof-top structure, and unauthorized alteration to exit route. In the past two decades, UBW proliferation led to many casualties in Hong Kong. During the period between 1990 and 2002, at least 21 people were killed and 135 were injured in building-related accidents involving UBWs in the city (Task Force on Building Safety and Preventive Maintenance, 2001; Leung & Yiu, 2004). In this light, proliferation of UBWs has long been a big issue to be addressed by the government of Hong Kong, and attracted widespread academic attention (Davison, 1990; Lai & Ho, 2001; Lai, 2003; Yiu et al., 2004; Yiu, 2005; Yiu & Yau, 2005). Yet, the topic was predominately studied qualitatively and relevant empirical research is lacking. Against this background, this study aims to straddle the research gap by empirically investigating the extent and the roots of the UBW problem in private multi-storey residential buildings in Hong Kong. Given that the urban areas of Hong Kong are dominated by these high-rises, UBW proliferation can seriously jeopardize the safety of the populace in the city. In this regard, empirical findings of the study can offer valuable insights to public administrators into government policies on building illegality.

Moreover, this paper is a response to a degree of frustration with the academic literature in the field of building control. While there has been a growing body of research throughout the world dedicated to the study on building control, focus of the literature tends to be one-sided. Previous studies regarding building control overseas (e.g. Meijer & Visscher, 1998; McAdam & O'Neill, 2002; Listokin & Hattis, 2005; Baiche et al., 2006; Imrie, 2004; 2007) usually put their focal points on building submissions and approvals for new developments, with the majority (Meijer et al., 2002; Meijer & Visscher, 2006; van der Heijden et al., 2007) concentrating their attention on privatization of plan checking duties of the local authorities. Also, more and more research (e.g. Wong et al., 2003; Yang & Xu, 2004) is dedicated to the applied research about automatic or computer-aided checking against building code compliance. However, only little academic effort (e.g. Burridge & Ormandy, 2007) has been paid to the control of existing buildings. Given that research on existing building stock is gaining its importance in the light of the rising public concerns about sustainability (Kohler & Hassler, 2002; Augenbroe & Park, 2005; Wood, 2005), there is no point to sustain such an imbalance of research focus between new and existing buildings. Indeed, a stand-alone research area should be warranted for the control of existing buildings for the reason that vigilant long-term management of the building stock is one of the key determinants of an urban city's sustainable development (Kohler & Yang, 2007). Therefore, this empirical study on UBWs constructed in existing buildings can contribute to the body of knowledge in that particular area.

This article is organized as follows. The regulation of UBWs in Hong Kong will be first introduced, followed by a brief account of the UBW problem in the city. Then, the methodology of the empirical study and the key findings will be detailed. What come next are the analysis results. The implications of the analysis results will be discussed before the empirical study is concluded.

THE PROLIFERATION OF UBWS IN HONG KONG

To commence, different types of building illegality in Hong Kong are discussed. This helps to give an account on why UBW, but not other types of building illegality, forms the subject of this study.

Housing Illegalities in Hong Kong

Broadly speaking, housing illegality can be classified into two categories according to their natures. The first category is the construction of building structures on government land without obtaining the legal title from the government. Informal housing or squatter structures of this type have been proliferating in the third world's big cities such as Johannesburg, Sao Paulo and Jakarta (Few *et al.*, 2004; Winayanti & Lang, 2004). In fact, similar squatter structures were once very common in Hong Kong before the 1990s, and their erection was ascribed to the population boom in the 1950s and 1960s due to the influx of Mainland immigrants and rising birth rate (Hopkin, 1972; Smart, 2006). Consequently, housing demand far exceeded the supply, and thus thousands of squatter huts were erected on undeveloped and unleased government land. An estimated 650,000 people squatted throughout Hong Kong in 1965 (Smart, 2006), and the number peaked at 750,000 in the early 1980s (Smart, 2002). With the continuous endeavours of the government in squatter control and clearance over the years, the number of squatters in the city urban dropped to 220,000 in the early 2000s (Smart, 2002).

The second category is the UBW constructed on leased land. From the legal viewpoint, all building works in Hong Kong are subject to statutory control under the Buildings Ordinance (Chapter 123 of *the Laws of Hong Kong*) and its subsidiary legislations. The Buildings Department executes and enforces these legislations. The term "building works" is very broadly defined under the Buildings Ordinance (Davison, 1990), and thus building works can range from construction of new buildings, demolition works to additions and alterations to existing buildings. To make sure the design and carrying out of a building work are up to minimum acceptable standard, approval and consent must be obtained from the Building Authority[1] before the building work can commence, unless the work is exempted from this requirement by the ordinance (Chan & Chan, 2003; Yiu & Yau, 2005). Building works that contravene this stipulation are generally regarded as UBWs (Yiu *et al.*, 2004).

Enforcement against UBWs

The Buildings Ordinance serves as a statutory weapon for the government to fight against UBWs in the city. The Building Authority is empowered by Section 24(1) of the ordinance to serve statutory orders on building owners to remove any UBW within a specified period of time. The UBWs that should be removed are explicitly specified in the orders. For their very purposes, these orders are commonly known as "removal orders" (Chan & Chan, 2003). However, the use of this term can be misleading because UBWs, by their natures, are not necessarily limited to unauthorized additions to a building, but also include unlawful alterations or removals of approved building elements. The subject of the statutory order, who can be an individual or owners' association, is then required to reinstate the parts of the building so affected as per the original approved building plans (Chan & Chan, 2003). In case the reinstatement work is substantial or involve structural work, the Building Authority may specifically require the order subject to appoint an authorized person or registered structural engineer to coordinate and supervise the reinstatement works on the subject's behalf. Concurrently, the subject is also required to appoint a registered general building contractor to carry out the reinstatement work under the supervision of the authorized person (Chan & Chan, 2003).

For an effective control against the UBWs, there must be sanctions for those who do not observe the legal requirements. In Hong Kong, both statutory and non-statutory means are employed. On the statutory side, Section 40(1BA) of the Buildings Ordinance stipulates that non-compliance with a statutory order served under Section 24(1) without any reasonable excuse is a criminal offence. The offenders are liable on conviction to a maximum fine of HK$200,000 and to a maximum imprisonment of one year. In addition, defaulted owners may be subject to a further fine of HK$20,000 for each day of continuation of the failure to comply with an order. To supplement the criminal punishment, the statutory orders issued are registered with the Land Registry against the titles of the properties concerned (Chan & Chan, 2003). Such a registration will only be discharged when the owners comply with the subject order to the satisfaction of the Building Authority (Buildings Department, 1997).[2] This non-statutory mechanism aims to create economic disincentives for carrying out UBW as properties with encumbrances in title are usually valued lower by the market, keeping other things constant.

---

[1] The Building Authority means, under the Buildings Ordinance, the Director of Building (i.e. the head of the Buildings Department).

[2] After the enactment of the Building (Amendment) Ordinance 2004 in December 2004, the Building Authority is authorized under the Buildings Ordnance to issue warning notices to owners of premises with UBWs, and to register the notices in the Land Registry if the UBWs are not removed within a specified period.

The Extent of the UBW Problem in Hong Kong

As aforementioned, the extent of the squatter problem in Hong Kong has been reducing in the past decade. A survey conducted in 2006 revealed that there were only 13,224 squatters structures in the urban areas of the city (Lau, 2007). Conversely, unlike the problem of squatter structures, the proliferation of UBWs has attracted the attention of Hong Kong's community and the public administrators since the mid-1970s. As illustrated in Figure 1, the number of reports on UBWs received by the Buildings Department rose from 12,427 in 1997 to 24,633 in 2007 (Buildings Department, 2001; 2004; 2008a). In other words, the figure has almost doubled over a decade. Meanwhile, the number of removal orders issued by the Building Authority fluctuated greatly. From 1997 to 2007, the number of orders issued each year for UBW removal increased more than tenfold from 3,103 to 32,898 (Buildings Department, 2001; 2004; 2008a). Yet, all these figures may not be able to reflect truly the order of magnitude of the number of UBWs in Hong Kong.

The former Planning and Lands Bureau estimated in 2001 the number of UBWs, including illegal rooftop structures, in Hong Kong at about 1.02 million (Planning and Lands Bureau, 2001). As estimated by the Buildings Department, the number of UBWs reduced to about 0.75 million in 2003 (Buildings Department, 2003). Hong Kong had approximately 39,000 private buildings in 2005, and 1.08 million private domestic units in the end of 2007 (Housing, Planning and Lands Bureau, 2005; Rating and Valuation Department, 2008). Assuming that the government's estimate in 2003 does not deviate a great deal from the actual figure and it has stood still for years, there were on average 19 UBWs present in every private building in Hong Kong, and 0.69 UBW in every domestic unit. Approximately, two UBWs can be spotted in every three domestic units. Given more frequent occurrence, UBW proliferation, rather than squatter problem, in the contemporary city of Hong Kong should warrant more academic attentions.

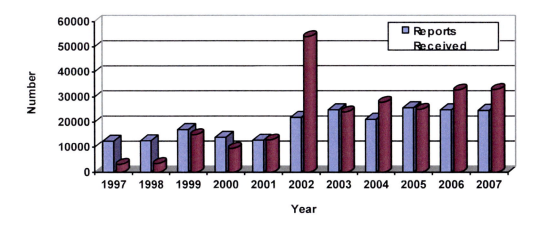

Figure 1. Number of reports on UBWs received and removal orders issued by the Buildings Department (Buildings Department, 2001; 2004; 2008a).

Reasons Underlying UBW Proliferation in Hong Kong

Notwithstanding the very extent of the problem, there still lacks a systematic exploration of the causes of UBW proliferation in Hong Kong. By and large, the roots are threefold. First, UBWs, particularly those unlawful additions, were carried out as a result of the scarcity of developable land in the territory (Lai, 2003). The dynamism from unremitting population growth and influx of immigrants from Mainland China in the 1950s and 1960s gave rise to a surge in the demand for low-cost housing in urban areas of the city (Buildings Department, 1998). UBWs then served as a way to create extra usable floor space for accommodation or other purposes like storage and home manufacturing.[3] On the other hand, the second explanation of UBW proliferation is originated from the inadequacy in building design. Perhaps for cost saving, building-integrated drying facilities and supports for air-conditioners were not provided in many residential buildings built before 1990. However, like the populaces in many other Asian countries, people in Hong Kong prefer drying clothes in open or airy areas to the use of clothes dryer (Wong *et al.*, 2005). Besides, the humid and hot summers in sub-tropical Hong Kong make air-conditioners a necessity for most households (Lam, 1996; 2000). For these reasons, drying facilities and supporting frames for air-conditioners have become essential amenity features for local residents. In the absence of these features in the original building designs, people may install metal frames on their buildings' external walls to support air-conditioners and for drying clothes, enhancing the amenities of their residences.

Apart from the hardware, UBW proliferation in Hong Kong can be ascribed to the meagreness in building management (Lai & Ho, 2001). The co-ownership arrangement, which has long been prevalently adopted in multi-storey buildings in the city, creates predicaments in building management. Under such an arrangement, an individual purchases only undivided shares of a multi-storey building, rather than the ownership of the physical unit to be occupied (Nield, 1997; Walters, 2002). Instead, the individual has the exclusive rights to use and occupy that unit, and he or she co-owns the whole building with the purchasers of other units in the same building. In this case, a substantial portion of a building, including corridors, staircases and external walls, are so-called common areas because they are co-owned by all unit purchasers and no one has the exclusive rights to use or occupy these areas. As a consequence of the ownership arrangement of this kind, the common areas in a building can be kept up and managed properly only with a high degree of coordination and cooperation among the co-owners, which is unlikely when the number of co-owners is large. That explains why many people take this advantage to abuse common areas in the a building by say constructing unauthorized appendages on the building's external wall.

Aside from civil proceedings, aggrieved parties can also lodge complaints to the Buildings Department against any UBWs. However, in spite of the provisions in the Buildings Ordinance, the government's enforcement actions against UBWs seem to be not very effective (Audit Commission, 2003). Moreover, building works that are exemptible from the requirements of obtaining prior approval and consent from the Building Authority are ambiguously defined in the Buildings Ordinance, and new UBWs are accordingly constructed

---

[3] In the 60s and 70s, it was very common in Hong Kong for the housewives and children to do embroidery, make plastic flowers or stitch dolls at home. That explains why there existed many home factories in the city in that period.

by some property owners as a result of their misinterpretation of the Buildings Ordinance (Yiu & Yau, 2005). Furthermore, the ambiguities in the Buildings Ordinance increase the enforcement costs, particularly litigation costs, impeding the government's efforts paid for UBW clearance (Yiu, 2005).

Consequences of UBW Proliferation

As aforementioned, UBWs, in common terms, are building works constructed without any prior approval or consent from the Building Authority. For this nature, the safety standard of these unauthorized works is not guaranteed because their designs and constituent materials have not been scrutinized and approved by the public officers. This explains why quite a number of UBW-related building accidents previously discussed were associated with collapses of the unlawful structures which eventually fell onto the street due to their poor construction and stability. Save for their uncertain structural soundness, UBWs affect building safety negatively because they may impose excessive loads on the buildings in which they are erected, thereby jeopardizing the latter's structural stability (Yiu & Yau, 2005). In addition, UBWs endanger fire safety of a building in many different ways. For instance, illegal additions and alterations inside the building may render the essential escape routes blocked, and unauthorized openings in the staircase enclosures may remove necessary protections of the escape routes against fire and smoke. When built on the external walls of the building, UBWs may hamper the firemen's access to the scene using fire ladders in case of fires.

In fact, the influences of UBWs are not only limited to the direct safety hazards discussed above. UBWs can affect property owners and occupants in numerous other ways. For example, unauthorized advertisement signs protruding from the external walls of buildings are sometimes very massive, as shown in Figure 2. They block the natural sunlight from entering the buildings nearby, and adversely affect the natural lighting and ventilation for the building occupants. Besides, UBWs built on a roof or external wall of a building inevitably hinder the repairs and maintenance carried out to the building (Chan, 2000). This will increase the risk for the building running into dilapidation, aggravating the problem urban decay in Hong Kong in the medium to long run. From another perspective, UBWs, especially those being the subjects of statutory removal orders, render the legal titles of affected properties unclean, encumbering property transactions. In the presence of these encumbrances against the property titles, the values of the properties concerned will diminish.

SURVEY ON UNAUTHORIZED APPENDAGES

In spite of the wide discussions of the problem of UBW proliferation in Hong Kong in the literature, nearly none of the previous studies looked into the issue in a rather empirical manner. To close the current research gap, this study aims to find out which unauthorized appendages are most commonly found in Hong Kong, and what types of multi-storey residential buildings are the most vulnerable to the proliferation of unauthorized appendages. The Building Quality Index (BQI) Research in Hong Kong, financially supported by the

Research Grants Council and the University of Hong Kong, provided us with the opportunity to investigate the UBW problem in the city empirically. The research findings described here are drawn from the dataset of that funded project in which the health and safety performance of 148, 175 and 106 multi-storey residential buildings located in Yau Tsim Mong (YTM), the Eastern District and Kowloon City respectively were assessed within the period between June 2004 and Oct 2006 using the BQI. The primary aim of the BQI research was to benchmark the health and safety performance of multi-storey residential buildings in Hong Kong through a theoretically sound and systematic assessment scheme (Wong *et al.*, 2006).

In the BQI research, information for assessing a building was obtained through a series of structured procedures, starting from desk studies (e.g. studying approved building plans and other relevant documents and records), on-site inspections and interviews with residents and/or management staff using a pre-designed questionnaire (Wong *et al.*, 2006; Ho *et al.*, 2008). One of the assessment criteria for building safety performance was the number of unauthorized appendages present in the subject building, and therefore such information was extractable from the dataset of the BQI research project. Other than such practicality, another ground for delimiting the scope of study to unauthorized appendages, but not other types of UBW, rests on the potential safety hazards posed on the general public. Unlike UBWs such as unauthorized removal of a smoke lobby or construction of an internal staircase which affect mainly the building occupants and visitors in case of failures, collapses of unauthorized appendages do not only make these parties suffer, but also endanger the safety of passers-by and the properties of others.

Figure 2. Sizable unauthorized advertisement signs in buildings along King's Road, North Point (photo by Y. Yau).

Characteristics of the Surveyed Buildings

The physical and management characteristics of the buildings under investigation are summarized in Tables 1 and 2. Surveyed buildings in YTM and the Eastern District were of similar mean ages (around 32 years) while the mean age of Kowloon City's buildings was a bit higher (39 years). The average flat sizes of buildings in the Eastern District were the highest (62.51 m^2) among the three districts, with those in Kowloon City being the smallest (44.21 m^2). Buildings from the Eastern District sample were comparatively taller (14.49 storeys on average) than those in the other two districts (11.70 storeys and 8.10 storeys respectively). Besides, the average development scale of the buildings, measured in terms of the total numbers of domestic units per building and per development, was the largest in the Eastern District (65.65 units per building and 216.35 units per development) because there were more estate-type developments (i.e. developments with more than one residential towers in each development) in the sample of buildings in the Eastern District.

Table 1. Physical characteristics of the buildings surveyed buildings

Attribute	District	Maximum	Mean	Minimum	Standard Deviation
Age (years)	Yau Tsim Mong	52.00	31.38	3.00	11.93
	Eastern District	60.00	33.52	4.00	12.59
	Kowloon City	57.00	39.06	8.00	12.18
	Overall	60.00	34.15	3.00	12.59
Average unit size (m^2)	Yau Tsim Mong	170.20	53.53	19.60	21.90
	Eastern District	148.03	62.51	26.19	23.99
	Kowloon City	245.85	44.21	11.44	35.00
	Overall	245.85	54.89	11.44	27.42
Number of storeys	Yau Tsim Mong	28.00	11.70	3.00	5.87
	Eastern District	40.00	14.49	2.00	8.42
	Kowloon City	21.00	8.10	3.00	3.83
	Overall	40.00	11.95	2.00	7.11
Number of units per building	Yau Tsim Mong	420.00	54.28	3.00	68.04
	Eastern District	500.00	65.65	4.00	77.25
	Kowloon City	198.00	40.35	3.00	42.38
	Overall	500.00	55.48	3.00	67.50
Number of units per development	Yau Tsim Mong	1,038.00	70.90	3.00	132.28
	Eastern District	12,896.00	216.35	4.00	1,041.88
	Kowloon City	1,820.00	77.91	3.00	206.42
	Overall	12,896.00	131.96	3.00	680.22

Table 2. Management characteristics of the buildings surveyed buildings

Management Regime	Yau Tsim Mong	Eastern District	Kowloon City	Overall
Managed by an external property management agent and statutory owners' association	58 (14%)	60 (14%)	25 (6%)	143 (33%)
Managed by a statutory owners' association only	32 (7%)	60 (14%)	23 (5%)	115 (27%)
Managed by an external property management agent but without any statutory owners' association	13 (3%)	14 (3%)	6 (1%)	33 (8%)
Not managed by an external property management agent nor a statutory owners' association	45 (10%)	41 (10%)	52 (12%)	138 (32%)

Among the 429 surveyed buildings, 143 (33 percent) were managed at the same time by a statutory owners' association, so-called owners' corporation,[4] and an external property management property. 115 buildings (27 percent) were managed by an owners' corporation only. Buildings managed by an external property management agent only, either with no owners' corporation formed or with a non-statutory owners' association such as owners' committee and mutual aid committee, accounted for only eight percent of the whole sample. The remaining 138 buildings (32 percent) were not unmanaged by any external property management agent or statutory owners' association. Overall speaking, a higher proportion of surveyed buildings in Kowloon City were unmanaged.

Number of Unauthorized Appendages by Type

Taken as a whole, each surveyed building had an average of 95.42 unauthorized appendages at the time of inspection. Yet, the variation in the figure was rather large, with the number ranging from zero to 869 and a standard deviation of 117.76. The average number of unauthorized appendages per domestic unit in the surveyed buildings was 2.78. The average per-unit figure was quite widely distributed, ranging from zero at the lower end to 24.50 at the upper end with a standard deviation of 2.52. Table 3 summarizes the statistics by type of unauthorized appendage. Generally speaking, air-conditioner supporting frame was the most common type of unauthorized structure, followed by drying rack. Figure 3 illustrates some UBWs spotted in the surveys.

---

[4] As an independent legal entity formed under the Building Management Ordinance (Chapter 344 of *the Laws of Hong Kong*), an owners' corporation mainly serves to represent all co-owners of a multi-storey building or development. By virtue of the ordinance, the owners' corporation is empowered to convene general meetings of owners, and to enforce resolutions arrived at the meetings, provided that the rulings and procedures of the

Table 3. Number of unauthorized appendages by type

Type	District	Maximum	Mean	Minimum	Standard Deviation
Solid canopy	Yau Tsim Mong	3.00	0.09	0.00	0.42
	Eastern District	2.00	0.05	0.00	0.27
	Kowloon City	0.00	0.00	0.00	0.00
	Overall	3.00	0.05	0.00	0.30
Light-weight canopy (> 0.5m)	Yau Tsim Mong	86.00	5.31	0.00	14.33
	Eastern District	7.00	0.16	0.00	0.71
	Kowloon City	0.00	0.00	0.00	0.00
	Overall	86.00	1.90	0.00	8.77
Light-weight canopy (≤ 0.5m)	Yau Tsim Mong	315.00	10.74	0.00	30.42
	Eastern District	179.00	15.82	0.00	29.78
	Kowloon City	46.00	10.91	2.00	7.31
	Overall	315.00	12.85	0.00	26.41
Solid extension	Yau Tsim Mong	2.00	0.05	0.00	0.28
	Eastern District	3.00	0.11	0.00	0.46
	Kowloon City	4.00	0.04	0.00	0.39
	Overall	4.00	0.07	0.00	0.39
Flower rack	Yau Tsim Mong	13.00	1.03	0.00	2.25
	Eastern District	37.00	1.22	0.00	4.65
	Kowloon City	13.00	0.69	0.00	1.67
	Overall	37.00	1.02	0.00	3.35
Drying rack	Yau Tsim Mong	156.00	21.39	0.00	26.17
	Eastern District	363.00	20.10	0.00	40.97
	Kowloon City	42.00	16.73	2.00	9.94
	Overall	363.00	19.71	0.00	30.74
Metal frame	Yau Tsim Mong	41.00	1.80	0.00	4.82
	Eastern District	44.00	1.26	0.00	4.45
	Kowloon City	6.00	0.40	0.00	0.99
	Overall	44.00	1.23	0.00	4.07
Air-conditioner supporting frame	Yau Tsim Mong	383.00	51.75	0.00	59.33
	Eastern District	549.00	80.57	0.00	102.17
	Kowloon City	233.00	28.28	7.00	25.42
	Overall	549.00	57.71	0.00	77.79
Metal cage	Yau Tsim Mong	23.00	0.53	0.00	2.15
	Eastern District	6.00	0.35	0.00	0.92
	Kowloon City	2.00	0.04	0.00	0.24
	Overall	23.00	0.33	0.00	1.40
Advertisement sign	Yau Tsim Mong	11.00	0.69	0.00	1.52
	Eastern District	12.00	0.53	0.00	1.43
	Kowloon City	3.00	0.37	0.00	0.84
	Overall	12.00	0.54	0.00	1.35
Overall	Yau Tsim Mong	869.00	93.37	0.00	104.16
	Eastern District	866.00	120.15	0.00	151.06
	Kowloon City	241.00	57.44	11.00	32.14
	Overall	869.00	95.42	0.00	117.76

meetings accord with those provisions set out in the ordinance. In addition, this legal entity is empowered to appoint, terminate, and monitor services provided by any external property management agent.

(a) Advertisement signs   (b) Lightweight canopy   (c) Drying rack

(d) Metal cage   (e) Flower rack   (f) Solid extension

Figure 3. Examples of unauthorized appendages spotted (photos by Y. Yau).

ANALYSIS RESULTS ANALYSES OF THE SURVEY FINDINGS

Based on the findings from the surveys, analyses were carried out to study for what purposes the unauthorized appendages were constructed, how risky the identified unauthorized appendages are, and in what types of buildings the unauthorized appendages proliferated most in Hong Kong.

Purposes of the Unauthorized Appendages: Space Creation or Amenity Enhancement?

As discussed earlier, literature suggested that lack of usable floor space and non-provisions of amenity facilitates have been the major drivers for the mushrooming of UBWs in Hong Kong. Built upon the premise, it is envisaged that UBWs are usually constructed to provide extra usable floor space for residents or to enhance the amenities the residents can enjoy. To verify the claims, the purposes of unauthorized appendage construction were studied. The types of unauthorized appendages listed in Table 3 were classified into two main groups according to their purposes, namely: i) for space creation and ii) for amenity enhancement. The former group includes solid extensions, metal cages and flower racks, and these appendages make more space for accommodation and storage purposes. Conversely, usable space created by the unauthorized appendages in the second group, including canopies, drying racks, metal frames and air-conditioner supporting frames, is minimal. These appendages serve for a variety of functions. For example, canopies protect rain from entering a dwelling unit; the support frames accommodate window-type air-conditioners or condensers of spilt-type air-conditioners; drying racks allow clothes to be hung for natural drying; metal frames in forms of security grilles to guard the premises against thieves. As for unlawful advertisement signs, they are not classified into either group owing to their frequent commercial uses.

Figure 4 shows that the vast majority of unauthorized appendages found in the buildings surveyed came from the amenity enhancement group. This group of unauthorized appendages accounted for over 98 percent of the total stock spotted in the surveyed buildings. Only less than two percent of the unauthorized appendages belong to either space creation or advertisement sign group. This extremely uneven distribution can be ascribed to the fact that air-conditioner supporting frames, drying racks and lightweight canopies protruding less than 0.5 metres from the external walls of buildings were the most commonly seen unauthorized appendages in the surveys, if one recalls the statistics in Table 3, and all these were erected for the purpose of amenity enhancement.

Risk Levels of the Unauthorized Appendages

As a matter of fact, classification of unauthorized appendages can be done not only based on their purposes of construction, but also on their levels of risk. The latter approach is usually adopted by the public authorities to prioritize enforcement actions. For example, it is rather impossible for the Buildings Department to tackle all types of unauthorized appendages at the same time in light of limited resources; therefore, enforcement actions against those illegal structures have to be prioritized according to the extent of the hazard the structures pose on the occupants and general public and the degree of severity in case of failures. Following the guidelines issued by the Buildings Department for the Blitz UBW Clearance Campaign,[5] unauthorized appendages in Hong Kong are classified into two categories, namely actionable and non-actionable UBWs (Buildings Department, 2005). The former are thought to create an imminent danger to building occupants and the public so they warrant priority removal. Examples of these high-risk UBWs include solid canopies, lightweight canopies projecting more than 0.5 metres from the external walls, solid extensions, flower racks, metal frames and metal cages. On the other hand, the non-actionable UBWs are those that are less hazardous. They include lightweight canopies projecting less than 0.5 metres, drying racks, air-conditioner supporting frames and advertisement signs. Owing to the comparatively lower safety risks associated and the amenities they offer to building occupants, these UBWs are tolerated by the Buildings Department under the current policy provided that they are in good condition and not too substantial.[6]

Perhaps, this prioritization of enforcement actions by the government helps to explain why unauthorized appendages for space creation were less frequently observed as discussed in last section. One may find that unauthorized appendages of this type have also been

---

[5] To effectively fight against the UBW problem in Hong Kong, the Buildings Department launched the largest-ever special 'blitz' operation in September 1999 to remove UBWs from the external walls of buildings.

[6] Despite these non-actionable unauthorized appendages are tolerated by the Hong Kong government, they are still unlawful, by nature, for being constructed without the approval and consent of the Building Authority. These unauthorized structures may be subject to enforcement actions once the government's policy on UBWs changes.

categorized as actionable UBWs so they are more likely to be the subjects of the statutory enforcement actions. Consequently, people are more hesitant to construct these appendages or these high-risk UBWs previously built have been removed in response to the statutory orders issued. As graphically illustrated in Figure 5, over 90 percent of the unauthorized appendages spotted were of the non-actionable type.

	Space Creation	Amenity Enhancement	Advertisement Sign
Overall	0.43%	99.01%	0.57%
Yau Tsim Mong	0.62%	98.64%	0.74%
Eastern District	0.38%	99.18%	0.44%
Kowloon City	0.13%	99.23%	0.64%

Figure 4. Distribution of unauthorized appendages by purpose of construction.

Building Characteristics and Number of Unauthorized Appendages

Main foci of the previous two sections were put on the characteristics of the unauthorized appendages spotted in the survey. Yet, these sections did not address the question on what types of building accommodated more unauthorized appendages. The answer to this question gives hints to the local government about types of building most vulnerable to UBW proliferation. To study the relationship of the number of unauthorized appendages per domestic unit in a building with other building-specific factors like building age, scale of development and average unit size, correlation tests were performed. The Pearson's correlation coefficients presented in Table 4 show that the number of unauthorized appendages per domestic unit of each surveyed building was positively correlated with the building's age and average unit size, but negatively correlated with its development scale (in terms of the numbers of residential units in the building and throughout the whole development). All these relationships were found to be statistically significant at the 1 percent level.

Regarding the management characteristics of the buildings, the average numbers of unauthorized appendages per unit found in buildings with an owners' corporation and without any owners' corporation were 2.42 and 3.32, respectively. The means were significantly different at the 1 percent level ($t = -3.70$). Similarly, buildings with an external property management agent were found to have fewer unauthorized appendages per unit (mean = 1.84) than those without any external property management agent (mean = 3.43), and the difference was statistically significant at the 1 percent level ($t = -6.76$). These results imply that the formation of owners' corporation and appointment of property management company resulted in a reduction in the number of unauthorized appendages per unit in a building.

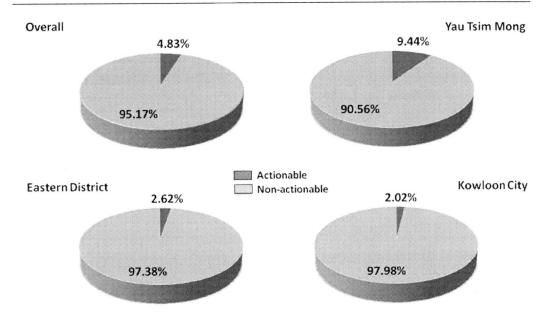

Figure 5. Distribution of unauthorized appendages by level of risk.

Table 4. Correlation coefficients between the number of unauthorized appendages per unit and other factors

	Pearson's Correlation Coefficient	t-statistic	
Building age (years)	0.4508	10.4352	*
Number of units per building	− 0.3471	7.6475	*
Number of units per development	− 0.1354	2.8244	*
Average unit size (m^2)	0.2752	5.9156	*
Note: * significant at the 1 percent level			

As discussed earlier in this article, drying racks and air-conditioner supporting frames were the most commonly seen unauthorized appendages in the surveyed buildings. However, if building-integrated drying facilities, such as approved drying racks and utility platforms, were provided in a building, the average per-unit number of unauthorized appendages was 1.17, compared with 3.24 for those without these drying facilities. The difference between the means were statistically significant at the 1 percent level (t = -7.55). Likewise, buildings equipped with building-integrated air-conditioner platforms, which were usually made of reinforced concrete, were found to have fewer unauthorized appendages per dwelling unit (mean = 0.95) than those without these provisions (mean = 3.32), and the difference was statistically significant at the 1 percent level (t = -8.94).

IMPLICATIONS OF THE ANALYSIS RESULTS

Despite the aforementioned results look factual, they contain far-reaching implications for the formulation of building safety policy in Hong Kong.

Promotion of Better Building Designs

The analysis results above vividly suggest that building design or configuration do have a significant impact on the extent of UBW problem in a multi-storey residential building. The incorporation of clothes-drying facilities and air-conditioner platforms in the original building design resulted in fewer unauthorized appendages on average. As a matter of fact, these provisions, particularly the utility platforms, do not only reduce the number of UBWs present in a building; they also help diminish the chance of accidents involving falling bodies or objects. For example, numerous cases in which individuals fell out the window when collecting clothes from a drying rack occur each year (e.g. South China Morning Post, 2004; 2005; 2008). Moreover, air-conditioner platforms usually come with built-in condensation pipes, so the nuisances created by air-conditioners can be minimized. The provision of these facilities or similar ones should thus be encouraged in new developments, perhaps by means of plot ratio concessions or mandatory stipulations.

Promoting of Better Building Management

Other than a building's hardware, its software, i.e. building management, also plays a momentous role in controlling the UBW problem. As shown by the analysis results, management characteristics did significantly affect the number of unauthorized appendages present in the surveyed buildings. Buildings with an owners' corporation and/or external property management agent were found to have fewer unauthorized appendages per unit, compared with those without any of these forms of management agents. These results provided strong empirical support to the government's initiative to address the UBW problem in Hong Kong by advocating the formation of owners' corporations and/or appointment of external property management agents in private residential buildings. In view of the effects of building management characteristics on the degree of UBW proliferation, the government should dedicate more resources to educate the public on the importance of proper building management; or it should make a decisive step to oblige all multi-storey residential buildings in Hong Kong to have an owners' corporation and a professional property management company through legislation.

Minor Works Control Regime

The vast majority of unauthorized appendages identified in this study were less risky ones erected for amenity enhancement purposes. Perhaps, one of the roots for the proliferation of these minor unauthorized works, including drying racks, small-scale light-weight canopies and air-conditioner supporting frame, lies in the disproportionate statutory submission requirements. Less stringent requirements or more streamlined process for the statutory submission should be warranted for these types of minor building works. Therefore, the government is on the track in putting forward a minor works control regime in which construction of the aforementioned amenity features can be self-certified by the prescribed registered contractors (Housing, Planning and Lands Bureau, 2004; Buildings Department, 2008b). Under the new arrangement, cost and time can be saved so the property owners are less reluctant to follow the statutory requirements in building submission. In addition, the construction of these amenity features will be carried out by prescribed registered contractors, rather than those "cowboy builders". As a result, enhancement is envisaged for the overall building safety in Hong Kong.

CONCLUSIONS

Undoubtedly, shelter for habitation or housing is a basic need for humans (Pryor, 1983). To serve its purpose, not only should housing be weatherproof and comfortable, but it should also be safe. Given the living and development patterns in the modernized cities, building safety is not a simple concern of the owners or occupants. On the account of the high-rise high-density built environment in urban areas, failures in housing structures do not only affect the these parties in many cases; it puts the general public at risk as well. This circumstance is particularly true in compact and highly urbanized cities with many residential high-rises in the Greater China such as Hong Kong, Beijing, Shanghai, Chongqing, and Taipei. However, safety of the urban built environment of Hong Kong has been jeopardized by the proliferation of UBWs in many residential buildings. With a view to the potentially calamitous threats posed by the unauthorized appendages in these buildings, public concern has been aroused over the issue. Further neglect of the prolonged problem of UBWs in the city is no longer acceptable, and thus it is necessary for the government to take valiant steps to tackle the problem. In this light, an in-depth empirical study is called for to provide valuable insights for the public administers.

While academic dialogues over UBW proliferation in Hong Kong abound, relevant empirical studies are still lacking. To straddle the existing research gap, the current research aims to establish a profile of unauthorized appendages in Hong Kong using the data obtainable from the BQI research project in which the health and safety performance of 429 multi-storey residential buildings in YTM, the Eastern District and Kowloon City was assessed through a structured and theoretically sound benchmarking protocol. The analysis results of the data on unauthorized appendages present in these surveyed buildings indicated that a vast majority of these unlawful building items were erected mainly for enhancing the amenities enjoyed by the residents, rather than increasing the floor areas of the dwelling units. Besides, building age, development scale and management characteristics were found to have a bearing on the degree of proliferation of unauthorized appendages. Even though the significance of this research is not quantifiable, it is certainly substantial. The research findings provide valuable insights for the public administers into causes and determinants of UBW proliferation in the Hong Kong so more informed decisions can be made in the formulation of building safety policy for the city.

ACKNOWLEDGMENTS

An earlier version of this manuscript was submitted to the Sixth China Urban Housing Conference held in Beijing, People's Republic of China in 2007. The authors would like to thank the delegates of the conference for their valuable comments. Besides, the authors gratefully acknowledge the financial support provided by the Research Grant Council of the Hong Kong Special Administrative Region (HKU 7107/04E and HKU 7131/05E), the Small Project Funding of The University of Hong Kong, and the HKU Research Group on Sustainable Cities Seed Grant, which made this research possible.

REFERENCES

Audit Commission. (2003). *A report no. 41 of the Director of Audit*. Hong Kong: Audit Commission.

Augenbroe, G. & Park, C. S. (2005). Quantification methods of technical building performance. *Building Research and Information, 33*, 159-172.

Baiche, B., Walliman, N. & Ogden, R. (2006). Compliance with building regulations in England and Wales. *Structural Survey, 24*, 279-299.

Buildings Department. (1997). *Maintaining your building*. Hong Kong: Buildings Department.

Buildings Department. (1998). *Building development and control in Hong Kong*. Hong Kong: Buildings Department.

Buildings Department. (2001). *Monthly digest, January 2001*. Hong Kong: Buildings Department.

Buildings Department. (2003). *Buildings Department's enforcement policy against unauthorized building works*. Hong Kong: Buildings Department.

Buildings Department. (2004). *Monthly digest, January 2004*. Hong Kong: Buildings Department.

Buildings Department. (2005). *Blitz UBW clearance 2005*, unpublished internal guideline, Hong Kong: Buildings Department.

Buildings Department. (2008a). *Monthly digest, January 2008*. Hong Kong: Buildings Department.

Buildings Department. (2008b). *Minor works control system*. Hong Kong: Buildings Department.

Burridge, R. & Ormandy, D. (2007). Health and safety at home: private and public responsibility for unsatisfactory housing conditions. *Journal of Law and Society, 34*, 544-566.

Chan, J. C. C. & Chan, W. T. (2003). Building safety and timely maintenance. *Proceedings of the HKIE Building Division 2nd Annual Seminar*, 28 March 2003, Hong Kong, 1-7.

Chan, K. J. K. (2000). Maintenance of old buildings. *The Hong Kong Surveyors, 11*, 4-7.

Davison, J. (1990). Illegal structures. In S. Nield, & J. Sihombing (Eds.) Multi-storey building management (43-58). *Hong Kong: Hong Kong Law Journal*.

Few, R., Gouveia, N., Mathee, A., Harpham, T., Cohn, A., Swart, A. & Coulson, N. (2004). Informal sub-division of residential and commercial buildings in Sao Paulo and Johannesburg: living conditions and policy implications. *Habitat International, 28*, 427-442.

Henderson, J. (2001). Heritage, identity and tourism in Hong Kong. *International Journal of Heritage Studies, 7*, 219-235.

Ho, D. C. W., Chau, K. W., Cheung, A. K. C., Yau, Y., Wong, S. K., Leung, H. F., Lau, S. Y. & Wong, W. S. (2008). A survey of the health and safety conditions of apartment buildings in Hong Kong. *Building and Environment, 43*, 764-775.

Hopkin, K. (1972). Public and private housing in Hong Kong. In D. J. Dwyer (Eds.) *The city as a centre of change in Asia* (200-215). Hong Kong: Hong Kong University Press.

Housing, Planning & Lands Bureau. (2004). *Minor works control regime: way forward*, paper no. CB(1)1578/03-04(02) submitted to the Bills Committee on Buildings (Amendment) Bill 2003. Hong Kong: Legislative Council.

Housing, Planning & Lands Bureau. (2005). *Building maintenance and management: public consultation on mandatory building inspection*. Hong Kong: Housing, Planning and Lands Bureau.

Imrie, R. (2004). The role of the building regulations in achieving housing quality. *Environment and Planning B, 31*, 419-437.

Imrie, R. (2007). The interrelationship between building regulations and architects' practices. *Environment and Planning B, 34*, 925-943.

Kohler, N. & Hassler, U. (2002). The building stock as a research object. *Building Research and Information, 30*, 226-236.

Kohler, N. & Yang, W. (2007). Long-term management of building stocks. *Building Research and Information, 35*, 351-362.

Lai, A. W. Y. (2003). Control on unauthorized building works in Hong Kong. In Division of Building Science and Technology, City University of Hong Kong (Eds.) *Building design and development in Hong Kong* (37-57). Hong Kong: City University of Hong Kong Press.

Lai, L. W. C. & Ho, D. C. W. (2001). Unauthorized structures in high-rise high-density environment. *Property Management, 19*, 112-123.

Lam, J. C. (1996). An analysis of residential sector energy use in Hong Kong. *Energy, 21*, 1-8.

Lam, J. C. (2000). Residential sector air conditioning loads and electricity use in Hong Kong. *Energy Conversion and Management, 41*, 1757-1768.

Lau, P. L. C. (2007). Controlling officer's reply to initial written question, no. HPLB(PL)070. http://www.landsd.gov.hk/en/images/doc/1912e.pdf.

Leung, A. Y. T. & Yiu, C. Y. (2004). A review of building conditions in Hong Kong. In A. Y. T. Leung, & C. Y. Yiu (Eds.) *Building dilapidation and rejuvenation in Hong Kong*. (11-34). Hong Kong: City University of Hong Kong and the Hong Kong Institute of Surveyors.

Listokin, D. & Hattis, D. B. (2005). Building codes and housing. *Cityscape, 8*, 21-67.

Macgregor, K. & Price, D. C. (2002). *New city, Hong Kong: a portrait of the city and its neon at night*. Hong Kong: Cameraman.

McAdam, R. & O'Neill, L. (2002). Evaluating best value through clustered benchmarking in UK local government: building control services. *The International Journal of Public Sector Management, 15*, 438-457.

Meijer, F. M. & Visscher, H. J. (1998). The deregulation of building controls: a comparison of Dutch and other European systems. *Environment and Planning B, 25*, 617-629.

Meijer, F. M. & Visscher, H. J. (2006). Deregulation and privatisation of European building-control systems. *Environment and Planning B, 33*, 491-501.

Meijer, F. M., Visscher, H. J. & Sheridan, L. (2002). *Building regulations in Europe, part I: a comparison of the systems of building control in eight European countries*. Delft: DUP Science.

Nield, S. (1997). *Hong Kong land law*. Hong Kong: Longman Group (Far East) Limited.

Planning and Lands Bureau. (2001). *For a culture of building care: a comprehensive strategy for building safety and timely maintenance: implementation plan*. Hong Kong: Planning and Lands Bureau.

Pryor, E. G. (1983). *Housing in Hong Kong*. Hong Kong: Oxford University Press.

Rating & Valuation Department. (2008). *Hong Kong property review*. Hong Kong: Rating and Valuation Department.

Smart, A. (2002). Agents of eviction: the squatter control and clearance division of Hong Kong's Housing Department. *Singapore Journal of Tropical Geography*, 23, 333-347.

Smart, A. (2006). *The Shek Kip Mei myth: squatters, fires and colonial rule in Hong Kong, 1950-1963*. Hong Kong: Hong Kong University Press.

South China Morning Post. (2004). Indonesian maid falls to her death. *South China Morning Post*, 24 July 2004, EDT4.

South China Morning Post. (2005). Wan chai woman survives. *South China Morning Post*, 10 November 2005, CITY4.

South China Morning Post. (2008). Maid falls from 2nd-floor flat. *South China Morning Post*, 11 November 2008, CITY4.

Task Force on Building Safety and Preventive Maintenance. (2001). Accidents related to building safety since 1990. http://www.devb-plb.gov.hk/taskforce/eng/info/arbs.htm.

van der Heijden, J., Meijer, F. M. & Visscher, H. J. (2007). Problems in enforcing Dutch building regulations. *Structural Survey*, 25, 319-329.

Walters, M. (2002). Transaction costs of collective action in Hong Kong high rise real estate. *International Journal of Social Economics*, 29, 299-314.

Winayanti, L. & Lang, H. C. (2004). Provision of urban services in an informal settlement: a case study of Kampung Penas Tanggul, Jakarta. *Habitat International*, 28, 41-65.

Wood, B. (2005). The role of existing buildings in the sustainability agenda. *Facilities*, 24, 61-67.

Wong, M. C., Li, H. & Shen, Q. (2003). Electronic building control: an IT-enabled re-engineering process. *International Journal of IT in Architecture, Engineering and Construction*, 1, 329-345.

Wong, S. K., Cheung, A. K. C., Yau, Y., Ho, D. C. W. & Chau, K. W. (2006). Are our residential buildings healthy and safe? A survey in Hong Kong. *Structural Survey*, 24, 77-86.

Wong, W. S. Lorne, F. & Chau, K. W. (2005). Solving externalities problems using innovative architectural design. In: Y. K. Cheung, & K. W. Chau (Eds.) *Tall buildings: from engineering to sustainability* (1107-1112). Singapore: World Scientific Publishing, 2005.

Yang, Q. Z. & Xu, X. (2004). Design knowledge modeling and software implementation for building code compliance checking. *Building and Environment*, 39, 689-698.

Yiu, C. Y. (2005). Institutional arrangement and unauthorized building works in Hong Kong. *Structural Survey*, 23, 22-29.

Yiu, C. Y., S. Kitipornchai, S. & Sing, C. P. (2004). Review of the status of unauthorized building works in Hong Kong. *Journal of Building Surveying*, 4, 28-34.

Yiu, C. Y. & Yau, Y. (2005). Exemption and illegality – the dividing line for building works in Hong Kong. *CIOB(HK) Quarterly Journal, October*, 16-19.

In: Built Environment: Design, Management and Applications   ISBN: 978-1-60876-915-5
Editor: Paul S. Geller, pp. 261-279   © 2010 Nova Science Publishers, Inc.

Chapter 11

UNDERSTANDING THE IMPACT OF THE ENVIRONMENT ON PHYSICAL ACTIVITY

Katherine S. Hall

Geriatric Research, Education, and Clinical Center,
Veterans Affairs Medical Center, Durham, NC

ABSTRACT

As the number of aged persons in the population grows, it is important to identify and understand those lifestyle components that contribute to an improved quality of life and maintenance of functional independence. Benefits of regular physical activity include reduced risk for diabetes, obesity, cardiovascular disease, cognitive dysfunction, and other physical and psychological health outcomes [Rejeski et al., 2002; Taylor et al., 2004; Weinstein et al., 2004; Kramer et al., 2003; McAuley, Kramer, & Colcombe, 2004]. Despite these benefits, a large proportion of adults fail to engage in any physical activity [United States Department of Health and Human Services (USDHHS), 2000] with women, the elderly, and minority populations reporting the greatest levels of inactivity.

Increasingly, links between elements of the built/physical environment and physical activity are being examined. One potentially useful approach to understanding physical activity behavior is through the use of ecologic models [Berrigan & Troiano, 2002]. An ecologic model examines both environmental and individual determinants of behavior, conceding both direct and indirect influences of each on behavior, and the tendency of their respective influences to shift over time. Individual characteristics interact with behavior, and both are influenced by the social and physical environments in a reciprocally determining manner; contributing to overall health. This triadic form of reciprocal determinism is central to Bandura's social cognitive theory [Bandura, 1986, 1997]. To date, very few studies have utilized transdisciplinary theories and research designs to determine the utility of individual and environmental interactions in predicting physical activity [Li, Fisher, Brownson, & Bosworth, 2005; Satariano & McAuley, 2003; King, Stokols, Talen, Brasington, & Killingsworth, 2002]. Consequently, there exist numerous conceptual, theoretical, and definitional ambiguities associated with this literature.

This chapter provides a critical review of the literature on physical activity and the environment in older adults. Specifically, attention is given to the large degree of variability that exists in the operationalization of common environmental characteristics (e.g., population density, access to recreational facilities, land-use mix) as well as physical activity behavior (e.g., step counts, self-report activity behavior, walking for transport) and how these different viewpoints influence the types of outcomes assessed and the measurement tools utilized. Moreover, the role of theory in the design and implementation of research into the environment-behavior relationship and directions for future research are discussed.

INTRODUCTION

The benefits of a physically active lifestyle on a variety of physical and psychological outcomes are well-established [Dishman, Washburn, & Heath, 2004]. However, despite efforts to increase physical activity in older adults, relatively few older adults engage in leisure-time physical activity of sufficient duration, frequency, and intensity to elicit health benefits [USDHHS, 2000]. There is a substantial literature examining factors associated with physical activity among older adults [Rhodes et al., 1999]. Factors such as self-efficacy and physical function have been consistently associated with physical activity levels. The relationship between the built environment and physical activity in older adults, however, is less clear. There has been an increasing level of interest in the role played by the built environment in determining physical activity patterns. Satariano and McAuley [2003] and Li and colleagues [2005] have recently proposed a multi-level relationship between the environment and health behavior. This ecological model proposes to examine variables and constructs at both the individual and the community level and their respective influences on each other. Bandura's [1997] social cognitive theory has emerged as the most comprehensive and applicable explanation of behavior at the individual level. The determinants and outcomes of behavior are examined within a triadic and reciprocally determining framework involving the context (e.g., built environment), the individual, and behavior; such that each component influences and is influenced by the other two factors.

Defining and Measuring the Environment

The importance of promoting more active lifestyles has lead researchers to consider more closely the role of the environment as a potential correlate of activity participation. The environment encompasses both the physical and social components and is believed to exert independent influences on behavior [Sanderson et al., 2003]. The physical environment is comprised of such characteristics as sidewalk conditions, presence of trails or paths for biking or walking, safety of surrounding areas, and adequate lighting [Humpel, Owen, & Leslie, 2002]. The social environment encompasses variables such as social support from family members and friends to be active, seeing comparative others being active, and affiliation with community or religious groups [Humpel et al., 2002].

Table 1. Tools commonly used to measure the built environment for physical activity in older adults

Instrument	Mode of data collection	Variables assessed	Notes
Neighborhood Environment Walkability Scale [NEWS; Saelens, Sallis, Black, & Chen, 2003]	Self-report (68 items)	Residential density Land-use mix density Land-use mix access Street connectivity Walking/cycling facilities Aesthetics Pedestrian/traffic safety Crime safety Neighborhood access Neighborhood characteristics	No cumulative score across all 8 subscales may be calculated.
Perceptions of Environmental Support Questionnaire [Kirtland et al., 2003]	Interview (26 items)	Neighborhood barriers Neighborhood social issues Neighborhood use Community acccss Community barriers Community social issues	Low levels of agreement between neighborhood and community items.
International Physical Activity Questionnaire (IPAQ) Environmental Module [Alexander, Bergman, Hagstromer, & Sjostrom, 2006]	Self-report (17 items)	Residential density Access to destinations Neighborhood infrastructures Aesthetic qualities Social environment Street connectivity Neighborhood safety	Measure designed to be relevant to all countries. Most validation studies have been done in U.S. and European samples.
St. Louis Environmental Instrument [Brownson et al., 2004]	Interview (30 items)	Walking trails Safe from crime Workplace incentives Workplace policy support Workplace safe stairways Walking/cycling infrastructure Neighborhood surroundings Neighborhood safety	Distinct item reliability differences noted between urban and rural respondents.
Irvine-Minnesota Inventory to Measure Built Environments [Boarnet, Day, Alfonzo, Forsyth, & Oakes, 2006; Day, Boarnet, Alfonzo, & Forsyth, 2006]	Environmental audit tool; coding of specific attributes (162 items)	Accessibility Pleasurability Perceived safety from traffic Perceived safety from crime	Audit tool that examines a broad range of built environment features related generally to active living. Requires extensive training.

Table 1. (Continued)

Instrument	Mode of data collection	Variables assessed	Notes
Geographic Information Systems (GIS)	Objective: integration of neighborhood- and community-level data	Built environment attributes Residential density Urban/rural setting Access to formal (e.g. gym, tennis court) recreation facilities Access to informal (e.g., streets, public open spaces) recreation facilities Street connectivity Land-use	Presence Yes/No Proximity to destination Total number Rating

Environmental researchers conceive of the environment as both a potential barrier and facilitator for physical activity, depending heavily on the presence or absence of specific qualities [Eyler et al., 2003b]. Environmental research has predominantly focused on the influence of the social environment variables, and has only recently begun to focus on the built environment as a factor in the study of physical activity. In Table 1, the techniques and tools most commonly used to assess the built environment are listed. As can be seen from this table, there exist a broad range of tools that include self-report, interview, and objective data collection techniques. These tools also vary widely based on the dimensions of the built environment that are assessed and the respective subject/interviewer burden. Finally, important points of consideration for each identified tool have been listed, and include limitations relative to measurement, generalizability, feasibility, and validity. All of these characteristics are important to consider when selecting a measure. Until recently, perceptions of the built environment were assessed using self-report or interview tools. However, the demand for objective measures of environment characteristics has precipitated borrowing techniques (e.g., GIS, audit tools) from other disciplines including city planning and urban design.

An objective of Healthy People 2010 [USDHHS, 2000], has been the incorporation of more objective collection techniques and the development of more comprehensive health data sets through the use of Geographical Information Systems (GIS). This initiative has been undertaken in the local, state, and national health departments and has lead to the increased utilization of this mapping and investigative software in other disciplines [Mullner, Chung, Croke, & Mensah, 2004]. Behavioral researchers have also made a call for the use of objective measures of the physical environment in conjunction with self-report measures [Wendel-Vos et al., 2004]. Although GIS have generally been used in the public health sector to study variables such as disease incidence, they also hold a great deal of potential in behavioral investigations of the role of individual and physical environment characteristics on health [Rushton, 2003].

The conceptualization of neighborhood and community boundaries in the literature varies, and is often defined by the scope of the geographic area to be analyzed and the nature of the research questions. Individual data points are identified through geocoding, or the input of addresses [Troped et al., 2001]. The use of addresses is advantageous for performing individual-level analyses as it allows for the identification and examination of various elements in relation to these specific points of interest. Neighborhood is typically defined as a 0.5-mile (1 km) radius or a 10 minute walk from the place of residence, and community extends to a 10-mile (15 km) radius or a 20-minute drive. The use of shorter distances to define these variables may result in increased correspondence between reported and measured characteristics of the environment, as individuals have a tendency to misinterpret these distances [Addy et al., 2004]. Other researchers utilize postal codes to determine community boundaries; however this may not be the best approach for studies measuring variables such as demographics and SES, as postal codes may not appropriately identify regions according to these characteristics [Mullner et al., 2004]. The absence of standardized definitions for neighborhood and community is a source of significant variation in the environment literature, and optimal values for use in GIS need to be identified [Addy et al., 2004]. Until these distances are quantified and validated, the rationale for using specified distances to denote neighborhood and community for GIS research must be presented.

Defining and Measuring Physical Activity

Physical activity is comprised of both recreational behaviors and leisure-time physical activity, thus appropriate measures need to be employed that best represent the population of interest. Physical activity measures that are commonly used are listed in Table 2. As the majority of energy expenditure in older adults comes from participation in leisure-time activities rather than from sport, several activity measures that include both forms of activity have been developed for use with older adults. As can be seen in Table 2, the self-report and interview tools vary widely along dimensions of length, types of activities assessed, and the specificity of results. Clearly, the selection of one measure over another depends on the size of the sample, the time available, preferred mode of data collection, characteristics of the sample, and the research question. As with any self-report measure, risk for participant bias and report error does exist, thus making objective measures more desirable.

Objective measures of physical activity such as activity monitors and pedometers have been widely utilized in physical activity research, and have seen great improvements over the last couple of years. Accelerometers are typically worn on the waist and measure activity using a piezoelectric sensor that captures 3-dimensional acceleration. However, these monitors record only those movements which fit within a prescribed bandwidth, excluding other data which is likely simulated vibrations such as from a car, lawn mower, or overzealous shaking by a participant. Accelerometers provide researchers with time-stamped estimates of cumulative activity (i.e., total activity counts, step counts) or rates of activity (i.e., counts/min, counts/day). Pedomcters are also typically waist-mounted and measure movement in the horizontal plane only, and provide daily physical activity volume (i.e., steps/day). The lower cost of pedometers and the feasibility of data collection and analysis make the pedometer a preferred method of activity monitoring, particularly in studies with large sample sizes. However, depending on which elements of the environment are being assessed, accelerometers may be preferable to pedometers or vice versa.

Table 2. Measure commonly used to assess physical activity behavior in older adults

Instrument	Mode of data collection	Activities assessed	Notes
Physical Activity Scale for the Elderly [PASE; Washburn, Smith, Jette, & Janney, 1993]	Self-report (10 items)	"Over the past 7 days, how often did you participate in…" Lifestyle/leisure activities Household activities Occupation activities	Includes a variety of activities, and collects frequency and duration of each activity. Validity has been demonstrated in older adults. Relatively quick to complete, and provides a total activity score.
Community Healthy Activities Program for Seniors Questionnaire [CHAMPS; Stewart et al., 2001]	Interview (41 items)	"In a typical week during the past 4 weeks, did you…" Sedentary/light-intensity activities (e.g., use a computer, do arts or crafts, shoot pool or billiards) Moderate-intensity activities (e.g., play golf, do heavy gardening or housework, walk briskly for exercise) Vigorous-intensity activities (e.g., jog or run, moderate or heavy strength training, singles tennis)	Caloric expenditure per week (requires body weight). Frequency per week. Requires 10-30 minutes to complete, and shows a tendency for over-reporting.
Leisure-Time Exercise Questionnaire [LTEQ; Godin & Shephard, 1985]	Self-report (4 items)	"During a typical week, how many times on the average do you do the following kinds of exercise for more than 15 minutes…" Strenuous exercise (i.e., heart beats rapidly) Moderate exercise (e.g., not exhausting) Mild exercise (e.g., minimal effort) "During a typical week, how often do you engage in any regular activity long enough to work up a sweat?" Often Sometimes Rarely	Calculate weekly leisure activity score. Ideal for large, population-based studies. General measure of activity; does not provide much detail relative to modes of activity etc.

Table 2.

Instrument	Mode of data collection	Activities assessed	Notes
Pedometer	Objective	Step counts-responds to vertical oscillations of the body	Very sensitive to movement; may record extraneous vibrations. Can be used to calculate energy expenditure; many now have a software-based interface to allow remote downloading. Validity varies across pedometers, populations, and gait speeds. Relatively inexpensive.
Accelerometer	Objective	Activity counts and/or step counts-responds to acceleration of the body	Placements include: wrist, hip, ankle. Validity of cut-points to measure energy expenditure still being assessed. Time spent in various intensities of activities. Expensive equipment and software. Records acceleration only within specified bandwidth.

Accelerometers have emerged as the superior method for objectively assessing activity levels and have demonstrated validity in studies of youth and middle-aged adults [Bassett et al., 2000; Tudor-Locke, Ainsworth, Thompson, & Matthews, 2002; Trost, 2001; Welk, Blair, Wood, Jones, & Thompson, 2000]. As the ratio of leisure-time physical activity to structured exercise is high in older adults, objective measures that adequately assess and account for this "lifestyle" physical activity are of great utility, as these behaviors are often under-reported in self-report physical activity measures [Welk et al., 2000]. However, it should be noted that a general indicator of activity volume (e.g., LTEQ, pedometer) may be sufficient when studying sedentary populations, who are not expected to engage in activities of higher intensity [Tudor-Locke & Myers, 2001].

The Built Environment and Physical Activity

In their examination of the role of the physical environment in a sample of African-American women, Ainsworth and colleagues [2003] found that a lack of recreation facilities, lack of sidewalks, presence of unattended dogs, and inadequate street lighting were cited as barriers to exercise. Having light traffic patterns in the neighborhood was identified as being conducive to meeting physical activity recommendations in this same sample. Corroborating other studies in which individuals who reported obtaining adequate levels of physical activity also reported being in closer proximity to exercise and recreational facilities [Booth, Owen, Bauman, Clavisi, & Leslie, 2000; Addy et al., 2004; Sanderson, Littleton, & Pulley, 2002]. Giles-Corti and Donovan [2002] also reported that access to facilities was correlated with activity levels, however, the social and individual factors determined how and if these facilities were used, highlighting the need to address all behavioral components related to physical activity.

Sanderson et al. [2003], however, reported no association between the physical environment and physical activity. Others have reported little or no association between environmental correlates and activity level [Eyler, 2003; Eyler et al., 2003a; Giles-Corti & Donovan, 2002; Thompson et al., 2002]. In fact, some of the studies showing significant associations reported associations that were complete contradictions of those reported in other publications; such that those women who reported inadequate/poor street lighting were more likely to be physically active than were women who reported having adequate/good street lighting in their neighborhood. One possible explanation for results such as these is that active individuals spend more time operating in their surrounding environment, and consequently are more aware of the environmental qualities. Inactive individuals however, are believed to spend less time being active in the environment and so perhaps are less likely to be aware of such environmental characteristics/deficiencies.

Although some studies report conflicting results, the general trend in the literature supports an effect of the built environment on activity levels. Booth and colleagues [2000] found that enhanced perceived access to facilities was associated with increased activity levels. Similar results were reported by Troped and colleagues [2003, 2001] in their study of physical activity and the environment using GIS data. Addy et al. [2004] reported that better street lighting and the use of private recreational facilities and community parks were associated with increased levels of physical activity. Moreover, increased walking behavior was associated with having sidewalks available in the neighborhood and using the mall for walking.

Brownson and colleagues [2001] examined the effects of the environment on physical activity among sex and income levels and found several instances of divergence. The presence of sidewalks, aesthetics, heavy traffic, and hills were all positively associated with physical activity in the sample as a whole. However, individuals with lower incomes reported fewer sidewalks in their neighborhood than those with higher incomes, highlighting a potential barrier to low-income groups. Access to facilities also differed based upon sex and income level, such that women with higher incomes reported greater access than low-income women. In men, however, the effect was reversed, such that men with higher incomes reported less access to facilities than did low-income men. No connection was made to physical activity however, so no conclusions can be drawn as to the effect of access on activity levels in these groups. However, Ball et al. [2001] reported greater physical activity in those individuals with more accessible environments, and those environments that are more aesthetically pleasing.

It is important to note that the majority of studies examining environmental influences on physical activity have largely employed cross-sectional designs and self-report measures. Future research employing prospective and longitudinal designs and objective assessments are necessary for a more comprehensive understanding of these relationships.

The Social Environment and Physical Activity

The built environment is but one variable in the assessment of the environment; the other facet is encompassed in the social environment. The role of social support, a variable commonly assessed in environmental research and a component of the social environment, is also addressed in social cognitive theory. The social environment includes variables such as seeing others exercise in the neighborhood, receiving support from others to exercise, and having another individual with whom to exercise. The evidence gathered for the enabling powers of a supportive social environment is extensive, with social support from family and friends shown to be consistently related to physical activity engagement [Oka, King, & Young, 1995; Turner, Rejeski, & Brawley, 1997].

According to social cognitive theory, social support is a source of efficacy information, and thus its effects on behavior are mediated by efficacy cognitions [Bandura, 1986]. Physical activity researchers have long measured social support in activity settings, assessing its role specific to exercise. The consensus from such studies is that its effects on behavior are not direct; instead they are mediated by self-efficacy [McAuley, Jerome, Elavsky, Marquez, & Ramsey, 2003; Duncan & McAuley, 1993; Duncan, McAuley, Stoolmiller, & Duncan, 1993]. It can therefore be reasonably argued that any correlations reported to exist between social support and physical activity behaviors are the result of a mediating relationship with self-efficacy.

In their examination of the longitudinal influences of social support, affect, and physical activity on self-efficacy in older adults, McAuley and colleagues [2003] reported higher levels of self-efficacy in those individuals receiving higher levels of social support. The strength of this association was second only to the relationship between previous activity behavior and self-efficacy. Moreover, those individuals displaying higher levels of efficacy at program end also reported higher levels of physical activity when contacted 18 months post-program completion. These findings suggest that characteristics of the individual, the environment, and past behavior all play a role in determining future activity engagement.

Another element of the social environment is the role of social modeling [Eyler et al., 2003a; Eyler, 2003]. Viewing others being active in the neighborhood has been associated with a greater likelihood to be active in African American, White, Native American, and Latina women. Knowing people who exercise was also significantly correlated with activity in these sub-populations examined by Eyler. Social-cognitive theory would hypothesize that knowing people who exercise and seeing referent others exercising both serve as social sources of efficacy information [Bandura, 1986]. Thus, the level of reported self-efficacy in this sample is influenced in part by the presence of others engaging in exercise behavior.

In their examination of African American women, Ainsworth et al. [2003] found that the majority of women who attended church or church services, knew people who exercised, but did not report seeing people exercise in the community. The latter finding is particularly disturbing given that seeing people exercise in the community was associated with being active oneself. Lower social strain and a more favorable rating of women who exercised were also associated with increased levels of activity in women from this sample.

Eyler [2003] reported similar findings with regard to knowing people who exercise and attending religious services. Again, those women with lower social strain also reported being more active than the women with higher social strain. Additionally, having friends who exercise and significant others reinforcing exercise behavior significantly influenced activity levels [Booth et al., 2000; Brownson et al., 2001]. These relationships have also been demonstrated in other minority populations [Ainsworth et al., 2003; Eyler et al., 2003a].

In a sub-sample of African American women, Eyler et al. [2003a] found that those women who saw people exercising were more likely to engage in any physical activity than those women who did not observe any such behavior. This finding highlights the need to target African American communities on a grander scale in order to encourage continued active behavior. Additionally, it was found that attending religious services once again emerged as a significant correlate of physical activity [Eyler et al.], as did lower social strain. The role of social strain is particularly salient in African American communities as it is often cited as a reason for not being more physically active [Richter, Wilcox, Greaney, Henderson, & Ainsworth, 2002; Wilcox, Richter, Henderson, Greaney, & Ainsworth, 2002]. It appears that interventions that highlight the need and ability to balance life's demands and responsibilities are much needed in this demographic.

Findings similar to those reported in previous sections were presented by Sanderson and colleagues [2003] in their examination of the role of the social environment in rural African American women. Specifically, active women were more likely to see people exercise in their neighborhood, to know people who exercise, to have higher perceptions of and feelings about physical activity, and to attend religious services. In one of the few studies to examine the role of the environment on activity levels in Native American women, Thompson et al. [2002] found that these women received little social support from family and community members to be physically active, and reported high social strain predominantly in the housekeeping and childcare realms [Evenson, Sarmiento, Tawney, Macon, & Ammerman, 2003]. Clearly, changes at both the community and the individual level are needed in order to encourage these sedentary individuals to become more active.

The mediational role of self-efficacy in the association between characteristics of the social environment and physical activity is often neglected in the environmental literature. Instead, social support is merely reported as a correlate of physical activity, providing no real explanation as to the mechanisms through which social support exerts an influence on behavior. Future work into the nature of these associations, particularly in the presence of an intervention to bolster social support, is needed.

Self-Efficacy as a Mediator

Self-efficacy is the strongest individual predictor of physical activity, and consistently demonstrates significant effects on behavior [McAuley et al., 2003; King et al., 2002; McAuley & Blissmer, 2000; Parks, Housemann, & Brownson, 2003; Wilbur, Chandler, Dancy, & Lee, 2003]. This situation-specific perception of one's abilities to successfully carry out a course of action is influenced by four primary sources of feedback: previous experience, modeling of others, persuasion by others, and physiological arousal [Bandura, 1997].

In their examination of the role of self-efficacy as a determinant and consequence of physical activity in older adults, self-efficacy was identified as a reliable predictor of exercise behavior during the early stages of adoption and adaptation, and less so during the maintenance stage [Oman & King, 1998; McAuley, 1992; McAuley, Lox, & Duncan, 1993]. Specifically, self-efficacy was identified as the single most reliable predictor of behavior in the early stages of an exercise program; suggesting that individual perceptions of the environment and ability to cope within that environment is of importance, not merely documenting the actual elements of that environment [Bandura, 2001]. Support for this position comes from Brownson and colleagues [2000] who reported that it was not the actual proximity of facilities, but the perceptions of access to those facilities, that influenced usage.

Other studies of physical activity and the environment identify self-efficacy as an important correlate of physical activity [Ainsworth et al., 2003; Booth et al., 2000; Eyler et al., 2003a; Sanderson et al., 2003]. Eyler et al. reported that self-efficacy was a stronger correlate of activity level than the social and physical environment variables in a sample of ethnically/racially diverse women. Similar results were reported by Ainsworth et al., such that highly efficacious women were twice as likely to achieve recommended activity levels as women low in self-efficacy. Moreover, these effects of self-efficacy were larger in number and magnitude than any reported correlations of physical environment variables (e.g. presence of sidewalks and having light traffic) with activity level. Eyler's study of Midwestern white women and Sanderson's sample of rural African-American women reported a significant association of physical activity with self-efficacy but not with built environment variables on activity.

Unfortunately most explanatory environmental studies fail to incorporate or account for the effects of social cognitive variables adequately in their theoretical models and analyses. Without consideration of variables such as self-efficacy and self-regulation in the structural models, conclusions cannot be drawn as to the motivation of individuals to engage in or cease active behavior. For example, it is not known why some individuals choose to use local facilities and why others do not. Perhaps this provides a preliminary explanation as to the variability in the findings of environment-behavioral research. Without assessing individual differences and one's propensity to act in a predictive manner, it is of little use to gather information about surrounding areas.

Brownson et al. [2001] conducted similar analyses, again in the absence of any individual factors, and found corroborating results: neighborhood characteristics were associated with activity levels, with social variables asserting the greatest influence. In their review of the environment-behavioral literature, Humpel and colleagues [2002] called for more studies that incorporate an ecological framework, addressing the physical environment, the social environment, and the individual, citing Bandura's social cognitive theory [1986] as a sound foundation for such research. Yet in their cross-sectional study of walking, Humpel et al. [2004] neglect to incorporate any measures of individual factors. Consequently, only those associations between weather, aesthetics, accessibility, and location with walking behavior were reported.

Booth et al. [2000] presented their research as a study looking at the effects of the social and physical environments in concert with social-cognitive variables on physical activity. In their sample of older adults, strong associations between perceived safety and access to facilities and activity levels were reported. A significant positive correlation among those who received support from friends and family to be physically active and those individuals who were active was also reported. Finally, a significant positive association between efficacy beliefs and activity levels was reported. The researchers in this study utilized logistic regression analyses to further examine these relationships between variables, and found that a greater number of the physical environment variables were associated with activity in a bivariate analysis than in the logistic regression analysis. Self-efficacy on the other hand was once again identified as a direct indicator of physical activity in the logistic regression analysis. In subsequent analyses however, self-efficacy was excluded, citing its already established association with physical activity.

The absence of such a critical individual component as self-efficacy in both the conceptualization and analysis of the environment-behavioral relationship is apparent in these studies. Perhaps it is only through a mediating mechanism of self-efficacy that the physical environment influences activity. Thus, instead of predicting direct influences of the environment on physical activity, the role of individual cognitions should be emphasized. This argument is a primary tenet of the social cognitive theory, such that the presence of a facilitative or restrictive environment is not the determining factor; it is the efficacy cognitions that result from perceptions of this environment and how they are interpreted by the individual that influences behavior [Bandura, 2001]. The power then does not rest in environmental factors to influence behavior; instead the real power lies in the beliefs of the individual acting in that environment.

Prodaniuk and colleagues [2004] examined this mediational relationship in their study of the assocations between outcome expectations, self-efficacy, and perceptions of the workplace environment using bivariate and multiple regression analyses of cross-sectional data. This study reported physical activity and self-efficacy to be weakly correlated with environmental perceptions. Subsequent mediational analyses demonstrated a decrease in the magnitude of the environmental effect on physical activity. The amount of variance in physical activity accounted for by the perceived environment also decreased as a result of including self-efficacy as a mediating variable in the model from 4% to 2%.

Further evidence of a mediatonal relationship is reported in a recent study conducted by Motl and colleagues [2007] using cross-sectional and longitudinal data in a sample of adolescent girls. Although this study found a cross-sectional relationship between equipment accessibility and self-reported physical activity, this relationship was accounted for by self-efficacy. However, there was not strong support for this pattern of relationships in the longitudinal analysis, as self-efficacy only weakly mediated the effects of equipment accessibility and neighborhood safety on physical activity. Importantly, there were a number of measurement shortcomings in this study including the use of a limited measure of the environment and only a self-report measure of physical activity.

A Comprehensive Study of the Environment-Physical Activity Relationship

In an effort to address a number of the aforementioned shortcomings of the current literature surrounding the environment and physical activity in older adults, we designed a study to examine a mediation model of environment and physical activity in older women

[Morris, McAuley, & Motl, 2008b]. 136 community-dwelling older women (M age = 69.6) were recruited to participate in a study of health and aging. All data was collected through the mail, and was comprised of self-report measures of functional limitations, self-efficacy, and perceptions of the environment (NEWS). In addition, participants were instructed to wear an accelerometer for 7 days to measure physical activity. These procedures were repeated again 6 months later. Objective environment data were collected using GIS software. A neighborhood boundary of 1 km was placed around each participant's home. The presence or absence of schools, recreational areas, parks, walking/cycling paths, and exercise facilities, were identified, along with the prevalence of each variable in the neighborhood of each participant. The objective environment data was aggregated through the summation of the prevalence scores for the measured variables; reflecting the existence of active areas at the neighborhood level.

At baseline, women who were more efficacious and had fewer functional limitations demonstrated significantly higher levels of physical activity. Of the 8 subscales contained in the NEWS questionnaire, only 3 subscales were significantly associated with physical activity: street connectivity, access to walking/cycling facilities, and satisfaction with neighborhood aesthetics. Surprisingly, GIS data was only significantly associated with functional limitations and 6 subscales of the NEWS: residential density, land use mix-diversity, access to services, street connectivity, walking/cycling facilities, and aesthetics. Contrary to our hypothesis, no significant association between GIS and physical activity was observed.

To determine the unique contributions of these variables on physical activity behavior at baseline, hierarchical multiple regression analyses were then conducted. The results of the regression analyses demonstrated that only self-efficacy, functional limitations, and street connectivity accounted for unique variance in physical activity. Thus, bivariate associations of aesthetics and access to walking/cycling facilities with physical activity were no longer significant when controlling for self-efficacy and functional limitations.

Using longitudinal data, the effects of the perceived environment, functional limitations, and self-efficacy on changes in physical activity over six months were examined [Morris, McAuley, & Motl, 2008a]. As the 8 subscale scores from the NEWS cannot be aggregated to form a total score for the perceived environment, one subscale (neighborhood satisfaction) was selected for use in these analyses. Alternatively, the model tested would need to be run separately for all 8 subscales; which we determined was neither feasible nor theoretically viable. The panel model tested includes: (a) paths from neighborhood satisfaction and functional limitations to self-efficacy at both baseline and 6 months; (b) a path from self-efficacy to physical activity at both baseline and 6 months; (c) and paths between the same variables measured across time (i.e., stability coefficients). Thus, this model tests the extent to which *changes* in neighborhood satisfaction and *changes* in functional limitations influence *changes* in self-efficacy, and consequently, *changes* in physical activity, controlling for baseline values and relationships.

Within the baseline data, self-efficacy had a direct effect on physical activity. That is, individuals with fewer limitations reported higher levels of self-efficacy and in turn, were more physically active. The relationship between neighborhood satisfaction and self-efficacy was non-significant. Relative to changes over time, changes in neighborhood satisfaction and functional limitations had significant direct effects on residual changes in self-efficacy, and changes in self-efficacy were significantly associated with changes in physical activity at

Month 6. In a second series of analyses, a direct pathway from neighborhood satisfaction to physical activity was specified; however, this path was not significant, suggesting that perceptions of the environment do not influence physical activity directly.

CONCLUSION

Recent proposals by environment behavioral researchers to improve public health have included recommendations to modify those characteristics of the environment which have been identified as barriers to physical activity [Humpel et al., 2002; Sallis et al., 2006]. Although recent reports suggest a significant association of the environment with physical activity, our results indicate that such findings may be the result of the exclusion of other influential variables in the models tested. Importantly, our results do not state that the environment is an unimportant variable to consider with respect to health behavior. Instead, our results suggest that other, individual-level factors such as self-efficacy operate as the predominant influence on behavior. These results reject the Skinnerian approach characteristic of much of the earlier work on this topic, in which it is solely the environment which drives behavior (i.e., if you build it they will come and use it). Clearly, our findings need to be replicated and, more importantly, prospective studies of longer duration as well as intervention studies are needed to determine how environment, individual, and functional factors influence physical activity behavior over time. The value of longitudinal research to inform future activity interventions and provide evidence as to the predictive power of social cognitive theory cannot be overstated. Social cognitive theory is one theoretical approach that allows one to integrate in a meaningful way the contributions of the environment, individual and social influences on behavior and has demonstrated its effectiveness for understanding physical activity behavior among older adults. The consideration of other theoretical approaches, however, is also warranted.

Additionally, this comprehensive review of the literature and the measures available to examine the built environment-physical activity relationship emphasize that this area of research is still in its early stages. Although the last 10 years or so have produced several measures of varying modalities, the quality of these measures and their utility in the areas of scientific research and public health remains to be determined. Indeed, a large degree of variability exists in the operationalization of common environment characteristics, particularly when applied to GIS. Finally, the role played by the environment on physical activity behavior across ethnic groups, and the differences in environment perceptions between individuals of different ethnicities, remains to be determined. Future efforts to address the issues raised in this chapter will do much to improve the science of understanding complex health behaviors such as physical activity.

REFERENCES

Addy, C. L., Wilson, D. K., Kirtland, K. A., Ainsworth, B. E., Sharpe, P. & Kimsey, D. (2004). Associations of perceived social and physical environmental supports with physical activity and walking behavior. *American Journal of Public Health, 94(3)*, 440-443.

Ainsworth, B. E., Wilcox, S., Thompson, W. W., Richter, D. L. & Henderson, K. A. (2003). Personal, social, and physical environmental correlates of physical activity in African-American women in South Carolina. *American Journal of Preventive Medicine, 25(Suppl 3)*, 23-29.

Alexander, A., Bergman, P., Hagstromer, M. & Sjostrom, M. (2006). IPAQ environmental module; reliability testing. *Journal of Public Health, 14(2)*, 76-80.

Ball, K., Bauman, A., Leslie, E. & Owen, N. (2001). Perceived environmental aesthetics and convenience and company are associated with walking for exercise among Australian adults. *Preventive Medicine, 33*, 434-440.

Bandura, A. (1986). Self-efficacy mechanism in physiological activation and health-promoting behavior. . In S. M. J. I. Madden, & J. Barchas, (Ed.), *Adaptation, Learning and Affect.* New York, New York: Raven Press.

Bandura, A. (1997). *Self-efficacy: The exercise of control.* New York, NY: W.H. Freeman and Company.

Bandura, A. (2001). Social cognitive theory: An agentic perspective. *Annual Reviews of Psychology, 52*, 1-26.

Bassett, D. R. J., Ainsworth, B. E., Swartz, A. M., Strath, S. J., O'Brien, W. L. & King, G. A. (2000). Validity of four motion sensors in measuring moderate intensity physical activity. *Medicine and Science in Sports and Exercise, 32(Suppl 9)*, 471-480.

Berrigan, D. & Troiano, R. P. (2002). The association between urban form and physical activity in U.S. adults. *American Journal of Preventive Medicine, 23(Suppl 2)*, 74-79.

Boarnet, M. G., Day, K., Alfonzo, M., Forsyth, A. & Oakes, M. (2006). The Irvine-Minnesota inventory to measure built environments: reliability tests. *American Journal of Preventive Medicine, 30(2)*, 153-159.

Booth, M. L., Owen, N., Bauman, A., Clavisi, O. & Leslie, E. (2000). Social-cognitive and perceived environment influences associated with physical activity in older Australians. *Preventive Medicine, 32*, 15-22.

Brownson, R. C., Baker, E. A., Housemann, R. A., Brennan, L. K. & Bacak, S. J. (2001). Environmental and policy determinants of physical activity in the United States. *American Journal of Public Health, 91(12)*, 1995-2003.

Brownson, R. C., Chang, J. J., Eyler, A. A., Ainsworth, B. E., Kirtland, K. A. & Saelens, B. E., et al. (2004). Measuring the environment for friendliness toward physical activity: a comparison of the reliability of 3 questionnaires. *American Journal of Public Health, 94(3)*, 473-483.

Brownson, R. C., Housemann, R. A., Brown, D. R., Jackson-Thompson, J., King, A. C. & Malone, B. R., et al. (2000). Promoting physical activity in rural communities. Walking trail access, use, and effects. *American Journal of Preventive Medicine, 18(3)*, 235-241.

Day, K., Boarnet, M. G., Alfonzo, M. & Forsyth, A. (2006). The Irvine-Minnesota inventory to measure built environments: development. *American Journal of Preventive Medicine, 30(2)*, 144-152.

Dishman, R. K., Washburn, R. A. & Heath, G. W. (2004). *Physical Activity Epidemiology*, (4th ed.). Champaign, IL: Human Kinetics.

Duncan, T. E. & McAuley, E. (1993). Social support and efficacy cognitions in exercise adherence: a latent growth curve analysis. *Journal of Behavioral Medicine, 16(2)*, 199-218.

Duncan, T. E., McAuley, E., Stoolmiller, M. & Duncan, S. (1993). Serial fluctuations in exercise behavior as a function of social support and efficacy cognitions. *Journal of Applied Social Psychology, 23*, 1498-1522.

Evenson, K. R., Sarmiento, O. L., Tawney, K. W., Macon, M. L. & Ammerman, A. S. (2003). Personal, social, and environmental correlates of physical activity in North Carolina Latina immigrants. *American Journal of Preventive Medicine, 25(Suppl 3)*, 77-85.

Eyler, A. A. (2003). Personal, social, and environmental correlates of physical activity in rural midwestern white women. *American Journal of Preventive Medicine, 25(Suppl 3)*, 86-92.

Eyler, A. A., Matson-Koffman, D., Young, D. R., Wilcox, S., Wilbur, J. & Thompson, J. L., et al. (2003a). Quantitative study of correlates of physical activity in women from diverse racial/ethnic groups. The women's cardiovascular health network project. Summary and conclusions. *American Journal of Preventive Medicine, 25(Suppl 3)*, 93-103.

Eyler, A. A., Matson-Koffman, D., Young, D. R., Wilcox, S., Wilbur, J. &Thompson, J. L., et al. (2003b). Quantitative study of correlates of physical activity in women from diverse racial/ethnic groups. Women's cardiovascular health network project. Introduction and methodology. *American Journal of Preventive Medicine, 25(Suppl 1)*, 5-14.

Giles-Corti, B. & Donovan, R. J. (2002). The relative influence of individual, social and physical environment determinants on physical activity. *Social Science and Medicine, 54(12)*, 1793-1812.

Godin, G. & Shephard, R. J. (1985). A simple method to assess exercise behavior in the community. *Canadian Journal of Applied Sport Sciences., 10(3)*, 141-146.

Humpel, N., Owen, N. & Leslie, E. (2002). Environmental factors associated with adults' participation in physical activity. *American Journal of Preventive Medicine, 22(3)*, 188-199.

King, A. C., Stokols, D., Talen, E., Brasington, G. S. & Killingsworth, R. (2002). Theoretical approaches to the promotion of physical activity: Forging a transdisciplinary paradigm. *American Journal of Preventive Medicine, 23*, 15-25.

Kirtland, K. A., Porter, D. E., Addy, C. L., Neet, M. J., Williams, J. E. & Sharpe, P. A., et al. (2003). Environmental measures of physical activity supports: perception versus reality. *American Journal of Preventive Medicine, 24(4)*, 323-331.

Kramer, A. F., Colcombe, S. J., McAuley, E., Eriksen, K. I., Scalf, P. & Jerome, G. J., et al. (2003). Enhancing brain and cognitive function of older adults through fitness training. *Journal of Molecular Neuroscience, 20(3)*, 213-221.

Li, F., Fisher, K. J., Brownson, R. C. & Bosworth, M. (2005). Multilevel modelling of built environment characteristics related to neighbourhood walking activity in older adults. *Journal of Epidemiology and Community Health, 59(7)*, 558-564.

McAuley, E. (1992). Self-efficacy and the maintenance of exercise participation in older adults. *Journal of Behavioral Medicine, 16*, 103-113.

McAuley, E. & Blissmer, B. (2000). Self-efficacy determinants and consequences of physical activity. *Exercise and Sport Sciences Reviews, 28(2)*, 85-88.

McAuley, E., Jerome, G. J., Elavsky, S., Marquez, D. X. & Ramsey, S. N. (2003). Predicting long-term maintenance of physical activity in older adults. *Preventive Medicine, 37*, 110-118.

McAuley, E., Kramer, A. F. & Colcombe, S. J. (2004). Cardiovascular fitness and neurocognitive function in older adults. *Brain, Behavior, and Immunity, 18(3)*, 214-220.

McAuley, E., Lox, C. L. & Duncan, T. (1993). Long-term maintenance of exercise, self-efficacy, and physiological change in older adults. *Journals of Gerontology: Psychological Sciences and Social Sciences, 54*, 283-292.

Morris, K. S., McAuley, E. & Motl, R. W. (2008a). Neighborhood satisfaction, functional limitations, and self-efficacy influences on physical activity in older women. *International Journal of Behavioral Nutrition and Physical Activity, 5*, 1-13.

Morris, K. S., McAuley, E. & Motl, R. W. (2008b). Self-efficacy and environmental correlates of physical activity among older women and women with multiple sclerosis. *Health Education Research, 23(4)*, 744-752.

Motl, R. W., Dishman, R. K., Saunders, R. P., Dowda, M. & Pate, R. R. (2007). Perceptions of physical and social environment variables and self-efficacy as correlates of self-reported physical activity among adolescent girls. *Journal of Pediatric Psychology, 32(1)*, 6-12.

Mullner, R. M., Chung, K., Croke, K. G. & Mensah, E. K. (2004). Geographic information systems in public health and medicine. *Journal of Medical Systems, 28(3)*, 215-221.

Oka, R. K., King, A. C. & Young, D. R. (1995). Sources of social support as predictors of exercise adherence in women and men ages 50 to 65 years. *Women's Health, 1(2)*, 161-175.

Oman, R. F. & King, A. C. (1998). Predicting the adoption and maintenance of exercise participation using self-efficacy and previous exercise participation rates. *American Journal of Health Promotion, 12*, 154-161.

Parks, S. E., Housemann, R. A. & Brownson, R. C. (2003). Differential correlates of physical activity in urban and rural adults of various socioeconomic backgrounds in the United States. *Journal of Epidemiology and Community Health, 57*, 29-35.

Prodaniuk, T. R., Plotnikoff, R. C., Spence, J. C. & Wilson, P. M. (2004). The influence of self-efficacy and outcome expectations on the relationship between perceived environment and physical activity in the workplace. *International Journal of Behavioral Nutrition and Physical Activity, 1(1)*, 1-7.

Rejeski, W. J., Focht, B. C., Messier, S. P., Morgan, T., Pahor, M. & Penninx, B. (2002). Obese, older adults with knee osteoarthritis: Weight loss, exercise, and quality of life. *Health Psychology, 21(5)*, 419-426.

Rhodes, R. E., Martin, A. D., Taunton, J. E., Rhodes, E. E., Donnelly, M. & Elliot, J. (1999). Factors associated with exercise adherence among older adults: An individual perspective. *Sports Medicine, 28*, 397-411.

Richter, D. L., Wilcox, S., Greaney, M. L., Henderson, K. A. & Ainsworth, B. E. (2002). Environmental, policy, and cultural factors related to physical activity in African American women. *Women and Health, 36(2)*, 91-109.

Rushton, G. (2003). Public health, GIS, and spatial analytic tools. *Annual Review of Public Health, 24*, 43-56.

Saelens, B. E., Sallis, J. F., Black, J. B. & Chen, D. (2003). Neighborhood-based differences in physical activity: An environment scale evaluation. *American Journal of Public Health, 93(9)*, 1552-1558.

Sallis, J. F., Cervero, R. B., Ascher, W., Henderson, K. A., Kraft, M. K. & Kerr, J. (2006). An ecological approach to creating active living communities. *Annual Review of Public Health, 27*, 297-322.

Sanderson, B., Littleton, M. & Pulley, L. (2002). Environmental, policy, and cultural factors related to physical activity among rural, African-American women. *Women and Health, 36*, 75-90.

Sanderson, B. K., Foushee, R., Bittner, V., Cornell, C. E., Stalker, V. & Shelton, S., et al. (2003). Personal, social, and physical environmental correlates of physical activity in rural African-American women in Alabama. *American Journal of Preventive Medicine, 25(Suppl 3)*, 30-37.

Satariano, W. A. & McAuley, E. (2003). Promoting physical activity among older adults: From ecology to the individual. *American Journal of Preventive Medicine, 25(Suppl 2)*, 184-192.

Stewart, A. L., Mills, K. M., King, A. C., Haskell, W. L., Gillis, D. & Ritter, P. L. (2001). CHAMPS physical activity questionnaire for older adults: Outcomes for interventions. *Medicine and Science in Sports and Exercise, 33(7)*, 1126-1141.

Taylor, A., Cable, N., Faulkner, G., Hillsdon, M., Narici, M. & Bij, A. V. (2004). Physical activity and older adults: a review of health benefits and the effectiveness of interventions. *Journal of Sports Science, 22(8)*, 703-725.

Thompson, J. L., Allen, P., Cunningham-Sabo, L., Yazzie, D. A., Curtis, M. & Davis, S. M. (2002). Environmental, policy, and cultural factors related to physical activity in sedentary American Indian women. *Women and Health, 36(2)*, 59-74.

Troped, P. J., Saunders, R. P., Pate, R. R., Reininger, B. & Addy, C. L. (2003). Correlates of recreational and transportation physical activity among adults in a New England community. *Preventive Medicine, 37(4)*, 304-310.

Troped, P. J., Saunders, R. P., Pate, R. R., Reininger, B., Ureda, J. R. & Thompson, S. J. (2001). Associations between self-reported and objective physical environmental factors and use of a community rail-trail. *Preventive Medicine, 32*, 191-200.

Trost, S. G. (2001). Objective measurement of physical activity in youth: Current issues, future directions. *Exercise and Sport Sciences Reviews, 29(1)*, 32-36.

Tudor-Locke, C., Ainsworth, B. E., Thompson, R. W. & Matthews, C. E. (2002). Comparison of pedometer and accelerometer measures of free-living physical activity. *Medicine and Science in Sports and Exercise, 34(12)*, 2045-2051.

Tudor-Locke, C. & Myers, A. M. (2001). Challenges and opportunities for measuring physical activity in sedentary adults. *Sports Medicine, 31*, 91-100.

Turner, E. E., Rejeski, W. J. & Brawley, L. R. (1997). Psychological benefits of physical activity are influenced by the social environment. *Journal of Sport and Exercise Psychology, 19*, 119-130.

U.S. Department of Health and Human Services (2000). *Healthy People 2010: Understanding and Improving Health.* Washington, D.C.

Washburn, R. A., Smith, K. W., Jette, A. M. & Janney, C. A. (1993). The physical activity scale for the elderly (PASE): Development and evaluation. *Journal of Clinical Epidemiology, 46(4)*, 153-162.

Weinstein, A. R., Sesso, H. D., Lee, I. M., Cook, N. R., Manson, J. E. & Buring, J. E., et al. (2004). Relationship of physical activity vs. body mass index with type 2 diabetes in women. *Journal of the American Medical Association, 292(10)*, 1232-1234.

Welk, G. J., Blair, S. N., Wood, K., Jones, S. & Thompson, R. W. (2000). A comparative evaluation of three accelerometry-based physical activity monitors. *Medicine and Science in Sports and Exercise, 32(Suppl 9)*, 489-497.

Wendel-Vos, G. C. W., Schuit, A. J., De Niet, R., Boshuizen, H. C., Saris, W. H. M. & Kromhout, D. (2004). Factors of the physical environment associated with walking and bicycling. *Medicine and Science in Sports and Exercise, 36(4)*, 725-730.

Wilbur, J., Chandler, P. J., Dancy, B. & Lee, H. (2003). Correlates of physical activity in urban midwestern African-American women. *American Journal of Preventive Medicine, 25(Suppl 3)*, 45-52.

Wilcox, S., Richter, D. L., Henderson, K. A., Greaney, M. L. & Ainsworth, B. E. (2002). Perceptions of physical activity and personal barriers and enablers in African-American women. *Ethnicity and Disease, 12(3)*, 316-319.

In: Built Environment: Design, Management and Applications   ISBN: 978-1-60876-915-5
Editor: Paul S. Geller, pp. 281-306   © 2010 Nova Science Publishers, Inc.

Chapter 12

POTENTIAL REUSE OF CONTAMINATED SOILS IN THE BUILT ENVIRONMENT

M. Pritchard and D. Allen*
Leeds Metropolitan University, School of the Built Environment,
Leeds, LS2 8AJ, UK.

ABSTRACT

The industrial heritage of the UK has given rise to around 100,000 sites, being classified as contaminated. In particular the former Avenue Coking Works, Chesterfield, UK in the 1990s was classified as one of the most contaminated sites in Western Europe. The site was left derelict after its working life and was never decommissioned. Chemical tanks, pipe work, sumps, gantries and contaminated lagoons were never drained down; the condition of these deteriorated. This coupled with site operation has caused severe ground pollution. Major contaminants identified included PAHs, phenols, mineral oil, BTEX, cyanides, tyres, metal, asbestos and spent oxide. In 2001 a series of remediation techniques were trialled to determine the superlative form of remediation. These trials included bioremediation, thermal desorption, soil washing, cement stabilisation and co-incineration. The enhanced thermal conduction (ETC) was the thermal desorption technology used which was found to be the most effective technology for treating this type of soil. Contamination levels such as PAHs (total 16 USEPA) were significant reduced from 76,000 mg/kg to 500 mg/kg at a cost of £60 per cubic metre.

In addition to the reduction in contamination levels the ETC process also significantly improved the geotechnical properties/parameters of the soil. The CBR value increased from 11% to 107%; this was related to the high uniformity coefficient of the soil where excellent particle interlock would have been obtained. Humus would have also been destroyed by the thermal process, which in turn would relate to a greater interlock between soil particles. The thermal process also enhanced the shearing resistance by a factor of two from 17° to 34°. Thus, the residual soil has the potential to be reused as a fill material (Grade 1A) under the Specification for Highway Works, UK (2007). The application for its reuse has been demonstrated in the chapter for an engineered embankment.

---

* Corresponding author: E-mail: m.pritchard@leedsmet.ac.uk

Keywords: Avenue coking works, bioremediation, contaminated land, remediation, thermal desorption.

1.0. INTRODUCTION

In the UK large-scale contamination began at the end of the 18th century with the start of the industrial revolution. Past industrial activities (as detailed in Table 1) have resulted in harmful substances being released on the land and into watercourses. Many of these industries have now ceased; however as a result of the industrial heritage of the land, the land use remains idle. Also, the expansion of the chemical and petroleum industries in the 20th century has resulted in the production of a vast array of chemical compounds. Fuels, solvents, food additives and other products have resulted in around eight million synthetic and naturally occurring compounds being produced (NAS, 2003). The manufacture and processing of these products has also resulted in a detrimental legacy of contaminated land, which can affect not only human health but also terrestrial ecosystems. There are approximately 100,000 sites covering between 50,000 and 200,000 hectares classified as contaminated (Dinsdale, 2005).

Table 1. Industrial activities

Large scale and heavy industrial groups	
Asbestos works	Pulp and paper manufacture
Brickworks (kilns, flues, coke, often lime kilns)	Radioactive materials
Ceramics, cement and asphalt manufacturing works	Railway land
Chemical works	Research laboratories
Coal preparation works	Scrap yards
Docks	Sewage works and farms
Engineering works	Tanneries
Garages and petrol filling stations	Textile works and dye works
Gasworks, coke works and coal carbonisation plants	Timber treatment works
Heavy engineering installations	Tyre manufacture
Landfills and waste treatment / disposal sites	Waste recycling and transfer stations
Metal manufacturing	*Small scale groups*
Metal processing	Automobile foam production works
Mineral workings - non-ferrous metals	Charcoal works
Mineral processing works	Dry-cleaners
Munitions production and testing sites	Fibreglass and resin manufacturing works
Oil refineries, storage of crude oil and petroleum	Glass manufacturing works
Paint manufacture	Photographic processing industry
Power stations	Printing and bookbinding works

1.1. Contamination

Figure 1 illustrates the most prevalent contaminants found in soils. Contaminants can be either organic or inorganic and may exist in solid, liquid, or gaseous forms. Many contaminants occur as a range of compounds, some of which are more soluble and toxic than others. Toxicity is also dependant on the oxidation state; however, chemical analyses of soil usually only provides an indication of the total soil concentration. Assumptions, therefore, have to be made as to which compounds exist and the risks arising from the level of soil concentration (Cairney and Hobson, 1998). Toxic effects can generally be divided into those resulting from short-term (acute exposure) and those administered over a long period of time (chronic exposure). Absorption, ingestion and inhalation are the main pathways through which contamination from soil can become toxic to humans.

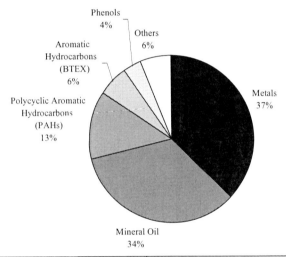

PAHs (USEPA 16)	Structure No. benzene (total) rings
Naphthalene	2
Acenaphthylene	2
Acenaphthene	2
Fluorene	2 (3)
Phenanthrene	3
Anthracene	3
Fluoranthene	3 (4)
Pyrene	4
Benzo(a)anthracene	4
Chrysene	4
Benzo(b)+(j)fluoranthene	4 (5)
Benzo(k)fluoranthene	4 (5)
Benzo(a)pyrene	5
Indeno(1,2,3-cd)pyrene	5 (6)
Dibenz(a,h)anthracene	5
Benzo(g,h,i)perylene	6

Figure 1. Overview of most prevalent contaminants (including the 16 most predominate PAHs) affecting European soils, adapted from data published by the European Environment Agency (2007).

1.1.1. Inorganic Contamination

Metal compounds are the most common, representing 9 of the top 25 hazardous inorganic compounds (NAS, 2003). This is the result of industrial activities such as mining, iron and steel works, electroplating and galvanising. Contaminants in this group can also be attributed to leaded petroleum, batteries, paint/wood preservatives, agricultural dips, fungicides and pesticides (NAS, 2003). Metallic compounds such as arsenic, cadmium, chromium, lead, mercury and selenium attack the central nervous system causing gradual paralysis of the body (Sarsby, 2000). In the context of contaminated land, excessive levels of copper, nickel and zinc are considered problematic, due to their toxicity effects on plant growth (Cairney, 1993).

Other inorganic contaminants such as sulphides, sulphates and cyanides generally arise as by-products from manufacturing industries (Sarsby, 2000). Problems associated with sulphate contamination are corrosion of materials, ingestion and water pollution. If sulphides or cyanides come into contact with waste acids they can rapidly liberate highly toxic hydrogen sulphide and hydrogen cyanide gases, which can cause asphyxiation (Cairney, 1993). Sulphates are also toxic in excessive concentrations. The microbial transformation of sulphates to sulphide in waterlogged soils also leads to the formation of hydrogen sulphide and toxic sulphide salts may also form.

1.1.2. Organic Contamination

Mineral oils are complex nonaqueous phase liquids (NAPLs), which can serve as long-term sources of groundwater contamination (Kong, 2004). Mineral oil is a by-product of the distillation of petroleum to produce gasoline. It is found on a large number of contaminated sites due the volumes of refined fuels and other products produced, approximately 25 billion barrels annually (Ward and Singh, 2004). Mineral oil is not a single compound but a mixture of different hydrocarbons with its composition depending on the type of mineral oil product. Hydrocarbons are generally grouped according to their carbon range. For example, petroleum range organics (PRO) and diesel range organics (DRO) are grouped as follows:

- PROs represent the lighter fraction carbon range C6 – C13, such as benzene, toluene, ethylbenzene, and xylene (BTEX), together with phenolic compounds.
- DROs represent groups of heavier hydrocarbon compounds based on the carbon range C10 – C36. These include polycyclic aromatic hydrocarbons (PAHs), ê.g. naphthalene and pyrene.

PAHs also represent a significant proportion of the most prevalent contaminants due to their wide dissemination and persistence in the soil environment (EUGRIS, 2008). PAHs are a range of over 100 organic chemical compounds that contain more than one fused benzene ring. They are commonly found in petroleum fuels, tars and coal products (EUGRIS, 2007). PAHs with more than three benzene rings are often referred to as high molecular weight (HMW) PAHs and this increase in molecular weight prolongs their environmental persistence (Kanaly and Harayama, 2000). Due to their tendency to sorb strongly with soil particles, PAHs are often found close to the soil surface. They also have high octanol/water partition coefficients (log Kow values) within the range 4 - 7 and, as a result, they are not very mobile in soil and groundwater. Most also biodegrade very slowly while some of them, especially HMWs, are persistent even under aerobic conditions (Pirika, 2004). Due to their global distribution and potential detrimental effect on human health 16 PAHs (Figure 1) have been considered a priority by the United States Environmental Protection Agency (USEPA) (National Pollutant Inventory, 2004).

Phenols are a specific class of compounds consisting of at least one hydroxyl group attached to an aromatic hydrocarbon ring. They are derived from benzene and are commonly used in resins, plastics, pharmaceuticals and in dilute form as a disinfectant and antiseptic (Farlex, 2008). A number of reproductive disorders in humans and wildlife are linked to these compounds as they are carcinogenic and have the capacity to disrupt the body's endocrine (hormone) system (Report LIFE 98, 2008).

Other persistent organic contaminants commonly found in soils and sediments are polychlorinated biphenyls (PCBs) and certain synthetic pesticides such as dichloro-diphenyl-trichloroethane (DDT). PCBs were once used in a variety of industrial materials, including electric transformers. They are almost completely indestructible in the natural environment and tend to accumulate in aqueous sediments (Procuro, 2006). DDTs, on the other hand, are persistent pollutants and are considered immobile in most soils due to their insolubility. Both PCB and DDT compounds are considered carconogenic with long-term exposure and can cause other problems in humans related to the nervous system (Health Canada, 2005).

1.1.3. Other Forms of Contamination

Other forms of contamination can also be present in soils and groundwater e.g. pathogenic organisms such as bacteria and viruses. The principal hazard with these is infection and the severity is dependant on the organisms involved. Bacteria such as *Bacillus anthracis* (the causative agent of anthrax), for example form spores that can remain dormant in soils for centuries and can be fatal if left untreated. Such a fatality due to anthrax was recorded in London on the 3rd of November, 2008 in London, UK (Anon, 2008a).

Gases and vapours can also be a hazard associated with contaminated sites. The main issue here is usually the build up of vapours within confined spaces. A list of the main gases and associated health affects is shown below (Cairney, 1993 and Sarsby, 2000):

- **Methane**: non toxic but in sufficient concentrations can cause asphyxiation.
- **Carbon dioxide**: a toxic asphyxiant gas, denser than air and when dissolved in water forms a highly corrosive liquid.
- **Hydrogen sulphide**: highly toxic, flammable gas. It is a well known contaminant of Kraft paper mill sites, oil refineries, coal carbonisation sites and chemical works.
- **Carbon monoxide**: highly toxic and flammable, produced during incomplete combustion of organic material.
- **Sulphur dioxide**: an irritating gas at concentrations of 6 to 12 ppm, typically released during the combustion of contaminated materials e.g. spent oxide.

1.2. Current Policies, Targets and Measures in the UK

The UK Government has identified three objectives relating to the treatment of contaminated land (DEFRA, 2006):

1. To identify and remove unacceptable risks to human health and the environment.
2. To seek to bring damaged land back to beneficial use.
3. To seek to ensure that the cost burdens faced by individuals, companies and society are proportionate, manageable and economically sustainable.

In order to achieve this, a "*suitable for use*" approach has been adopted. This seeks to ensure that land is suitable for its current/existing use, or is made suitable for any new use. This makes certain that the level of remediation is proportionate to the proposed use (Dti, 2002). Risk is considered greatest for land whose end use will be domestic and declines with reducing human contact. Least risk is associated with commercial use (Sarsby, 2000).

There are two principal legislative controls within the UK to regulate redevelopment and remediation of contaminated sites. These are the Town and Country Planning Act 1990 and Part IIA of the Environmental Protection Act 1990 (EPA). The latter, inserted in Section 57 of the 1995 Environmental Act, specifically deals with existing contamination and came into force over the period April, 2000 to July, 2001 (Dti, 2002). Where contamination is deemed to be a sufficient risk Part IIA of the EPA places a duty on local authorities to inspect their areas in order to identify contaminated land. In the UK the majority of local authorities have a contaminated land strategy and potential contaminated sites are being investigated. The EPA has also introduced a definition for contaminated land given in Section 78a (2). This states that:

> "*any such land which appears to the local authority in whose area it is situated to be in such a condition, by reason of substances in, on or under the land that –*
> *(a) significant harm is being caused or there is a significant possibility of such harm being caused; or*
> *(b) pollution of controlled waters is being, or is likely to be caused: and, in determining whether any land appears to be such land, a local authority shall...act in accordance with guidance by the secretary of state...*" *(DEFRA, 2008).*

Once formally classified as 'contaminated', a process of risk management is then implemented. The purpose of this approach is to determine the significance of the risks and decide on whether further action, including remediation, is needed. This approach led to the Contaminated Land Exposure Assessment (CLEA) model being developed by the Environment Agency. This was carried out on behalf of the Department of Environment, Food and Rural Affairs (DEFRA) and has led to the compilation of the contaminated land report (CLR) series of documents. This model superseded the previous Interdepartmental Committee for the Redevelopment of Contaminated Land (ICRCL) trigger values and enables site specific action values to be developed for specific contaminants. The information covered by the new CLRs relates primarily to the assessment of the impact of land contamination on human health and uses Monte Carlo simulations to model potential pathways dependant on health criteria values, age and land use.

Separate guidance is provided for situations relating to short-term exposure to high contaminant concentrations (EA, 2002). In some cases, the assessment process will also need to consider whether other targets of concern, such as controlled waters, ecosystems, buildings, crops or livestock, are at risk. Where such risks are identified, other good practice guidance should be used to evaluate the extent of the risk to these targets and to determine appropriate actions. The guidance requires an identifiable contaminant, pathway and specified receptor which is suffering harm or pollution - referred to as a significant pollution linkage (Figure 2). Where this linkage is identified and the land meets the statutory definition the local authority can make a formal determination that the land is contaminated. For land where the criteria of a special site (described in Regulation 2 of the Contaminated Land (England) Regulations 2006) are also met, the Environment Agency becomes the enforcing authority to ensure that the pollution linkage is broken.

When contaminated soil is unsuitable for the current or proposed use and requires disposal to landfill, the material then becomes waste. Waste management licensing controls originate from the Control of Pollution Act 1974 (Anon, 2008b). This evolved into the Environmental Protection Act 1990, which greatly tightened up the licensing system to ensure that all sites are run by competent people and do not harm the environment. More recently the EU landfill directive, transposed into domestic legislation by the Landfill (England and Wales) Regulations 2002, has been introduced. These regulations require landfill sites to be classified into one of three categories; hazardous, non-hazardous/non-inert and inert according to the type of waste they receive. The EU directive also sets demanding targets to reduce the amount of biodegradable waste landfilled to 75% of 1995 levels by 2010, with further staged reductions to 35% by 2020.

These latest regulations have impacted on the remediation of contaminated land in the following ways:

- Certain hazardous and other wastes, including liquids, will be prohibited from landfills.
- Biodegradable waste will be progressively diverted away from landfills, and upper limits on organic carbon content have been set for all categories.
- Pre-treatment of all wastes prior to land filling will become a requirement.
- Co-disposal of hazardous and non-hazardous wastes will be banned.

Figure 2. Illustration of the 'Source-Pathway-Receptor' linkage.

The traditional remediation approach of excavation and disposal is directly affected with the requirement to pre-treat all wastes. Treated contaminated soil that cannot be reused will need to be disposed of, either to hazardous or non-hazardous sites depending on the concentrations of dangerous substances in the waste soil. For example, soil containing hydrocarbons (e.g. total petroleum hydrocarbons, diesel range organics, gasoline range organics) in excess of 1000 mg/kg would be subjected to a waste acceptance criteria (WAC) leachability test (Strange and Langdon, 2008). The results would determine the category of landfill that can accept the waste. This requirement has caused the number of licensed hazardous waste sites in the UK to be reduced from 67 to 24 between 2004 and 2007 (ICE, 2004; EA, 2007). The WAC has also increased the cost of disposal as both transportation and landfill costs have risen and current landfill space is estimated to run out within nine years (BBC, 2007).

1.3. Soil Remediation

Over the last decade significant advances have been made with regard to the science and engineering of contaminated sites and many different techniques have been developed to remediate land (Madabhushi and O'Neill, 2007). Remediation can be carried out both *in situ* and *ex situ*. *In situ* methods seek to treat the contamination without removing the soils whereas *ex situ* methods involve excavation and subsequent treatment on the surface. Both methods have advantages and disadvantages. *Ex situ* processes, for example, make contaminants more accessible and allow for greater process optimisation, whilst *in situ* techniques result in lower air emissions and waste regeneration (Barr *et al.*, 2003).

Table 2. Proven remediation techniques

Technique	Organic	Inorganic	Suitability	Limitations
Excavation and disposal	✓	✓	Contaminants are completely removed from site	Requires pre-treatment and a licence landfill
Thermal	✓	✓	Widely proven, effective for heavy organic contaminants	High energy costs & alters the consistency of the soil
Bioremediation	✓		Contaminants are destroyed	Treatment can take a long time
Soil Washing	✓	✓	Suitable for all soils & reduces volume of contaminated material	Extracted 'cake' and treatment liquid require further treatment
Capping	✓	✓	Easy to install	Does not treat the contamination - covers it
Vapour extraction	✓	✓	In situ technique for volatile contaminants	Non volatiles cannot be treated
Stabilisation/Solidification	✓	✓	Effective for heavy metals improves the engineering properties	Alters the consistency of the soil

Remediation techniques can be grouped under three main headings by the way in which they deal with contaminants namely containment, separation and destruction (Figure 3). Containment techniques operate by containing the contamination. This breaks the pathway between contaminant and receptor without removing the contaminants. Separation techniques on the other hand are physical processes that remove the contamination from the soil matrix. This concentrates the contaminants by creating a lesser quantity of higher contaminated material that subsequently requires processing. Finally, destruction techniques utilise either a physical, biological or chemical process to destroy contaminates.

Within the UK, physical processes have traditionally dominated the land remediation market with excavation and disposal to landfill being the predominant choice for developers since the 1970s (Anon, 2008c). The 2005 changes in the landfill regulations led to a decline in the amount of contaminated material being disposed of to landfill. Prior to 2005, 78% of contaminated soil had been disposed of in this manner and after 2005 only 76% of total contaminated soil went to landfill (Randall, 2007). Techniques such as thermal desorption, bioremediation, soil washing and stabilisation have since emerged to offset this decline and fulfil both remediation and landfill requirements.

Various thermal techniques can be applied to deal with soil contaminants. Incineration involves the use of high temperature (>1,000°C) to destroy contaminants. Thermal desorption is another technique in which temperatures between 450 and 600°C are used to physically separate moisture and contaminants from the soil. The volatised contaminants are then passed through a catalytic or thermal oxidiser to incinerate the contaminants prior to discharge to atmosphere. This technique is very effective at removing volatile organic compounds (VOCs) but is very energy intensive.

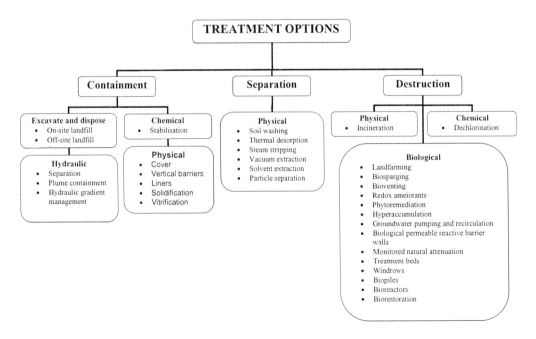

Figure 3. Hierarchy of treatment options (Barr *et al.*, 2003).

1.4. Potential Reuse of Remediated Soils

Bioremediation, which has been available in the UK for many years, breaks down organic contaminants. Over recent years very large sums of money have been invested in the biological remediation of contaminated land. Figures published in 2007, for example, estimated that the total UK remediation market was worth £600 million of with £36 million attributed to bioremediation (Randell, 2007). Various agencies, including the Environment Agency and Construction Industry Research and Information Association (CIRIA), have produced guidelines on the use of bioremediation and the conditions that optimise its efficiency. Bioremediation has many advantages over other forms of treatment as it requires little energy and does not affect the remediated soil as the contamination is removed. Its main disadvantage is the speed of treatment which is considered slow. Until recently therefore the application of this technology has been somewhat limited.

Soil washing is another technique that can remove a wide range of both organic and inorganic contaminants. This utilises the fact that contaminants adhere to the fine-fraction of the soil, such as organic matter and clay/silt particles. The method utilises the difference in grain-size and density of the materials and separates the different fractions by means of screens, hydrocyclones and upstream classification. Pollutants in the contaminated soil are separated and transferred from the sand and gravel fractions to a filter-cake residue, thereby minimising the volume needing further treatment or disposal.

Techniques that exploit stabilisation/solidification technologies can also be used to immobilise a range of organic and inorganic contaminants. Stabilisation techniques such as cement stabilisation use chemical reagents that react with the contaminants present to transform them into an immobile form. Solidification differs from stabilisation in that it uses chemicals to turn the contaminated soil into a monolithic mass, the contaminants then being physically encapsulated. A further alternative technique is vitrification which uses high temperatures (>1,000°C) to form a solid glassy mass from the soil. The high temperatures incinerate organic contaminants whilst the inorganic compounds are incorporated into the ceramic-like matrix. One drawback with this technology, as with other thermal techniques, is that they are relatively expensive due to the energy required to heat the soil to these temperatures. This is also influenced by the moisture content of the soils as the energy involved increases with increasing moisture content.

The selection of the most suitable remediation technique is site specific to ensure that the remediation target is met for its designated use. Not all techniques remove contaminants from the soil, for example the use of a barrier material to break the source/receptor pathway link. Although this reduces the current risks associated with the contaminants, any future change in land use may require further remediation. Other techniques change the structure of the soil matrix such as thermal desorption. This technique destroys the organic compounds, leaving a residual soil with the consistency of ash. Table 2 shows the more commonly applied UK techniques. It can be seen that they all have limitations, which will affect their success. Some techniques for instance can only treat organic contaminants such as bioremediation, whereas others can also treat inorganic compounds such as vitrification. Site specific trials would therefore normally be required to determine the most appropriate technique. In some cases a combination of techniques can also be employed to optimise efficiency and meet remediation targets.

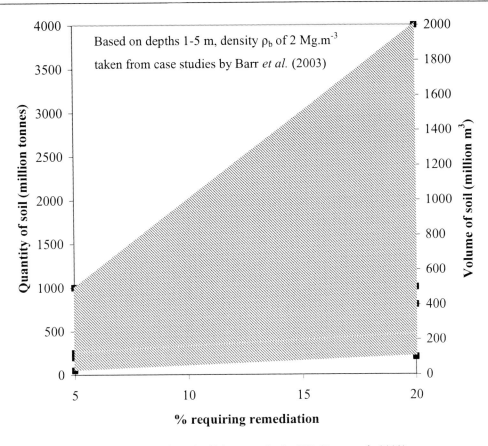

Figure 4. Potential amount of remediated soil for reuse in the UK (Barr *et al.*, 2003).

1.4. Potential Reuse of Remediated Soils

To date, the majority of remediation has been directed towards the treatment of soil prior to disposal. However, technological advances in remediation techniques have resulted in significant reductions in residual levels of contamination after treatment. Figure 4 illustrates the estimated quantity of soil requiring remediation in the UK. Based on the lower estimate that 5% of 50,000 hectares to a depth of one metre, 50 million tonnes of soil would require treatment. At the other extreme, based on 20% of 200,000 hectares to a depth of five metres, four billion tonnes would require treatment. Due to the problems in defining what constitutes land contamination, precise estimates of its presence are therefore difficult to make.

In practice there is a vast quantity of remediated soil that can potentially be reused within the UK. When coupled with the limitations placed on landfill, a new source of material namely 'the remediated soil' is at the hands of the engineer to innovatively reuse. To investigate potential reuse of remediated soils, soil samples from the remediation trials undertaken at the former Avenue Coking Works were investigated, particularly from the thermal desorption process.

Figure 5. The former Avenue Coking Works – Plant Area.

2.0. THE FORMER AVENUE COKING WORKS

The former Avenue Coking Works, Chesterfield, UK (referred to as the Avenue – Figure 5), in the late 1990s, was classified as one of the most contaminated sites in Western Europe. Between 1956 and 1992 the Avenue produced smokeless fuels and associated by-products (such as, sulphuric acid, ammonium sulphate, benzene, toluene, xylene and naphthalene) through the carbonisation of coal. The site was left derelict after its working life and was never decommissioned. Chemical tanks, pipe work, sumps, gantries and contaminated lagoons were never drained down and the condition of these has deteriorated over time. This, coupled with site operation, has caused severe ground pollution. In 1999 the site was governed under the Control of Major Accident Hazards (COMAH) Regulations due to the potential risk to human health and the environment. The site is now owned by East Midlands Development Agency (EMDA) and is being regenerated as part of the national coalfields regeneration programme. As part of this programme, all above ground structures and tanks were emptied and demolished. The site is currently classified under part IIA of the Environmental Protection Act 1990 as a special site due to the severe ground pollution causing ongoing pollution of the adjacent River Rother.

To identify the nature and extent of soil and groundwater contamination at the site, a detailed ground investigation was carried out over and beyond the 98 hectare site between August 2000 and August 2001. The Avenue was split into three distinct areas all requiring remediation; a solid waste tip; a tar lagoon and the plant area. Major contaminants identified included PAHs, phenols, mineral oil, BTEX, cyanides, tyres, metal, asbestos and spent oxide (Blue Billy). The spent oxide was used as a purifying agent for coal gas (Cairney, 1987) and was also found to contain free and complex cyanides, elemental sulphur, sulphate and sulphide.

2.1. Remediation Trials

Based on the results of the ground investigation and historic site information, three locations of contaminated material were chosen for the remediation trials by EMDA. These locations were chosen as they where the most contaminated and offered examples of the different soil matrices encountered at the site. Material was taken from the following areas (Figure 6):

Waste Tip containing spent oxide and covering an area of approximately three hectares, located in the western end of the site. The waste tip was licensed in 1977 for the receipt of waste, but aerial photographs suggest that it was in use from 1960. The licence was revoked under the Control of Pollution Act (CoPA) 1974 in March 1993 due to the environmental hazards posed.

Plant Area - Zone B within the former by-product processing area, containing hard pitch, tar bases, tar distillation and tar acid plant. Contaminants associated with this type of material were considered to be the most refractory and problematical for most remediation technologies to treat (Hughes and Bollermann, 2004).

Lagoon Two is a settlement lagoon located at the north end of the site containing leachate from the waste tip. It is separated into three compartments by clayey silt earth bunds. The clayey silt material was saturated with tar, with a moisture content and pH of 28-78% and 11-13 respectively. Contaminants in the tar and groundwater include PAHs, phenols and BTEX, which are migrating through the bunds towards the River Rother.

The ground investigation trials were carried out both in the laboratory and onsite to determine the most appropriate forms of remediation. These included bioremediation, thermal desorption, soil washing and cement stabilisation (Figure 7). From these tests (Table 3), thermal desorption was found to be the most effective technology for treating the contaminated soils from the Avenue (80% reduction PAHs). A drawback was, however, the fact that this was the most expensive of the technologies evaluated at £60 per m^3 (Hughes and Bollermann, 2004). Both bioremediation and advanced soil washing technologies were found to reduce contaminant levels but the residual levels of PAHs were 8,000 and 1,400 mg/kg respectively. As such this may preclude reuse of these treated soils. Cement stabilisation was not found to be an effective treatment option for the contaminants encountered at this site.

Table 3. Cost evaluation for treatment technologies at the Avenue site (2001)

Process	Reduction in levels of PAH for PA soil mg/kg	Limitations identified	Cost £/m^3
Thermal desorption (enhanced thermal conduction)	6430 to 290	Not suitable to treat liquids, heavy metals or asbestos	55 - 60
Soil washing	9,841 to 1,400	4% will need disposal	25 - 35
Bioremediation (solid phase)	13,000 to 8,000	<4 ring PAHs. Inhibited by cyanide	35 - 45
Cement stabilisation	N/A	Effective on cyanide and ammonium only	20 - 25

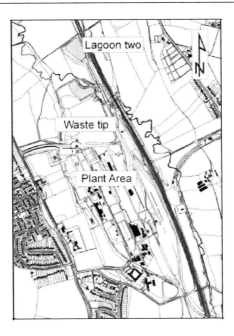

Figure 6. Illustration of the Avenue identifying the main areas.

Figure 7. Remediation trial (techniques) carried out at the Avenue site.

2.1.1. Enhanced thermal conduction

The specific type of thermal desorption undertaken at the Avenue was Enhanced Thermal Conduction (ETC). This uses burners to provide heat to deplete the levels of contaminates within the soil. The burners deliver heated air in the range of 450 - 600°C for a period of 5 - 14 days through a system of interlaced perforated pipes and manifolds. A three-layered soil cell is often used with each layer consisted of a series of small perforated pipes perpendicularly joining a large steel manifold running the length of the cell. The cell is encapsulated within a heat shield/cover (Figure 8). The heat caused the contaminants to volatilise and migrate to a low-pressure area, i.e. the gap between the soil cell and heat shield. The volatilised contaminants are then passed through a vapour system (either Catalytic or Thermal Oxidiser) under a vacuum to destroy the contaminated gases prior to the remaining vapour being discharge to atmosphere.

Catalytic oxidiser

Thermal oxidiser

Figure 8. Thermal desorption remediation trial.

3.0. PHYSICAL/CHEMICAL PROPERTIES OF THERMAL DESORBED SOIL

Two samples of soil from the Avenue were used in this research programme. These were taken from the same geographical location namely: National Grid Ref 438994,367888. This was within Zone B of the Plant Area. The first soil, referred to as TD soil, was originally excavated from a trench in the former by-product processing area. This soil was subsequently heat treated in the ETC remediation trial. Following treatment it was then stored under a tarpaulin on the high level stocking area until collection for use in this programme. The second soil, referred to as Plant Area (PA), was removed using a back acting excavator directly from the end of the original remediation trial trench in 2004. As such the soil had not been subjected to any form of remediation. This enabled its pre and post treatment geotechnical and chemical properties to be compared (Table 4). These tests confirmed that the levels of contamination were significantly lower in the thermally desorbed soil. For example, levels of contamination such as naphthalene, phenanthrene and fluoranthene were reduced from 34,000, 10,000 and 9,000 mg/kg to 30, 100 and 80 mg/kg respectively.

For the physical properties test programme the PA and TD soil samples were double bagged using shovels and the tops loosely tied using tie wraps and kept cool in the dark in accordance with BS 7755-2.6: 1994. A third soil (LB) was selected as a control soil. This consisted of uncrushed silica sand obtained from Leighton Buzzard, Bedfordshire, UK; it was washed, dried and graded. It is free from silt, clay and organic matter.

Table 4. Chemical profile of PA and TD soils

Parameter	PA soil	TD soil	Units	Method	Level of Detection
pH	6.8	7.6	pH units	GLpH pH Meter	N/A
Moisture Content	13.4	12.4	%	Modified BS 1377:1990	0.1%
Total Organic Carbon	11.7	4.0	%	Colorimetry	0.1%
Phenols Total 8 speciated	<0.1	<0.1	mg/kg	HPLC-EC	0.1 mg/kg
Arsenic	17	19	mg/kg	IRIS Emission Spectrometer	0.1 mg/kg
Chromium	33	74	mg/kg	IRIS Emission Spectrometer	0.1 mg/kg
Copper	85	147	mg/kg	IRIS Emission Spectrometer	0.1 mg/kg
Lead	152	210	mg/kg	IRIS Emission Spectrometer	0.1 mg/kg
Mercury	5	6	mg/kg	IRIS Emission Spectrometer	0.1 mg/kg
Nickel	45	60	mg/kg	IRIS Emission Spectrometer	0.1 mg/kg
Zinc	406	387	mg/kg	IRIS Emission Spectrometer	0.1 mg/kg
PAHs (sum of EPA 16)	76334	548	mg/kg	GC-MS	0.1 mg/kg
Naphthalene mg/kg	34137	35	mg/kg	GC-MS	0.1 mg/kg
Acenaphthylene	184	0	mg/kg	GC-MS	0.1 mg/kg
Acenaphthene	1447	2	mg/kg	GC-MS	0.1 mg/kg
Fluorene	1747	2	mg/kg	GC-MS	0.1 mg/kg
Phenanthrene	10459	98	mg/kg	GC-MS	0.1 mg/kg
Anthracene	3642	34	mg/kg	GC-MS	0.1 mg/kg
Fluoranthene	9393	83	mg/kg	GC-MS	0.1 mg/kg
Pyrene	6352	51	mg/kg	GC-MS	0.1 mg/kg
Benzo(a)anthracene	2307	38	mg/kg	GC-MS	0.1 mg/kg
Chrysene	2770	55	mg/kg	GC-MS	0.1 mg/kg
Benzo(b)+(j)fluoranthene	1500	61	mg/kg	GC-MS	0.1 mg/kg
Benzo(k)fluoranthene	550	19	mg/kg	GC-MS	0.1 mg/kg
Benzo(a)pyrene	940	25	mg/kg	GC-MS	0.1 mg/kg
Indeno(1,2,3-cd)pyrene	405	20	mg/kg	GC-MS	0.1 mg/kg
Dibenz(a,h)anthracene	96	5	mg/kg	GC-MS	0.1 mg/kg
Benzo(g,h,i)perylene	404	20	mg/kg	GC-MS	0.1 mg/kg

* All results expressed on a dry weight basis

Table 5. Geotechnical properties/parameters of the soils

Property/Parameters		PA soil	TD soil	LB sand	Units
Moisture content		20.3	15.9	0.1	%
Uniformity coefficient C_u		3.5	61.1	1.6	-
Particle density		1.68	2.11	2.67	Mg/m^3
Maximum density		1.19	1.61	1.79	Mg/m^3
Minimum density		0.69	1.22	1.47	Mg/m^3
CBR @ OMC		11	107	n/a	%
Coefficient of permeability, k		2.38×10^{-5} ($@1.30Mg/m^3$)	7.64×10^{-6} ($@1.40Mg/m^3$)	9.85×10^{-3} ($@1.70Mg/m^3$)	m/s
Total stress	c_u	11	14	0	kN/m^2
	ϕ_u	17	34	44	°
Effective stress	c'	-	0	0	kN/m^2
	ϕ'	-	42	44	°

Table 5 summarises the range of geotechnical properties/parameters of the three soils that were established. The laboratory tests were carried out in accordance with BS 1377: 1990 'Methods of test for soils for civil engineering purposes' in conjunction with BS 5930: 1999 'Code of practice for site investigations' and was based on disturbed material.

The grading curve for both the PA and TD soils show that they have a similar distribution of particle sizes over the range 10 and 75 mm (Figure 9). This is to be expected as the feedstock for the TD trial was extracted from the same geographical location as the PA soil. At lower particle sizes the shape of the grading curve indicates that the PA soil contains less fine particles. This can be attributed to the contaminants present in the soil bonding these particles to others.

After thermal desorption and the removal of a majority of the contamination these fine particles are then separable and the percentage of fine material increases in the TD soil. The presence of contamination also reduces the permeability of the soil. This is illustrated by a reduction of one order of magnitude in the value of k between the PA and TD soils (Table 5). The very high CBR value for TD soil (Figure 10) could be due to the high uniformity coefficient of the soil, in that it is classified as a well graded material and, as such, excellent particle interlock would have been obtained. Humus would also have been destroyed by the thermal process. This, in turn, would lead to a greater interlock between soil particles and higher strength (CBR value). The thermal process significantly improved the engineering parameters of the soil, for example a 90% increase was obtained in CBR value at optimum and the shearing resistance improved by a factor of two from 17 to 34° (Table 5).

From Figure 9 the soils were classified in accordance with BS 5930 (1999) as follows:

PA   Black well graded slightly sandy GRAVEL containing high quantities of tar
TD   Black well graded very sandy GRAVEL
LB   Light brown/pale silver to brown uniformly graded slightly gravelly SAND

In terms of their geotechnical properties the TD soil has a classification under the Specification for Highway Works (2007) of Grade 1A – well graded granular material suitable for general granular fill, with reference to Table 6/1: Acceptable earthworks materials.

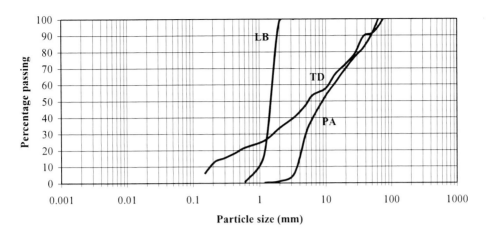

Figure 9. Particle size distribution curves for the three soil types.

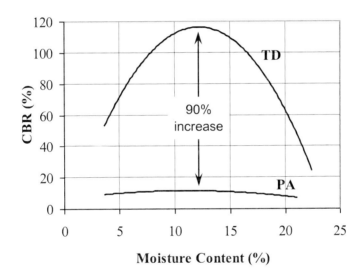

Figure 10. Optimum CBR plot for PA and TD soils.

4.0. THERMAL DESORBED SOIL AS EMBANKMENT FILL

A potential application for thermally desorbed soil for example is for embankment fill as the soil is well graded and non cohesive. In practice, these types of soils have traditionally been used for the building of embankments (TFHRC, 2008). During embankment construction soil is placed and compacted to a predetermined density, typically >95% maximum dry density. Embankments are often constructed quickly over soft compressible ground (Tandjiria *et al.*, 2002; Varuso *et al.*, 2005), and failure is most likely to occur towards the end, or immediately after, construction (Tandjiria, 2002). This is when the excess pore water pressure in the underlying soil is at its highest due to the self-weight of the embankment fill. The embankment will then become progressively more stable over time as the excess pore water pressure dissipates, resulting in an increase in the effective stress. This process is dependant on a number of factors including soil permeability and the length of the drainage path. To prevent failure a tensile inclusion, such as a geotextile, can be inserted into the soil. This provides tensile strength to the soil (Borges and Cardoso, 2002; Zhang *et al.*, 2006). When the inclusion is used between the underlying soil and the embankment fill, i.e. at the base, the term 'basal reinforcement' is used. Its function is to prevent the embankment from splitting or to provide a stabilising moment to resist rotation (Figure 11).

4.1. Coefficient of Interaction of Thermally Desorbed Soil

To further establish the suitability of the TD soil as an embankment material it was pertinent to consider its shearing interactive characteristics of thermally desorbed soil along a typical geotextile interface. The coefficient of interaction $(\propto)$ i.e. the ratio of the friction coefficient between soil and fabric (tan δ) and the friction coefficient for soil sliding on soil (tan ϕ) (BS 6909:8, 1991) indicates the shearing resistance at the soil/geotextile interface. There are a number of different ways to use the direct shear box, all of which stem from the method employed to clamp the geotextile to the shear box and the area of shear. Richards and Scott (1985) presented a review of the five main types of shear box, namely: fixed, partially fixed, free, large base and central base together with advantages and disadvantages from Ingold (1982) and Brant (1985). In the main, the fixed and partially fixed shear boxes give similar results and are the simplest, least expensive pieces of test equipment available.

The fixed method was employed during this testing stage, where a mid-range commercially available synthetic geotextile was carefully glued to a small wooden block in the lower section of the shear box. The upper section of the box contained the appropriate soil for the different tests. Nominal normal stresses of 122.6, 95.4, 68.1 and 40.9 kN/m² were applied to the samples. The relative horizontal displacement of the two halves of the shear box, the change in sample height during shearing of the upper half of the shear box were monitored by linear dial gauges reading directly to 0.01 mm. The tests were conducted at a strain rate of 1.25 mm/minute.

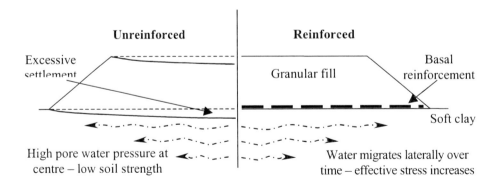

Figure 11. Diagrammatic illustration of a rapidly constructed embankment over soft compressible clay.

Table 6. Geotextile shearing interactive properties

Geotextile	TD soil		LB soil	
	δ'_{max}	$\propto'_{max}$	δ'_{max}	$\propto'_{max}$
Synthetic warp knitted polyester	41	0.98	35	0.79

Values of the shearing angle (δ') and coefficient of interaction ($\propto$) of the geotextile sheared in TD and LB soils, are shown in Table 6. The shearing interactive properties of the TD soil were superior to those of LB soil. The TD soil yielded a coefficient of interaction of around one, indicating that the additional of the geotextile in this soil type did not introduced a weak plane.

4.2. Tensile Strength Properties of Geotextile

A tensile test was conducted on the geotextile, at the standard testing temperature and relative humidity of $20 \pm 2°C$ and $65 \pm 2\%$ respectively. The geotextile was also conditioned, at this temperature and humidity, for at least 24 hours before testing. Due to the testing equipment not being able to accommodate the large clamp size recommended by BS EN ISO 10319: 2008 (wide width) tensile test, a gauge length of 200 x 50 mm was tested at a constant strain rate of 100 ± 10 mm/minute as in BS EN ISO 13934-1: 1999 (strip method). The warp knitted geotextile is essentially a collection of parallel yarns in one direction, thus the ground structure would not be affected by necking and thus the use of latter standard was adequate.

Figure 12 details the breakage characteristics for the geotextile. The geotextile curve undergoes relative large amounts of strain before rupture, which is an intrinsic characteristic of polyester.

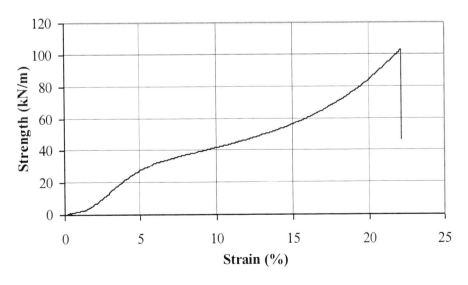

Figure 12. Stress-strain plot for synthetic warp knitted polyester.

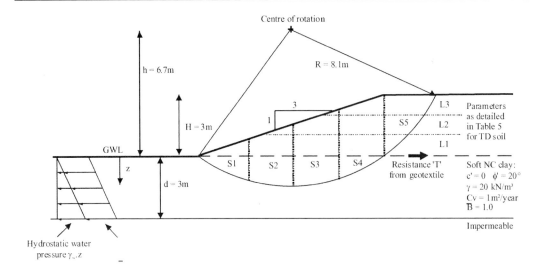

Figure 13. Illustrative example of basal reinforced embankment.

4.3. Stability Analysis of Thermal Desorbed Soil as Embankment Fill

To illustrate the geotechnical aspects using TD soil as an embankment fill containing basal reinforcement the example shown in Figure 13 has been analysed. The example has a typical configuration of an embankment over soft ground. Characteristic values have been assigned to the relevant parameters for the underlying soil. However, for the embankment fill the actual TD soil parameters obtained in this study were used. The embankment is assumed to be constructed in three layers (each taking 4 months to construct) on soft normally-consolidated clay with the water table at ground level. For the analysis the actual loading is taken as being applied step-wise, with each increment of load being applied instantaneously at the mid-time/point of each layer. The stability of the embankment slope during and after construction, until all the excess pore water pressure has dissipated, has been analysed for the typical circular failure surface indicated in Figure 13. Analysis has been by an extended Fellenius method of slices, in effective stresses with the additional force provided by the geotextile being incorporated as a resisting moment.

Figure 14 illustrates the variation in the factor of safety (FOS) during and after construction, for an unreinforced embankment. This demonstrates that the embankment would fail when the final (3rd) layer of the TD embankment is constructed, but post construction the FOS would increase with time to above unity. A typical acceptable FOS for a 3:1 slope would be 1.3. Figure 15 illustrates the force required by the geotextile in the reinforced embankment to maintain the necessary FOS throughout the life of the TD embankment. The stability of the embankment is at its most critical at the end of construction as demonstrated by the force required to be provided by the geotextile to be 45 kN/m. This required strength can easily be met by the synthetic warp knitted polyester geotextile used in this testing programme.

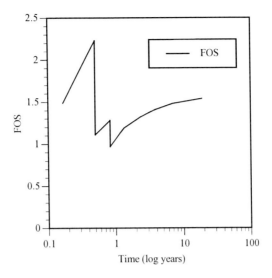

Figure 14. Variation with time (unreinforced embankment).

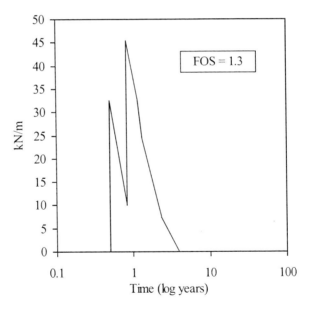

Figure 15. Force required by the geotextile for basal embankment stability.

5.0. CONCLUSIONS

- There are vast quantities of contaminated soil in the UK that has the potential to be remediated.
- Remediation technologies have significantly developed over recent years and at optimum conditions the remediate soil only contains trace levels of residual contamination.

- In some instances (e.g. thermal desorption) not only does the remediation activity improve the chemical properties, but an improvement in physical properties, such as CBR value, shear strength, density and grading is also achieved.
- It has been shown that the geotechnical properties/parameters of TD soil can be classified as suitable fill material for a reinforced embankment construction.
- It is now up to the engineer to innovatively reuse the remediated soil in engineering applications.

6.0. RECOMMENDATIONS

- Only one source of thermal desorbed soil has been tested in this research study; it is recommended that other sources of thermal desorbed soil are tested to ensure that the results can be replicated from soil taken from different sites.
- *In situ* design implications are needed to be assessed e.g. the leachability on any residual contamination.
- A life cycle assessment/life cycle cost analysis is needed to assess the reuse of thermal desorbed soil on a practical, financial and environmental scale.

ACKNOWLEDGMENT

Thank you to Engineering and Physical Sciences Research Council (EPSRC) and Leeds Metropolitan University for their financial support together with EMDA and Jacob (Babtie) for allow the soil from the former Avenue Coking Works, Chesterfield, UK to be tested during this research project.

REFERENCES

Anon. (2008a). *Anthrax spores kill drum maker.* Metro 3rd November. p2.
Anon. (2008b). *Legislation affecting waste.* Available at: <http://www.wasteonline.org.uk/topic.aspx?id=24> [Accessed 10th July 2009].
Anon. (2008c). *Remediation technologies.* Available at: <http://en.wikipedia.org/wiki/Remediation> [Accessed 12th May 2008].
Barr, D., Bardos, R. & Nathanail, C. (2003). *Non Biological methods for the assessment and remediation of contaminated land – case studies.* CIRIA C588, London.
BBC. (2007). *UK landfill dustbin of Europe* Available at: <http://news.bbc.co.uk/2/hi/uk_news/7089963.stm> [Accessed 5th July 2008].
Borges, J. L. & Cardoso, A. S. (2002). *Overall stability of geosynthetic-reinforcedembankments on soft soils.* Geotextiles and Geomembranes, 20, 395-421.
Brant, J. R. T. (1985). *Behaviour of soil-concrete interface,* PhD Thesis, Department of Civil Engineering, University of Alberta, 376.
BS EN ISO 10319 (2008) *Geosynthetics – Widewidth tensile test.* British Standards Institution, Milton Keynes.

BS EN ISO 13934-1 (1999) *Determination of maximum force and elongation at maximum force using the strip method*. British Standards Institution, Milton Keynes.

Bsi. (1990). *BS 1377-2: 1990 - Soil classification for civil engineering purposes*. British Standards Institute, Milton Keynes.

Bsi. (1991). *BS 6909-8: 1991 - Determination of Sand-Geotextile Frictional Behaviour by Direct Shear*. British Standards Institution, Milton Keynes.

Bsi. (1994). *BS 7755-2.6: 1994 / ISO 10381-6:1994 Soil quality - Part 2: Sampling – Section 2.6. Guidance on the collection, handling and storage of soil for the assessment of aerobic microbial processes in soi*. British Standards Institution, Milton Keynes.

Bsi. (1996). *BS EN ISO 10319: 1996 – Geotextiles wide width tensile test*. British Standards Institute, Milton Keynes.

Bsi. (1999). *BS 5930: 1999 – Code of practice for site investigations*. British Standards Institute, Milton Keynes.

Cairney, T. (1993). *Contaminated Land problems and solutions*. Blackie Academic press.

Cairney, T. & Hobson, D. (1998). *Contaminated Land*. 2nd Edition. London. E & FN Spon.

DEFRA. (2006). *Circular 01/2006 Environmental Protection Act 1990: Part 2A Contaminated Land*. Department for Environment, Food and Rural Affairs, London.

DEFRA. (2008). *Guidance on the Legal definition of contaminated land*. Department for the Environment, Food and Rural Affairs, Crown Copyright 2008.

Dinsdale, J. (2005). *How much contaminated land is there?* [Internet] Environment Agency. Available at: <http://www.environment-agency.gov.uk> [Accessed 22nd Feb 2006].

Dti - Department of Trade and Industry. (2002). *Brownfields – managing the development of previously developed land - a client's guide*. London. CIRIA.

EA. (2002). *Dealing with contaminated land in England – progress in 2002 with implementing the Part IIA regime*. Environment Agency.

EA. (2007). *Hazardous waste data 2007*. Available at: <http://www.environment-agency.gov.uk/research/library/data/97801.aspx> Environment Agency [Accessed 2nd June 2009].

European Environment Agency. (2007). *Overview of contaminants affecting soil in Europe* Available at: <http://dataservice.eea.europa.eu/atlas> [Accessed 14th Oct 2007].

EUGRIS. (2007). *Poly Aromatic Hydrocarbon. European Groundwater and contaminated land Remediation Information System*. Available at: <http://www.eugris.info/ EUGRISmain.asp?e=7&Ca=2&Cy=0> [Accessed 25th March 2007].

EUGRIS. (2008). *Further description - Poly-Aromatic Hydrocarbons*. Portal for soil and water contamination in Europe. Available at: <http://www.eugris.info/Further Description.asp?Ca=2&Cy=0&T=Poly-Aromatic%20Hydrocarbons&e=7> [Accessed 2nd July 2008].

Farlex. (2008). *Phenol – the free online dictionary*. Available at: <http://www.thefree dictionary.com/Phenols> [Accessed 26th Sept 2008].

Health Canada. (2005). *It'sYour Health PCBs*. Available at: <http://www.hc-sc.gc.ca/hl-vs/alt_formats/pacrb-dgapcr/pdf/iyh-vsv/environ/pcb-bpc-eng.pdf> [Accessed 26th Sept 2008].

Hughes, D. & Bollermann, A. (2004). *Avenue Coking Works remediation trials report*. Unpublished report for East Midlands Development Agency. Babtie, Manchester.

ICE - Institution of Civil Engineers. (2004). *The state of the nation 2004*. New Civil Engineer 17, June, 20-22.

Ingold, T. S. (1982). *Some observations on the laboratory measurement of soil-geotextile bond. American society for testing and materials geotechnical testing journal*, 5, 3-4, 57-67.

Kanaly, R. A. & Harayama, S. (2000). Biodegradation of High-Molecular-Weight Polycyclic Aromatic Hydrocarbons by Bacteria, *Journal of Bacteriology*, April, 182, *8*, 2059-2067.

Kong, L. (2004). *Characterization of mineral oil, coal tar and soil properties and investigation of mechanisms that affect coal tar entrapment in and removal from porous media*. PhD Thesis. Georgia Institute of Technology.

Madabhushi, S. & O'Neill, T. (2007). Remediation process optimization: A status report. *Remediation Journal*. *17*, *3*, 47-66.

NAS - National Academy of Sciences. (2003). *Bioavailability of contaminants in soils and sediments*. Washington. National Academic Press.

National Pollutant Inventory. (2004). *Poly Aromatic Hydrocarbons - substance factsheet*. Austrailian Government, Department of the Environment and Water Resources. Available at: <http://www.npi.gov.au/database/substance-info/profiles/74.html> [Accessed 26th March 2007].

Pirika. (2004). *Octanol-Water Partition Coefficient Estimation*. Available at: <http://www.pirika.com/chem/TCPEE/LOGKOW/ourlogKow.htm> [Accessed 25th March 2007].

Procuro. (2006). *PCBs the facts*. Available at: <http://www.procuro.co.uk/About%20Us. html> [Accessed 4th November 2008].

Randall, J. (2007). *The EIC guide to the UK environmental industry Contaminated land*. The Land remediation industry.

Report LIFE 98. (2008) *Detection and elimination of human exposure to environmental hormone disrupting substances*. Available at: <http://www.andrology.be/life98/ index.php> [Accessed 26th Sept 2008).

Richards, E. A. & Scott, J. D. (1985). *Soil geotextile frictional properties*, 2nd Canadian conference on geotextiles and geomembranes, 13-24.

Sarsby, R. (2000). *Environmental geotechnics*. Thomas Telford Publishing, London Chapter 13.

Specification for Highway Works. (2007). *Manual of contract documents for highway works*. *Vol.1*, 0600 Earthworks.

Strange, J. & Langdon, N. (2008). *ICE design and practice guides: contaminated land investigation, assessment and remediation*. 2nd Ed. Thomas Telford Publishing, London.

Tandjiria, V., Low, B. K. & Teh, C. I. (2002). *Effect of reinforcement force distribution on stability of embankments* Geotextiles and Geomembranes 20, 423-443.

TFHRC – Turner-Fairbank Highway Research Centre (2008) *Embankment or fill materials soils*. Available at: <http://www.tfhrc.gov/hnr20/recycle/waste/app4.htm> [Accessed 10th November 2008].

Varuso, R. J., Grieshaber, J. B. & Nataraj, M. S. (2005). *Geosynthetic reinforced levee test section on soft normally consolidated clays*. Geotextiles and geomembranes 23, 362-383

Ward, O. P. & Singh, A. (2004). *Soil Bioremediation and Phytoremediation –An Overview*. Soil Biology, *Vol. 1*. Applied Bioremediation and Phytoremediation.

Zhang, Y., Tao, L. Zou, Z. & Wang, F. (2006). *Modelling and performance analysis of a reinforcing loess embankment* Proceedings of the 8th international conference on geosynthetics, Yokohoma, 4, 1339-1342.

In: Built Environment: Design, Management and Applications   ISBN: 978-1-60876-915-5
Editor: Paul S. Geller, pp. 307-318

Chapter 13

INTERGENERATIONAL HOUSING AS A NEW VECTOR OF SOCIAL COHESION

*Nicolas Bernard**

Professor at Facultés Universitaires Saint-Louis (Brussels).

ABSTRACT

Today, when the prevailing current of individualism is coming in for well-warranted criticism, one expanding phenomenon deserves to be singled out for attention. This is intergenerational housing, which, as its name intimates, brings a couple of senior citizens and a younger household under the same roof. It differs from formulae such as "Béguine convents" or residential communities (for example, the Abbeyfield "Entre Voisins" house in Etterbeek) that bring together senior citizens *only* (Dossogne & Simon). Intergenerational housing appears to be a particularly stimulating initiative in the galaxy of collective housing, in any event, as it illustrates the fact that, beyond providing physical shelter in the strict sense of the word, housing also helps to *(re)forge social ties*. Although the occupants of such houses do not necessarily share a common space, they nevertheless take part in a lifestyle based on mutual assistance. The philosophy behind the founding of intergenerational housing, in which informal solidarity is the driving force, consists in offering a minimum degree of availability to the other household with a view to providing one or the other service. Yet for all that, this living arrangement does not require the occupants to modify their lifestyles substantially. For example, the younger members are not at all expected to serve as home nurses or family aides. Nor are the occupants required to live as a community. Intergenerational housing (also called "kangaroo houses" in reference to the kangaroo's sheltering and nurturing pouch for its young), which was devised in a spirit of pooling assistance and support, is in any event coming in for more and more attention from the population and authorities alike (Bonvalet). That is why in this article the spotlight is trained on this emerging alternative form of accommodation, which appears to be heavy with promise, despite the many legal obstacles to its official recognition.

---

* Corresponding author: E-mail: nbernard@fusl.ac.be

I. BACKGROUND – THE EMERGENCE OF COLLECTIVE HOUSING

a) A Spreading Phenomenon...

For a long time, housing, whether in the property rental or ownership sector, has been defined largely as an individual matter. Currently, one buys or rents essentially alone (or by household). This is paralleled by the prevalence of housing schemes governed by ordinary law: a lease on or ownership of the main residence. Yet, almost a half century after the communalism of the Sixties, we see collective living once again breaking through the surface in today's housing sector. Alternative housing seems to be making a comeback. Appearing with still fuzzy contours, this phenomenon, referred to under the common banner of "collective living", regroups in one place, in any event, people who may or may not be unified by a life plan. In fact, no one model for collective living exists. "Béguine Convents" for senior citizens (see the "Cité Jouet-Rey" or Abbeyfield "Entre Voisins" residence, both in Etterbeek), artists' workshops of the Parisian "Bateau Lavoir" type (see the Mommen workshops at Saint-Josse), squatters' sites (e.g., Tagawa on Avenue Louise, Ministry of Housing crisis on Rue Royale), intergenerational housing (Dar El Amal in Molenbeek), permanent camping settlements (Ourthe-Amblève), collective purchasing grants (City of Brussels), housing for troubled youth, etc.: These are some of the numerous applications of this "fertile" theme in the Brussels region and Wallonia. Furthermore, collective housing can concern rentals as well as purchases, be deliberate or incidental, and may or may not be accompanied by social service support. Whilst, finally, certain structures are created – and managed – by an institutional agent, the majority of collective housing is the result of private initiative (in a bottom-up approach).

Although the phenomenon is rather diffracted, there is one fundamental constant: The residents all felt that, at a given moment in their life, a collective arrangement was the best way truly to appropriate a living space and enjoy their dwellings fully. In any case, the theme is just as attractive to senior citizens (concerned about ending their isolation) as to ex-convicts who lack lodging upon their release from prison and whose past prison stays often put off owners), social welfare recipients (who wish to share certain particularly high expenses: rental security deposits, telephone, gas and electricity accounts, etc.), members of the upper middle class who want to band together in order to acquire lofts in downtown areas, workers at the beginning of their careers (who, before starting a family, wish to prolong their student lives for a bit longer and invest with their friends in a more or less fashionable neighbourhood that would be financially inaccessible to each individually[1]), and so on.

As we see, the motivations are diverse. However, unlike the communal living of the 1960's, these groups do not seem to be ideologically motivated. Nowadays we are no longer trying to change the world by our residential mode. People are currently being pushed towards group living above all out of material necessity. At the height of the housing crisis, the hour is one of firm pragmatism (which does not mean that more doctrinal objectives may not replace the initial purely financial concerns once the situation stabilises).

As regards those people in specifically precarious states, collective living seems all the more interesting in that poverty, contrary to current belief, is not just a matter of limited financial resources. First and foremost, it is a corollary of the erosion of social ties and the weakening of the informal network of mutual aid and solidarity. In this framework, collective living helpts to reforge social ties to great advantage (Noël & Perdrix).

Alas! Locked into a categorical logic that progressively borders on stubbornness, and finally on obsolescence, public authorities tend to penalise certain forms of mutual support and solidarity. Thus, welfare and unemployment beneficiaries who decide to live under the same roof for financial reasons run a high risk of losing their status as one-person households and "falling" into the category of "cohabitant" and this applies even when they would not have formed in any way a household in the traditional sense of the term. As a consequence, the parties concerned have to rent post boxes or even doorbells, naturally at prohibitively high costs (which endanger their budgets), for the sole purpose of simulating residence apart and thus maintaining their "isolated individual" rate. Furthermore, is this collective living movement properly understood by the social housing sector? Unfortunately, nothing is less certain. If indeed there exists one or another effort in this area (modern residences, for instance, in the capital), it is regrettable, for example, that the Project for Future Housing in Brussels, which aims to construct 3,500 new social housing units, does not dedicate even an infinitely small fraction of this new housing to collective arrangements. And yet by designing a layout that would dispel all confusion (independent kitchens, separate entries, etc), the promoter could even overturn the presumption of "cohabitation" that weighs upon the occupants. It is therefore time for building companies, amongst others, to take account of the full measure of this social evolution in order to offer housing that corresponds better to the needs of a population that is itself changing radically.

b) ...That Raises Questions

Although still emerging, the collective living phenomenon is no longer a marginal phenomenon in 2009. The individual housing matrix seems fairly well worn and the return swing of the pendulum has already begun. A sufficiently critical mass of collective housing experiences has been reached, it seems, so that one could begin to speak about a true alternative. And yet, this type of living arrangement does not occur without raising a battery of question; it calls our relationship with housing and, therefore, our relationship with the norm, into question. First of all, let's raise the following paradox. Those people who decide to invest in a collective residence – and who, in so doing, claim fierce independence from existing structures – need at the same time pronounced social support (whether financially, legally or again in terms of relationships), as the path to collective housing is strewn with a great many obstacles. If, in the same vein, one wants to ensure the persistence of collective housing, drawing up a relatively strict set of "house rules" (or even a "charter" from which experience shows that can have a strong self-disciplining power) is advisable. This would unavoidably limit the freedom of residents who are often convinced that such arrangements will finally enable them to live without constraints.

Furthermore, the encouragement of collective living formulas should not lead to the multiplication of social "isolates" containing a smattering of small collectives as a marginal phenomenon. We should ensure that alternative housing maintains the possibility of reintegration, in one manner or another, into the sphere of ordinary civil law. If collective

---

[1] This is the *Friends* effect, from the name of the US television series.

living, incontestably and by its nature, challenges the dominant norm, it must at the same time be able to return to such a norm. This is not the easiest of the challenges that it must take up. Good norms are those that carry off the feat of taking account of the characteristic situations of poverty whilst detaching themselves in such a way as to be able to encompass a broader and fully liberating horizon. Effectively, the risk for the lawmakers, if they want to satisfy these – numerous – demands to recognise collective living fully, is to lose sight of the structural elimination of the causes of socio-economic fragility and thereby to multiply differentiated sub-statuses that are incompatible with the promotion of social progress according to a unifying hypothesis, for an authentic democracy has a duty to propose a social plan likely to be shared by all citizens, *i.e.*, the possibility to be just like everyone else yet have the right to cultivate one's differences. Following on from that, the public authorities would not be able to find, through the possible institutionalisation of private housing initiatives, a convenient way to divest (financially in particular) from the problem. In other words, the State's involvement should not be merely to sound its retreat from the housing field.

When it comes to the normative context, the current ambivalence should be emphasised. Depending on the subject covered, the regulatory arsenal appears nowadays to be either overflowing or sickly. Too complicated or exaggeratedly discreet, the *corpus* of rules in any event does not always contribute to the development of a right to housing. We find, on the one hand, a veritable profusion of laws, as much in the building sector as in the sales and even the rental sectors (just think of the – draconian – quality standards for property or the regulations that hamstring the renovation lease). The standards can, therefore, be perceived as oppressive, even penalising, by some. On the other hand, different stimulus initiatives suffer cruelly from a lack of legal recognition; here, the legislator adopts an astonishingly low profile. It is now time to make certain that the law remains abreast with such changes in society.

Altogether, alternative housing (necessarily collective when the predominant model is of individual housing) restores a bit of liberty to the resident, who is in a largely disempowering "real estate world". Traditionally, housing is handed over as "turn key projects", one might say. Low-income renters, to use only this example, hardly have the right to configure their residences in their own image by doing their own renovations, for example. Their contributions are too often limited to choosing the colour of the carpeting. *Homo habitans* currently remains a pure consumer. As a result, the lodging, which has been completed without their help, is not truly appropriated by its occupant.[2]. In contrast, collective housing makes each citizen a party to their own housing solution. Their creative potential is thus developed and the likelihood of identifying with their living space – both being vectors of self-esteem – is reinforced commensurately (Bernard, 2005).

The issue here is, therefore, to "liberate" other forms of housing that have undeniably proven themselves in the face of the current housing crisis. We must innovate, with it being up to the public authorities, guided by a healthy principle of subsidies that has been renewed in this way, to frame – legally and financially – these initiatives before, if need be, stabilising

---

[2] Thus, in social housing, North African women, for example, tend to abandon the American kitchen, which does not leave them any intimacy, whilst the bathtub – which some judge to be redundant in relation to the shower - sometimes serves as a makeshift receptacle in which to soak motorbike parts in oil. Generally, many homeless people prefer to continue sleeping in the street because they often consider the housing offered to be too regimented, sterile, etc.

them and, ultimately, amplifying or multiplying them. After all, the social housing agencies that are so praised today are not created in any other way. Doesn't the housing crisis, which is wreaking such devastating social havoc, deserve a similar "paradigm shift"? To ask the question is to answer it already.

II. A Fundamental Question: Where to House the *Grandpappy Boomers* Adequately?

Having set the stage regarding collective housing, let's return to intergenerational housing in particular. We know that the baby boomers of the 1960s will reach retirement age before the middle of this century. So, around 2050, senior citizens (*i.e.*, people over 60) will constitute close to one third of the population in our country, and, even today, one of five Belgians is grey haired (Leleu) Also, in less than two generations, the life expectancy of women will brush up against 90 (Hochart). So, a "grey revolution" is well underway, one that must be accompanied and carefully framed when it comes to housing, amongst other things.

At a time – the present – when lodging the elderly in their children's homes is a thing of the past, senior citizens find themselves faced with an alternative housing situation for which neither of the options (staying at home or moving to a nursing home) is fully satisfying. The elderly can first choose to remain in their homes, but with the risk of a declining quality of life. As time passes the house becomes too large and inconvenient and more complicated – and onerous – to run. Their isolation, furthermore, is likely ultimately to have a negative effect on their mental health, especially in the case of widowhood. Coupled with the decrease in income, the increasing physical difficulties encountered by elderly people in maintaining a residence lead to a large number of them leaving the upper floors of a building unoccupied whilst they continue to live on the lower floors. This "temptation" is further strengthened by the fear of all sorts of problems if part of the building is rented out. In any event, 20% of the seniors in Flanders are ready to move to a residence better adapted to their needs (Vanderhaegen, 2003).

In light of these different inconveniences, the rest home naturally appears to be a potentially attractive solution[3]. However, it is not yet a panacea. The idea of residing in a senior residence continues to trigger much apprehension amongst the elderly. Besides which, most of the time, falling back on a home is submitted to (particularly under the more or less friendly pressure of the children) rather than freely consented to or chosen. We have here essentially a choice under duress. And all of this without taking account of the fact that moving to a residence, which is inexorably more impersonal than one's own home and located most often in a different neighbourhood, can be a truly traumatic experience for a senior, coupled with considerable emotional uprooting. One's entire past disappears at the same time that one leaves one's home. That is why in particular no fewer than 95% of the elderly prefer to remain in their own homes (Braeckman), although this proportion decreases with age, as fewer than one out of five elderly people manage to live at home beyond the age of 95 (Observatoire de la Santé et du Social).

---

[3] In Brussels, the sector has just finally been given an organic legal framework (see the ordinance of the Joint Community Commission's joint assembly of 24 April 2008 on centres and housing establishments for the elderly.

As we can see, an optimal housing situation for the elderly does not exist at the present time. This is a major problem that the authorities, however, do not seem to have gauged correctly. Moreover, given our particular demographic situation, this need is expected to grow during the next decades, and by a more than substantial rate. That being so, it is more than time to begin devising novel, creative residential formulas that are likely to respond to this decisive societal deal. Of course, one could always conceive of this "riposte" from a macro-economic angle and thus embark on vast housing construction projects. However, faced with the fairly elevated costs incurred by such an operation, it would seem preferable to begin by using the existing – unoccupied – premises as far as this is possible. In other words, it would be suitable, when possible, to keep the elderly in their own homes, *in situ,* whilst trying to arrive at a solution for the more or less abandoned upper floors. The current housing stock actually has strong hidden potential that it would be judicious to exploit intelligently, especially from the sustainable development standpoint. What if, concretely, it was decided to reallocate the empty floors within the buildings occupied by the elderly in order to make them available to younger households, for example? Such is the heart of the philosophy underlying this form of alternative housing known as intergenerational housing, the advantages of which will now be described in detail.[4].

III. THE VIRTUES OF INTERGENERATIONAL HOUSING

Intergenerational housing emblematically embodies the virtues of a "win-win" formula in which *all* of the parties (in this case, the elderly owner, the young couple, and even the public authorities) benefit. The mechanism is fully to the mutual advantage of the different interests involved.

First of all, the benefits are obvious from the senior's point of view. The habitual presence under the same roof of one or several more people is clearly reassuring to an elderly person, who traditionally runs a greater risk in the event of a fall, for example. This same presence can also induce an appreciable feeling of security, whilst helping to relieve the solitude that weighs upon the elderly in particular. In addition, the young couple so lodged can perform several small services, the small work load of which will be nothing in comparison with the pleasure that the older party will derive. A bit of repair here, a little errand there: The elderly person can be expected to welcome these little helping hands (which do not "cost" the person who provides them very much) as a blessing. We must point out, moreover, that from a strictly pecuniary viewpoint, accommodating another household *intra muros* in a property left partially abandoned beforehand is a fairly advantageous operation. Not only does the building, which was previously unproductive, generate a rental revenue, but occupying the house as a whole also protects the building from deteriorating too rapidly. Last but not least, setting up this type of arrangement allows the senior to stay in his or her own home.

---

[4] To start off, it should be noted that kangaroo housing can also be set up directly by the public authorities. The latter can very well install elderly and younger tenants simultaneously in a building that they own, provided, naturally, that some system for exchanging services with variable degrees of formality is organised if the scheme is to have the "kangaroo" label.

The advantages for the lodgers are no less numerous. First of all, in exchange for the small services referred to above, they will get a chance to live at low cost, which appears to be especially difficult on today's tight real estate market, where decent, affordable properties are hardly legion. In additon, the increased housing supply due to the insertion of a certain number of intergenerational residences into the rental market should logically reduce the current pressure on rentals. This beneficial deflationary effect, which will be obtained only once a critical mass of new housing units is reached, should be observed not only in the housing of the intergenerational type but also more generally through the entire residential housing sector, according to the law of supply and demand. Finally and more specifically, according to the logic of informal, *bilateral* solidarity, the lodger's household can in turn justifiably expect some help from its (elderly) co-contractor. The latter will be able to receive packages from the postal service for his or her renter during working hours, for example, as well as possibly stepping in as a providential emergency *baby sitter.*

Whilst it is advantageous for the two parties to the lease, the intergenerational housing formula is not without very substantial benefits for the public authorities as well. Faced with a cruel lack of adequate housing for the elderly, and at the same time, being drastically powerless to deal with the phenomena of vacant dwellings (particularly when the property is not entirely abandoned, as shown by the nagging problem of commercial properties with empty upper storeys), the authorities will find in the intergenerational housing mechanism a way to work on these two fronts simultaneously. They will "strike two birds with one stone" in a way. Besides this, the much-vaunted social mixity that the state bodies so dearly desire will be achieved within each kangaroo building. In the same vein, it may be opportune to note that 50% of the students who come from outside of Brussels to study – and to live – in Brussels choose to continue to reside in the city once they have finished their studies. Here are the ready-made "clients" for intergenerational housing, who would thus also bolster the Brussels-Capital Region's tax base in the medium term (whereas, currently, the local authorities rather tend to overtax student rooms). More generally, the blossoming of such islands of "hot solidarity" cannot but help, even if only marginally, to repair the social fabric and reintroduce the collective model in an area – that of housing – that is defined almost exclusively by inter-individual relations. Incidentally, a drop in rental rates could only relieve public authorities that gave up on trying to regulate rents long ago. Finally, providing a framework for the kangaroo house formula by providing tax incentives, for example, hypothetically costs the public authorities less than to subsidise old age homes (or finance home care aides). Now, if we extrapolate from the current data, more than 15,000 extra beds would be needed in these homes in the medium term (above the 125,000 that already exist) if Brussels wanted to accommodate the current demographic trend (Van de Putte, Leroux, & Inghels), whilst another, linear, study even advocates the construction of eighty old age homes per year from now until 2050 (Vanderhaegen, 2006).

IV. LEGAL OBSTACLES

Whilst intergenerational housing has many advantages, it is nevertheless, impeded by serious legal limitations. In other words, the mechanism can run up against many different legal obstacles that not only prevent all spreading of the principle but are also likely to imperil the existence of several initiatives already in place. In any case, the straitjacket of existing law, which is in no way configured for this type of alternative lodging, we must add, is capable of dissuading households interested in this formula by penalising them in several ways. Let's begin by mentioning that an intergenerational residence that is a bit too integrated, which provides one or more common spaces for example, risks pushing its social welfare beneficiary residents down the rather unfortunate slope from the individual tax rate to that of the cohabitant's rate, with a total loss oscillating, on average and according to the situation, between 200 and 400 euros per month. This is the case even when the two couples do not form any sort of socio-economic unit in the traditional sense of the term. Whether according to the legislation on integration income or that governing unemployment benefits, cohabitation officially exists when "several people live under the same roof and mainly handle their household issues in common "[5]. Even if, by virtue of legal definitions, the occupants of an intergenerational residence could theoretically lay claim to their separate statuses as separate or one-person households, there still remains the risk of an examination to prove their claims. Effectively, if a public social service centre or the National Employment Office infers a presumed cohabitation of the collective housing type, the beneficiary of integration income or an unemployment benefit must therefore provide proof that he or she does not actually share any expenses related to the household. This often turns out to be difficult in practice, as providing negative proof is more than a difficult task.

However, this is not all. Apart from financial considerations, intergenerational housing also faces legal vicissitudes tied to town planning regulations. For example, the senior couple who would like to re-organise the upper floors of their house for rental (either a kangaroo house or some other type) would very probably be forced to undertake construction in order to create one or more autonomous housing units within a previously one-family house. Moreover, if the construction is major (which it most probably will be as soon as the use of the building is modified), it will definitely require a building permit, which, however, is obtained from the municipal administration only with great difficulty. It is true that many excesses were committed in the past by fairly unscrupulous landlords who did not hesitate to subdivide buildings into as many flats as possible in order to derive the greatest profit from their investment. The negative consequence has been, on the one hand, to jeopardise the tenants (often subjected to deplorably insalubrious conditions in these veritable "rabbit hutches".) and, on the other hand, to overpopulate a neighbourhood in an unchecked manner (and creating massive mobility problems as a result as well). Keen to make a clean break from this era, which is understood to be long past, the municipal bodies are now showing extreme firmness when it comes to authorising the subdivision of buildings. Yet for all that, and in order to avoid falling into the opposite extreme, this severity should not result in nipping innovative housing projects supported by a fertile social philosophy in the bud. More flexibility would consequently be welcome in these cases, and only in these cases. In any event, the issue will not be easily resolved in so far as the Brussels Junior Minister in charge of urban planning just a few years ago felt that "the danger of shared residences is currently greater than the advantages offered by this system of intergenerational mixity " (Dupuis, 2006). Let us stress once more, still regarding construction and renovations, that home owners in the Brussels Region who rent out their property have no longer been eligible for the renovation premium for the past few years.[6].

---

[5] See Articles 14 of the law of 26 May 2002 and 59(1) of the Ministerial Order of 26 November 1991 as amended by the Ministerial Order of 6 February 2003, respectively.

[6] Unless, as we shall see, the property is entrusted to a social real estate agency, which many owners are highly reluctant to do.

Another legal deterrent on the road to intergenerational housing is the real estate tax scheme. Whoever leases a property automatically sees a 40% increase in their *cadastral income* assessment — which is the base from which the real estate tax for physical persons is calculated in particular. Although this is perfectly logical in general (as an attempt to adjust the cadastral income assessment to reality, given that it has been largely undervalued, based on assessments last made 30 years ago), this rule nevertheless puts a damper on the elderly's keenness and ability to convert their empty floors for intergenerational housing. In the name of social ties and today's extremely critical housing crisis, isn't there a way to change the norm somewhat? In the same vein of taxation, and even farther upstream, we can also point to the fact that every sort of building repair that requires a building permit leads to a site visit by the federal officers responsible for bringing the cadastral income assessment in line with current values. (This, amongst other things, explains the massive tendency to engage in undeclared construction projects.) Again, although understandable in a "normal" context, this regulation is likely, however, to be overly severe in the case of intergenerational housing.

Now, as regards the purely leasing aspects, and provided that intergenerational housing is institutionally consecrated one day, it is imperative that the lease joining elderly landlords with their young renters should come under the federal law of 20 February 1991 rather than the specific laws governing social housing. This is because in ordinary law the contract effectively *does not end* upon the death of the lessor (which, however, causes some difficulties for the heirs), whereas the social leasing covenant, which is drawn up in direct consideration of the lessee (*intuitu personae*), ceases to exist at the same time as the beneficiary. Still regarding the rental relationship *per se*, we note that certain kinds of property (furnished, less than an area of 28m2) may be rented only subject to previous approval by the authorities in the form of a "rental permit" attesting that the property meets certain quality requirements. Now, obtaining this official document can create considerable administrative headaches for the senior citizen as well as often revealing more or less severe problems of insalubriousness. Thus there is the risk that the senior will be dissuaded from pursuing the idea of a kangaroo house.

Generally, any rental relationship is certain to spawn a host of troubles for the lessor (late or unpaid rent, property damage, etc), all manner of vicissitudes that senior citizens could decide, with good reason, that they cannot face, even to the point of depriving themselves of rental income. To cope with this state of affairs in particular, in which certain people prefer to leave their properties empty rather than "take the risk" of renting it out, the public authorities decided some ten years ago to subsidise social real estate agencies ("A.I.S." in French). Their mission consists of convincing the owner of a vacant property to turn over its management to them in order to rent it out to people with a "needy" profile. In exchange, the A.I.S. promises to cover any possible rental arrears on the part of the lessee as well as to guarantee a continuous rental payment during the inevitable periods of vacancy between two leases. Moreover, this new type of real estate agency is bound to restore the property to its initial condition at the end of each lease. In exchange for these various services, the owner accepts a reduced (but regular) rent. As we can see, the social real estate agency seems to be eminently suited to serve as an intermediary in intergenerational housing, all the more so as its purpose is precisely to provide renters with social services and guidance. Within the context of a potentially tense or conflict-ridden cohabitation, in which two generations live under the same roof, such guidance would seem to be more than welcome if we want to create the trust without which any rental relationship is inevitably bound to deteriorate. The recourse to the

A.I.S. is all the more necessary as often the more senior party is fairly apprehensive about the younger party. Unfortunately, the social real estate agencies balk at managing lodgings situated within buildings in which the owner continues to reside for fear that the latter will constantly interfere in the rental relationship over which the agency is supposed to have sole control. Furthermore, the regional subsidies granted to the A.I.S are not to be used for bringing the property up to standard prior to the rental. These costs, in theory, are the prior responsibility of the property owner. Yet such renovations quickly will prove to be indispensable when one aims to transform a partially unused one-family house into a building for intergenerational housing[7].

As we see, some legal adaptations (and financial incentives) are needed to allow social real estate agencies to take on this definitely leading role in the development of kangaroo housing. In this regard, the lack of enthusiasm shown by Brussels's Junior Minister for Housing a few years ago, for example, is regrettable. For her, "[I]t does not seem advisable to provide a specific allowance for the elderly who wish to rent part of their house to young couples" as "this could lead to discrimination against the elderly who do not own their homes " (Dupuis, 2005). The argument, in any case, seems inadmissible to us, in that it favours a certain type of levelling down that ultimately results in the *status quo*.

CONCLUSIONS

Far from being utopian, the concept of intergenerational housing is already solidly implanted abroad, whether in Scandinavia, the Netherlands, or Australia ("Grannyflats") in particular. Furthermore, this idea is spreading throughout Belgium, in the form of actual projects in the Brussels Region such as the Jouet-Rey housing estate in Etterbeek (housing managed by the non-profit association "Les trois pommiers" that combines an old age home, a shelter for young mothers in need, and a residence for the people with mild disabilities) or in Molenbeek (where for the past twenty years the association"Dar al amal" has continued to refine the scheme since, in addition to mixing generations and socio-economic levels, it also combines different cultures by putting a Belgian household and an immigrant family under the same roof)[8] or again in Laeken, where a social housing company plans to transform part of its housing stock into an intergenerational housing project ("Versailles Seniors" project (Cherbonier). As we see, integration via housing is resolutely multiform (Fondation Roi Baudouin). Wallonia, for its part, is not to be outdone, (rural areas are threatened with decreasing social interactions even more so than urban areas) with, amongst others, experiences in Hannut (fourteen residences) and Opprebais (twenty-two residences) (Fondation Rurale de Wallonie). As a sign in any event of its fertility, the idea behind intergenerational housing has been applied at times to sectors other than housing alone. For example, the Saint Ignatius Home (Maison Saint-Ignace) in Laeken decided to pair a senior residence with a day care centre. In addition, a survey conducted in 2009 in the Walloon Region revealed that no fewer less than 20% of the people older than 55 would "probably" or

---

[7] Provided, of course, that the services of an A.I.S. are used, the landlord is authorised to claim a renovation allowance, but experience shows that this forced alliance puts off some owners.

[8] We note as well that in Mediterranean cultures for example, the elderly often enjoy more prestige than in Western civilisation.

"certainly" migrate towards group housing (Dedicated Research), which indicates the potential for this type of housing. In passing, the presumed reticence of seniors to share their living space takes quite a beating.

As overwhelming as they seem for the time being, the legal obstacles should not obscure the fact that *other hurdles*, especially architectural ones, loom on the road to intergenerational housing. Effectively, one must develop dwelling units that are both autonomous (this is not collective housing strictly speaking) and preserve sufficient connecting areas between the households so as to make the exchange of services possible. Furthermore, architects should learn to think more about the building as a progression, so that a house can be adapted to family variations and take on a different configuration during each phase of the family life cycle. In other words, housing would benefit from looking ahead to the foreseeable future upheavals that the nuclear family may go through. Therefore, the building should be designed in a modular spirit *from the very beginning*, at the moment of construction: moveable partitions, preliminary plumbing and electrical hook-ups in certain rooms, etc., (Friedman). Besides which, at this time, when environmental demands (in terms of mobility, pollution, and energy consumption) require a certain increasing housing density, it is truly time to rationalise, that is to say, optimise, the current use of residential buildings. By maintaining the proper balance between isolation (traditional in individual housing) and lack of privacy (inevitable in an integrated housing community), the intergenerational housing formula offers a system of freedom and availability that the elderly as well as young households plainly seem to value. The former manage to stay in their own home for a much longer time than if they were alone, whereas the latter find the means to reconcile work and family life more easily. Paradoxically, one must therefore accept limitations on one's independence in order to strengthen one's autonomy (Bernard, 2008).

Be that as it may, it is vital that this budding solidarity brought about by intergenerational housing should not be penalised. Whilst we certainly cannot demand everything from the public authorities in terms of social intervention, at the very least we must expect that they will not destroy the informal mechanisms of mutual aid that are emerging at the core of civil society. Besides, this would be especially unfortunate on their part, as they remain powerless to eliminate poverty structurally. In line, then, with the first principle of the Hippocratic oath ("Above all, do no harm"), they must avoid ruining collective housing initiatives by enacting untoward regulations . This is not at all a passive approach for all that, as it presumes that the authorities will have to take positive action to modify the different regulations that currently hobble the launching of interdependent housing schemes. Of course, solidarity cannot be decreed, but when a symmetrical supply of and demand for intergenerational housing exists, as they do today, the politician's first duty is to have them meet (Bernard, 2007).

REFERENCES

Bernard, N. (2005). J'habite donc je suis. Pour un nouveau rapport au logement. *Brussels: Labor.*

Bernard, N., MIGNOLET, D., THYS, P., & VAN RUYMBEKE, M. (2007). L'habitat groupé pour personnes en précarité sociale : et si on arrêtait de pénaliser la solidarité ? *Échos du logement*, 5, 1-16.

Bernard, N. (2008). Le logement intergénérationnel : quand l'habitat (re)crée du lien social. *La Revue nouvelle*, 2, 67-76.

Bonvalet, C., Ogg, J., Drosso, F., Benguigui, F. & Mai Huynh, P. (2007). Vieillissement de la population et logement. Les stratégies résidentielles et patrimoniales, Population ageing and housing. *Residential strategies and asset management.* Paris: La Documentation française.

Braeckman, D. (2004). Les aînés dans la ville. Vieillir à Bruxelles. Brussels: *Groupe Ecolo du Parlement bruxellois.*

Charlot, V. & Guffens, C. (2006). Où vivre mieux ? Le choix de l'habitat groupé pour personnes âgées . *Brussels: Fondation Roi Baudouin.*

Cherbonnier, A. (2007). L'habitat groupé. *Bruxelles Santé*, 48, 9-16.

Dedicated Research. (2009). Étude du parcours résidentiel des personnes âgées de plus de 55 ans en Région wallonne. *Namur: Ministère de la Région wallonne.*

Dossogne, I. & Simon, C. (2007). Au temps de la retraite, emménager dans un habitat groupé. *Brussels: Éducation permanente.*

Dupuis, F. (2005). Bulletin des interpellations et des questions. *Documents du Parlement de la Région de Bruxelles-Capitale*, 86, 14-21.

Dupuis, F. (2006). Compte-rendu intégral. *Documents du Parlement de la Région de Bruxelles-Capitale*, 68, 19-26.

Fondation Rurale De Wallonie. (2006). Le logement intergénérationnel. Une solution pour améliorer la qualité de vie des seniors et des jeunes ménages. *Cahiers de la Fondation rurale de Wallonie*, 3, 1-8.

Friedman, A. (2006). La maison évolutive et abordable. *Échos du logement*, 5, 37-41.

Hochart, I. (2005), Vieillir à Bruxelles, un défi d'avenir. *Bruxelles en mouvements*, 141, 2-4.

Leleu, M. (2000). Les nouveaux vieux. Portrait des personnes âgées de plus de 50 ans vivant dans la Région bruxelloise. Brussels : Commission communautaire française de la Région de Bruxelles-Capitale.

Noël, J., & Perdrix, M.-C. (2007). L'habitat groupé, un mode de vie attirant. In Habitat et âges de la vie. L'habitat groupé, d'une aspiration personnelle à une politique publique ? Grenoble : Les ateliers de la citoyenneté.

Observatoire de la Santé et du Social (2007). Vivre chez soi après 65 ans. Atlas des besoins et des acteurs à Bruxelles. Brussels : Commission communautaire commune de la Région de Bruxelles-Capitale.

Van de Putte, F., Leroux, V., & Inghels, E. (2007). Nursing homes Belgium. Brussels : DTZ Research.

Vanderhaegen, J.-Chr. (2003). Handicaps et vieillissement démographique: des défis pour la ville, Bruxelles : Confédération de la construction.

Vanderhaegen, J.-Chr. (2006). L'avenir de nos seniors et la création de leurs logements: défis et opportunités. Bruxelles: Confédération de la construction de Bruxelles.

In: Built Environment: Design, Management and Applications   ISBN: 978-1-60876-915-5
Editor: Paul S. Geller, pp. 319-333   © 2010 Nova Science Publishers, Inc.

Chapter 14

AUGMENTED REALITY: A NOVEL APPROACH TO INTERFACING BUILDING INFORMATION MODELLING SERVERS FOR DESIGN COLLABORATION

Ning Gu and *Xiangyu Wang***

¹School of Architecture and Built Environment, The University of Newcastle, Australia
²Design Lab. Faculty of Architecture, Design and Planning,
The University of Sydney, Australia

ABSTRACT

Building Information Modelling (BIM) enables intelligent use and management of building information embedded in object-oriented CAD (Computer-aided Design) models. The widespread use of object-oriented CAD tools such as *ArchiCAD*, *Revit* and *Bentley* in practice has generated greater interest in BIM. A number of BIM-compliant applications ranging from analysis tools, product libraries, model checkers to facilities management applications have been developed to enhance the BIM capabilities across disciplinary boundaries, generating research interests in BIM servers as tools to integrate, share and manage the model developed in distributed collaborative teams. This chapter presents a computational model for applying Augmented Reality (AR) as a novel approach to interfacing BIM servers for design collaboration in the architecture, engineering and construction (AEC) industries. The chapter firstly elicits technical requirements for a BIM server as a collaboration platform based on: (1) a case study conducted with a state-of-the-art BIM server to identify its technical capabilities and limitations, and; (2) an analysis of features of current collaboration platforms used in the AEC industries. The findings are classified into three main categories including: (1) BIM model management related requirements; (2) design review related requirements; and (3) data security related requirements. An integration framework is then proposed for adopting AR as the primary interface for a BIM server, supporting the above technical requirements for design collaboration in the AEC industries. The adoption of AR opens

---

* Corresponding author: E-mail: ning.gu@newcastle.edu.au
** E-mail: x.wang@arch.usyd.edu.au

up new opportunities for exploring alternative approaches to data representation, organisation and interaction for supporting seamless collaboration in BIM servers.

Keywords: BIM, Augmented Reality, Design Collaboration, Technical Requirements.

1. INTRODUCTION

Building Information Modelling (BIM) is an advanced approach to object-oriented CAD (Computer-aided Design), which extends the capability of the traditional CAD model by defining and applying intelligent relationships between building elements in the model. BIM models include both geometric and non-geometric data such as object attributes and specifications. The built-in intelligence allows automated extraction of 2D drawings, documentation and other building information directly from BIM models. This built-in intelligence also provides constraints that can reduce modelling errors and prevent potential technical flaws in the design, based on the rules encoded in the software (Ibrahim et al. 2003, Lee et al. 2006). Most CAD packages have adopted the object-oriented approach with certain BIM capabilities. A number of supporting applications have emerged that can exploit the information embedded in the BIM model for model integration, design analysis, error checks, facilities management (FM), and so on (Khemlani 2007a). The emergence of multiple applications with the ability to directly use and exchange building information between them provides opportunities for enhanced collaboration and distributed project development. BIM has been increasingly considered as an IT (information technology) enabled approach that allows design integrity, virtual prototyping, simulations, distributed access, retrieval and maintenance of building data. Hence, the scope of BIM is expanding from the current intra-disciplinary collaboration through specific BIM applications to multi-disciplinary collaboration through a BIM server that provide a platform for direct integration, storage and exchange of data from multiple disciplines. A BIM server is a collaboration platform that maintains a repository of the building data, and allows native applications to import and export files from the database for viewing, checking, updating and modifying the data. In general, a BIM server by itself has limited built-in applications. BIM servers are expected to allow exchange of information between all applications involved in a building project life cycle including design tools, analysis tools, FM tools, document management systems (DMS), and so on. In principle, BIM servers aim to provide collaboration capabilities similar to DMS. However, while DMS is meant for collaboration through exchange of 2D drawings and documents, BIM servers provide a platform for the integration and exchange of 3D building data with embedded intelligence.

While the potential benefits of BIM servers in design collaboration may seem evident, the adoption rate of the technologies has been rather lethargic. The literature attributes such slow uptakes to both technological and cultural issues. With the aims to identify and address these issues, this chapter presents an integration framework for applying Augmented Reality (AR) as a novel approach to interfacing BIM servers for design collaboration in the architecture, engineering and construction (AEC) industries. The chapter firstly elicits technical requirements for a BIM server as a collaboration platform based on: (1) a case study conducted with a state-of-art BIM server to identify its technical capabilities and limitations, and; (2) an analysis of features of current collaboration platforms used in the AEC industries.

The findings are classified into three categories including (1) BIM model management related requirements; (2) design review related requirements; and (3) data security related requirements. An integration framework is then proposed for adopting AR as the primary interface for a BIM server, supporting the above technical requirements for design collaboration in the AEC industries. The proposed integration framework will enhance BIM server research and development to better facilitate the adoption of the technologies, supporting greater design collaboration supports in the AEC industries.

2. BACKGROUND

2.1 BIM Adoption

Based on the current literature (Johnson and Laepple 2003, Bernstein and Pittman 2004, Holzer 2007, Khemlani 2007b, Howard and Bjork 2008) and focus group interviews (FGIs) with key industry players, we have identified the main factors that affect BIM adoption (Gu et al. 2008, 2009). The perception and expectation of BIM against the industry's current practice are summarised in terms of the following three main aspects: tools, processes and people.

1. Tools: The expectations of BIM vary across disciplines. For the design discipline, BIM is mainly an extension to CAD, while for non-design disciplines such as contractors and project managers, BIM is more like intelligent DMS that can take-off data directly from CAD packages. While there are evident overlaps, BIM application vendors seem to be aiming to integrate the two requirements. The existing BIM applications are not yet mature for either purpose. Users with CAD backgrounds, such as designers, expect BIM servers to support integrated visualisation and navigation that is comparable to the native applications they use. Users with DMS backgrounds, such as contractors and project managers, expect visualisation and navigation to be the important features of BIM servers that are missing in existing DMS solutions.

2. Processes: BIM adoption would require a change in the existing work practice. An integrated model development needs greater collaboration and communication across disciplines. A different approach to model development is necessary in a collaborative setting where multiple parties contribute to a centralised model. Standard processes and agreed protocols are required to assign responsibilities and conduct design reviews and validation. Experience from DMS will be useful for data organisation and management, but organisations will need to develop their own data management practices to suit their team structure and project requirements. Different business models will be required to suit varied industry needs. A BIM model can be maintained in-house or outsourced to service providers. In the latter case, additional legal measures and agreements will be required to ensure data security and user confidence.

3. People: New roles and relationships within the project teams are emerging. Dedicated roles, such as BIM model manager and BIM server manager will be inevitable for large scale projects. Team members need appropriate training and

information in order to be able to contribute and participate in the changing work environment.

In summary, as BIM matures it is likely to integrate the existing CAD packages and DMS into a single product. For BIM to succeed and be adopted widely in the industry all stakeholders have to be informed about the potential benefits to their disciplines. Earlier research shows that (1) the lack of awareness, (2) the over-focus on BIM as advancement of CAD packages only, and (3) the relative downplaying of BIM's document management capabilities have inhibited the interest of non-design disciplines of the AEC industries in BIM adoption. A user-centric BIM research has to be more inclusive, since the success of BIM adoption lies in collective participation and contribution from all the stakeholders in a building project. For our research, the above understanding of BIM adoption highlights that building data representation and interaction is one of the most important aspects of BIM. Like general CAD applications, current BIM servers lack intuitive and flexible interfaces for supporting customised needs from different disciplinary users and the effective collaboration among them. This has become most critical as BIM servers utilise much larger and more complex data sets to be shared by a wider range of users across all AEC disciplines, compared to a general CAD application that often serves a single discipline only.

2.2 Augmented Reality in Design Collaboration

In building projects, the difficulties in information sharing between different disciplines often result in delays in decision-making as well as errors in project outcomes. This section introduces the current applications of AR that aim to improve interdisciplinary design collaboration. Several attempts have been made to address this issue by implementing knowledge integration and sharing from different disciplines in AR-based environments. These AR-based environments can allow multiple disciplines to contribute and interact with the digital building information in a seamless and intuitive manner during design collaboration. For instance, the FingARtips project (Buchmann et al. 2004) applies AR technologies. The users can then wear head-mounted displays or virtual glasses to view digital information that is overlaid onto physical objects in the real world. However, this application is limited by the number of concurrent users the system could technically support at a given time. The high cost of equipments for individual users also make the application infeasible to be used by large-size interdisciplinary design teams. Furthermore, the tracking of the system is inaccurate - it is very difficult to have the system thoroughly evaluated for accuracy and precision. Another example of such is a MetaDESK called Tangible Geospace (Ishii and Ullmer 2006) which is essentially a Tangible User Interface (TUI) system. It has been applied in the field of urban planning. This application aims to support synchronous design collaborations in a face-to-face setting. In this system, several physical objects are used to support designers' tangible interactions with the site. These objects are placed on a translucent holding tray on the metaDESK's surface. During the collaboration process, by placing a miniature model (phicon), such as the Great Dome of the MIT campus onto the metaDESK, a 2D digital map will appear underneath the metaDESK bound to the Great Dome miniature at its actual location on the map. Simultaneously, the arm-mounted active lens will display a 3D view of the MIT campus with the Great Dome and its surrounding buildings in perspective. The main function here is for exploring existing designs, the prospects for supporting interdisciplinary collaboration has not been examined.

3. TECHNICAL REQUIREMENTS FOR A BIM SERVER

The lack of industry experience in the use of BIM servers means there is limited feedback from industry on their technical requirements (Gu et al. 2009). The technical requirements for a BIM server presented in this chapter have been developed by adopting a combined research method that involves the following two steps:

- Case study using a state-of-the-art BIM server: A state-of-the-art BIM server is tested on a real-world project with the following research objectives: (1) to test the current functionalities, usability and limitations of the BIM server as a collaboration platform; and (2) to use the results of (1) as a benchmark to propose features and technical requirements for a BIM server that will address the adoption issues as discussed above.
- Analysis of document-based collaboration platforms in the AEC industries: A review of current document-based collaboration platforms and their use in the industry is conducted to identify features and technical requirements that users may expect in a BIM server. This enables us to extract essential and relevant functions of collaborative technologies that are currently in use, to be integrated into the ideal BIM server environment.

3.1 Case Study Using a State-of-the-Art BIM Server

For the case study, a renovation project within an Australian landmark building is chosen. A BIM server: *EDMmodelServer™* and relevant BIM applications are used for space renovation and re-functioning of a service chamber in the building. Therefore, the existing building data, such as the original design drawings, the existing infrastructures in the renovating space and its spatial relationships with other surrounding spaces become important and increase the complexity of the project. These factors have been considered and respected in the case study.

The two main tasks of this case study are (1) the construction of different discipline-specific models and (2) the integration of the models as an integrated BIM model using *EDMmodelServer™*. Three disciplines involved in the case study are architecture, hydraulics, and lighting. Applications used for constructing these discipline-specific models include *ArchiCAD* for the architectural model and *DDS-CAD* for the hydraulic and lighting models. Other applications used within the case study include: *Solibri Model Checker, Solibri Model Viewer, DDS-CAD Viewer* and *Octaga Modeller* (a plug-in for *EDMmodelServer™* for 3D model viewing).

The case study tests a wide variety of issues including building data visualisation, analysis and collaboration, specific features tested are (1) object attributes in the discipline-specific models; (2) intelligent relationships within a discipline-specific model and between different models in the integrated BIM model; (3) data representation, visualisation and access functions; (4) analysis functions that focus on design analysis and model evaluation; (5) project collaboration and communication functions.

In general, the BIM server has been found useful for design collaboration in the case study. The BIM server has well developed features for data upload, model integration and information extraction such as documentation and report generation. However, some technological and implementation issues are identified during the case study, which are likely to scale up and develop into a major roadblock in adopting the server for more complex projects. These issues are primarily related to:

- **Set-up and access to the BIM server:** The model server allows defining access rights and permissions based on participant roles and responsibilities. In the case study few participants were involved. Even in the simplified scenario, interventions were required midway through the project and personal meetings were organised to coordinate activities. In a full-scale project, roles and responsibilities not only are likely to increase in complexity, but also overlap. Thus, it will be useful to methodologically identify role dependencies and responsibilities, which are critical to the set-up and access to the integrated BIM model.

- **User interface:** The user interface (UI) of the BIM server is reported to be complex and confusing. Users, who have extended experience with native disciplinary applications (e.g. CAD tools), find the BIM server interface non-intuitive. Users from different disciplinary backgrounds may have different usage patterns, which will require different standard interfaces. Currently, different data and information are shown and limited in one single window. A flexible, user-friendly UI that allows customisation to suit different user profiles is required. 3D visualisation of the plug-ins for viewing the BIM model is useful. However, further tests are required for project with larger data sets.

- **Data management and modification:** In the BIM server, objects and its operations are well defined. However, during model integration, object duplication or model conflict may occur especially when the same object is created in parallel in different discipline-specific models.

- **Extended functionalities for project communication:** Some other features and technical requirements reported as wish-lists in an earlier study (Gu et al. 2009) were reiterated, with most suggestions related to improving project communication. For example, although the BIM server allows object tagging and change notifications, if an object is tagged on the BIM server, and the tagged model is downloaded, the tagged message is not downloaded. The user needs to manually check and update the tagged object separately in the native disciplinary model. Flexibility to download the appended project communication records along with the model may be useful. It was also suggested that the BIM server should provide more synchronous communication features such as an embedded instant messenger for the easy and clarity of project communication.

- **Help function and tutorials:** Although the participants in the case studies had some training and familiarity with the BIM server, during the case study some difficulties were reported in tool usage and technical support. Participants emphasise the need for improved help functions and tutorials. There are suggestions for open-source training materials where users can learn form each others' experience.

3.2 Analysis of Document-Based Collaboration Platforms in the AEC Industries

The analysis includes the examination of existing online collaboration platforms, primarily DMS, such as *Incite*, *Aconex*, *TeamBinder* and *Project Centre*, along with inquiries and interviews with industry partners who have adopted the technologies in their practices. The review was conducted to understand the implementation and application of web-based collaboration platforms in the AEC industries. Such collaborative practices exist within the industry and therefore, they may act as a gauge for using BIM servers as CPs. The analysis suggests the following issues that should be considered when implementing BIM servers as collaboration platforms in the AEC industries.

- The initial collaboration platform set-up is complex. Various dependencies within the process and between activities and people need to be identified before its operation. The complexity in setting up a BIM server as a collaboration platform for building project development may be even greater because model-based data exchange will require greater coordination owing to larger size of the data set, varied file formats and tool compatibility issues.

- There are various levels of DMS usage. In some projects, DMS is used across the entire project lifecycle, involving most of the project participants. In other cases, only some of the project participants coordinate their activities through DMS, or the DMS is used in specific stages only. This scoping of DMS as a collaboration platform is usually conducted at the initial phases of the project. Such scoping is also critical for BIM server adoption.

- In the current practice, a customised project instruction document is generally developed to serve as a guide for the project operation. This ensures that as the project develops and the team dynamics change, key terminologies and standard procedures are agreed and complied by all participants.

- In general, DMS automates the process of uploading, validating, approving and distributing documents. A series of business rules encoded within the applications at the start of the project automates the decision making such as which folder to upload documents to, how to validate the document, who to distribute documents to, and so on. The inbuilt intelligence in form of business rules and distribution matrix would require knowledge elicitation from project partners.

- In practice, once the protocols are established, the DMS is configured for the project. The administration team and project partners are trained, before the system goes online.

- The project administrator requires in-depth knowledge of the required document flow process as well as basic configurations and user requirements on the DMS. In general, a company administrator is appointed to coordinate with the project administrator and has access rights similar to the project administrator.

- Communication is a critical part of all DMS. Most DMS provide multiple modes of communication including instant messages, SMS texts, emails and voice communications. They support both synchronous and asynchronous communications, and are used for multiple purposes, such as direct project communication, documentation, as well as sending notifications, reminders and clarifications.
- Training programs are available for all functional levels, from the standard tools that all project teams will use, to the more advanced construction project management tools such as workflows and tender modules. Help is provided in form of manuals, technical support and video demonstrations.

3.3 Technical Requirements for a BIM Server

Based on the case study and the analysis above, we have broadly grouped the features and technical requirements for a BIM server into three categories as shown in Table 1.

Table 1. Features and Technical Requirements for a BIM Server

Features and Technical Requirements
(a) BIM model management related requirements
BIM model organisation • Model repository • Sub-models, and objects with different levels of details • Public and private model spaces • Globally Unique Identifier (GUID) for all object data • Information Delivery Manuals (IDM) based specifications
Model access and usability • Hierarchical model administration structure • Download/upload model, and check- in/ check-out/ check-out with lock • Version lock and archiving • Model viewing options • Documentation and reports
User Interface (UI) • Customisable interface • Online real-time viewing, printing and markups • On-click object property check and modification
(b) Design review related requirements
Design visualisation and navigation
Team communication and interaction
(c) Data security related requirements
• Features supporting confidentiality, integrity, and availability • System security • User authentication • Data security • Access control • Encryption

- **BIM model management related requirements:** These features and technical requirements are directly related to the storage, operation and maintenance of the BIM model that contains 3D geometries, 2D documents, and other related building information.

- **Design review related requirements:** These features and technical requirements are specifically related to design review activities, including various functions needed for design visualisation and navigation, as well as team communication and interaction.

- **Data security related requirements:** These features and technical requirements are related to network security and the prevention of unauthorised access into the system.

Our research focuses on the features and technical requirements needed during the usage of the BIM server in direct support for a collaborative design project. Therefore, various support technical requirements that are expected to facilitate and assist the set-up, implementation and usage of the BIM server, are not considered in the chapter.

3.3.1 BIM model management related requirements

BIM model organisation: Features and technical requirements for model management and organisation should include the following.

- **Model repository:** A BIM server should provide a centralised data repository for the building project. This data repository can be linked to other federated data repositories to increase data capacity and efficiency of the server.

- Hierarchical model structure: A BIM model on a server is organised in a hierarchical structure. For example, at present the model-tree in *EDMmodelServer™* has the following hierarchy: project > site > building > building storey. However, users may want a different model structure based on their requirements, e.g. a client may want to group projects within a site rather than the other way round, i.e. site > project > building > building storey, and so on. Thus, BIM servers should support the flexibility to customise the model structure.

- **Sub-models, and objects with different levels of details:** the BIM server should provide the ability to *map objects with different levels of details*. For example, if level 1 detail only shows a rectangular volume for a room, and level 2 detail of same volume may show all openings and doors and windows. Users should be able to navigate and switch across different views through simple functions (toolbars) and shortcut keys. In order to support such functionalities, mapping of objects with different levels of details will be required.

- BIM server should support the ability to store and present objects of the model as text-based information in repositories, and link 3D objects with the model viewer. Choosing object in one window (text-based model tree or 3D models in the model viewer) should highlight the corresponding data in the other window.

- Object and model history, such as ownership and modification records, should be maintained in the data repository.

- Object property: The BIM server should provide the ability to *overlay additional object properties*, if customised object property is desired. For example, Quality of survey may not be a default object property, if this is the case, in the BIM server this can be an overlayed property linked with each object. Technical issues may arise if the data is downloaded and uploaded again. Additional technical measures will be required to deal with such issues.

- **Public and private model spaces:** BIM server should allow differentiation and management of *public and private model*. Public model is accessible to all users with access rights. Private model could be model in progress, but not ready to be shared with others. However private models may be shared with a select group of users.
- **GUID and IDM:** Globally Unique Identifier (GUID) allows each object to be uniquely identified, preventing duplication. Information Delivery Manuals (IDM) specifications and approaches for connecting the BIM approach with business processes. BIM servers should enable efficient integration of GUID and IDM to deal with problems encountered in merging different discipline-specific models.

Model access and usability features: Features and technical requirements related to model usability are explored below.

- **Hierarchical model administration structure:** BIM server administration deals with management and allocation of model access rights, data control and security. Typically, hierarchical administrative structures exist across distributed teams and in large organisations to manage day-to-day local and global issues. Thus, BIM sever should allow administrative structures that reflect and support existing organisational practices.
- **Download / Upload model:** Various modes of interaction for model download (upload) are possible to include download (upload) buttons as well as drag and drop options. Ability to download (upload) data straight to (from) an email account, which is possible in existing DMS when dealing with documents, will be useful.
- **Check-in / check-out and version lock:** *Check-in* options should allow addition of new partial model or merging with existing model. Again, different modes of interaction are possible to include buttons and drag and drop capabilities. Similarly, *check-out* options should allow download of complete model or partial model using different modes of interaction. A *check-out with lock* feature should be provided to notify other users that the checked-out data has been locked and deemed not usable. A *version lock* feature should be provided to lock version of the model after sign-off, as a form of archiving.
- **Model viewing options:** BIM servers should support the ability to *capture and save screen shots, which is a standard functionality provided across CAD packages*. Other features, such as the option to choose the *level of detail for viewing* should be available through toolbars and shortcut keys, i.e. sub-sets should be managed such that users can choose the level of detail for viewing by selecting options on a checklist, e.g. conceptual block model, space layout model, etc.
- **Documentation and reports:** When downloading a part model from the BIM server, there should be options to *generate reports* on parametric, linked, and external information for selected objects and other objects in the rest of the model. This information can be in form of a checklist, where users can choose to get details of only those objects they intend to modify, delete or replace. Ideally, a facility to

append this information to objects will be helpful, but that would be useless until the native applications can receive those additional data.

- BIM servers should support the ability to *generate and export PDF* or other document formats. This capability also allows direct offloading of ready to use information to DMS, in which case some users may not need to access the BIM server. They can continue interacting with DMS as they have been practising at present.
- BIM servers should support the ability to *integrate information from product libraries*. It should be possible to create a comparison report for alternative product options.
- Features should be provided to *validate rules* while uploading the files on the BIM server. Users should have the option to switch validation check on or off.
- Technical provisions are required for *data ownership transfer* and handover in a BIM server environment. These functionalities should account for security measures to deal with such change of hands, log-ins and passwords.

User Interface (UI): Other than the standard UI features, the BIM server interface should include: (1) model tree view position and 3D viewer position; (2) support for online real-time viewing, printing and mark-ups; (3) the ability to click on an object, and check what sub-sets it belongs to, and (4) the ability to click on an object, and switch between the sub-sets it belongs to, for another sub-set selection. Users should also be able to customise and choose the available UI functionalities.

3.3.2 Design review related requirements
The following features and requirements are related to distributed design review.

- **Team communication and interaction:** Distributed design reviews may require parallel video conferencing, virtual environments and other similar interactive media. BIM servers should be compatible with such technologies and some basic integration of BIM servers with these technologies will be useful for model navigation and viewing. Organisations may want to maintain a record of the design review interactions in the data repository. Thus, BIM servers should provide the ability to capture real time interaction data from meetings and online reviews. Some of these interaction platforms, such as instant document/ message exchange window, may be directly hosted on the BIM server environment.
- **Design visualisation and navigation:** Building projects often result in large data files, which reduce the online navigation and viewing capabilities. Hence, for effective design review across distributed teams, capabilities to create lightweight 3D data are essential. 3D Model viewers supported (provided) by BIM servers should have high data compression capabilities while maintaining the image quality. It will be useful to provide technical features that allow instant, online mark-up, and tagging on a shared document or model being viewed by design reviewers and users.

3.3.3 Data security related requirements:

Security of data on BIM server should account for confidentiality, integrity and availability of data. These features and related technical requirements are discussed below.

- **Confidentiality:** The data stored on the BIM server should be available to authorised users only, and on the need-to-know basis. This service is crucial to secure sensitive data from malicious intruders.
- **Integrity:** All BIM data should be only created, modified and deleted by authorised users having an authorised access, and should be a subject to integrity cross-checking.
- **Availability:** Data and services provided by a BIM server should be available to users when they need them. As BIM servers have a role of a central data repository serving simultaneously a range of users involved in a project, availability requirements are of greater consequence than in specialised systems.
- **System security:** BIM servers should have *user authentication* facilities to ensure that only authorised users can access it. The current BIM servers only partially satisfy the security requirements. While they would typically provide for internal user authentication in form of log-in and password, they fail to provide other forms of authentication. Moreover, if sufficient password management is not put in place, by the system and users alike, such authentication procedures may be regarded as inadequate.
- **Data security:** BIM server should provide effective *data access control*, with access privileges to individual pieces of data, including create, delete, read, write and execute. While access control is typically implemented by the tested BIM server, it does not provide for fine granularity. It would be desirable for BIM servers to adopt the Role Based Access Control (RBAC) (Sandhu et al. 1996), which is already used by leading DMS. RBAC allocates access privileges to *roles*, rather than users, which greatly simplifies the privilege management task, especially in such a dynamic environment as BIM. The BIM data should be protected by means of *encryption*, both when stored on the system and transferred over a network. This feature is available in major DMS, it should be adopted by BIM servers.

4. AN INTEGRATION FRAMEWORK FOR INTERFACING BIM SERVERS USING AR

The technical requirements for a BIM server as discussed in Section 3 highlights the urgent need for a new interface that can utilise the list of the requirements. Among the list of these technical requirements, the core is the enhanced support for interdisciplinary building data representation, organisation and interaction. We propose the following integration framework for interfacing BIM servers using AR. AR enables a mixed-reality design environment that supports both face-to-face and distant collaboration. In an AR environment, building information can be generated and represented as 3D virtual models; the models can then be viewed and interacted virtually, or extracted and inserted into the multi-disciplinary users' views of a real-world scene (i.e. the actual building site). With TUI, AR can also support tangible interactions with digital building information beyond the use of computer screen, mouse and keyboard, to allow more intuitive interactions such as gestures and body movements that are commonly used in traditional collaboration scenarios. The coupling of BIM and AR has opened up new opportunities for exploring alternative approaches to data representation, organisation and interaction for supporting seamless collaboration in a BIM server.

A BIM server supports multi-disciplinary collaboration by providing a platform for direct integration, storage and exchange of information from all payers involved in a building project. A BIM server is a collaboration platform that maintains a repository of the building information, and supports the viewing, checking, updating and modifying of both the geometric and non-geometric data. The integration framework as shown in Figure 1 highlight the main elements and relationships when interfacing a BIM server using AR.

The first aim of the integration is to support a wide range of activities occurred during building project collaboration including the viewing, checking, updating and modifying of building information whenever required, accurately, easily and securely. As shown in Figure 1, the BIM model server therefore will need to satisfy the three sets of technical requirements developed earlier in this study including BIM model management related requirements, design review related requirements, and data security related requirements. The second aim of the integration is to facilitate different levels of collaborations within a building project, through an intuitive interface. As shown in Figure 1, the AR-interfaced BIM server will need to provide a *private space* for intra-disciplinary collaboration where users can interact with disciplinary- specific building information. The BIM server will also need to provide a *public space* for multi-disciplinary collaboration where users can interact with integrated building information across different disciplines.

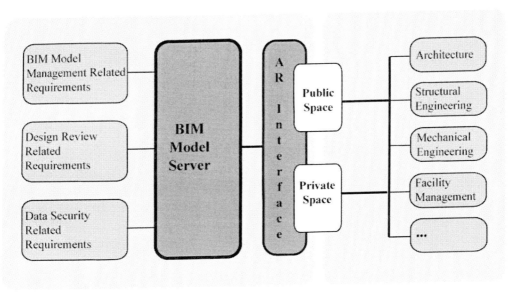

Figure 1. The relationship between the elements.

5. SUMMARY

This chapter firstly presents the technical requirements for a BIM server as a collaboration platform based on: (1) a case study conducted with a state-of-the-art BIM server to identify its technical capabilities and limitations, and; (2) analysis of features of current collaboration platforms used in the AEC industries. The findings are classified into three main categories including (1) BIM model management related requirements; (2) design review related requirements; and (3) data security related requirements. An integration framework is then proposed for adopting AR as the primary interface for a BIM server, supporting the above technical requirements for design collaboration in the AEC industries. The proposed framework provides new opportunities in BIM research that aims to explore alternative approaches to data representation, organisation and interaction for supporting seamless collaboration in the AEC industries. Through case studies and direct on-site observations, the future research directions can include (1) technical development to test and refine the framework for better integrating the AR technologies in BIM servers; (2) articulation and examination of suitable business cases for adopting the AR-interfaced BIM servers.

6. REFERENCES

Aranda-Mena, G. & Wakefield, R. (2006). "Interoperability of building information - myth of reality?", In: Martinez & Scherer (Eds.) *eWork and eBusiness in Architecture, Engineering and Construction*, London: 127-133.

Bernstein, P. G. & Pittman, J. H. (2004). "Barriers to the adoption of Building Information Modeling in the building industry", *Autodesk Building Solutions*.

Buchmann, V., Violich, S., Billinghurst, M. & Cockburn, A. (2004). "FingARtips: Gesture-based direct manipulation in Augmented Reality", *Proceedings of the 2nd International Conference on Computer Graphics and Interactive Techniques*, 212-221.

Eastman, C., Sacks, R. & Lee, G. (2004). "Functional modelling in parametric CAD systems", *Proceedings of G-CADS 2004*, Carnegie Mellon University, USA.

Fischer, M. & Kunz, J. (2004). "The scope and role of Information Technology in construction", *Proceedings of JSCE 763*, 1-8.

Gallello, D. (2008). "The new 'Must Have' - the BIM manager", *AECbytes Viewpoint 34*.

Gu, N., Singh, V., Taylor, C., London, K. & Brankovic, L. (2008). "BIM: Expectations and a reality check, *Proceedings of ICCCBEX11 & INCITE 2008,* Tsinghua University, China.

Gu, N., Singh, V., Taylor, C., London, K. & Brankovic, L. (2009). "BIM adoption: expectations across disciplines." In: Underwood & Isikdag (Eds.) *Handbook of Research on Building Information Modelling and Construction Informatics: Concepts and Technologies*, IGI Publishing.

Holzer, D. (2007). "Are you talking to me? Why BIM alone is not the answer", *Proceedings of AASA 2007*, University of Technology, Sydney, Australia.

Howard, R. & Bjork, B. (2008). "Building Information Modelling: Experts' views on standardisation and industry deployment", *Advanced Engineering Informatics*, 22, 271-280.

Ibrahim, M., Krawczyk, R. & Schipporiet, G. (2003). "CAD smart objects: potentials and limitations," *Proceedings of eCAADe 2003*, 547-552.

Ishii, H. & Ullmer B. (2006). "Tangible bits: towards seamless interfaces between people, bits and atoms", *Proceedings of CHI97 2006*.

Johnson, R. E. & Laepple, E. S. (2003). "Digital innovation and organizational change in design practice", *Working Paper*, CRS Center, Texas A&M University, USA.

Kalny, O. (2007). "Enterprise Wiki: An emerging technology to be considered by the AEC industry", *AECbytes Viewpoint 31*.

Khemlani, L. (2007a). "Supporting technologies for BIM exhibited at AIA 2007", *AECbytes, Building the Future*.

Khemlani, L. (2007b). "Top criteria for BIM solutions: AECbytes survey results", *AECbytes Special Report*.

Lee, G., Sacks, R. & Eastman, C. M. (2006). "Specifying parametric Building Object Behavior (BOB) for a Building Information Modeling system", *Automation in Construction*, 15, 758-776.

Sandhu, R. S., Coyne, E. J., Feinstein, H. L. & Youman, C. E. (1996). "Role-based access control models", *Computer*, 29 (2), 38-47.

Yassine, A. & Braha, D. (2003). "Complex concurrent engineering and the design structure matrix method", *Concurrent Engineering*, 11, 165-176.

In: Built Environment: Design, Management and Applications   ISBN: 978-1-60876-915-5
Editor: Paul S. Geller, pp. 335-339   © 2010 Nova Science Publishers, Inc.

Chapter 15

MAYORS IN GHANA: 'PUBLIC' SERVANTS OR AGENTS OF CAPITAL?[1]

Franklin Obeng-Odoom

Department of Political Economy, University of Sydney, Australia

ABSTRACT

The mayors in the newly elected Mills government have now been appointed to help the government to deliver on its promise of making Ghana 'better'. What is the strategy of these mayors? How does this strategy impact different groups in the city? Does this distribution matter? This brief paper attempts to answer these questions.

Keywords: Africa, Ghana, Urban Economy, Mayors, Poverty

INTRODUCTION

With 70 per cent of the urban population in slums, 20 per cent in poverty, 13 per cent officially classified as unemployed (Obeng-Odoom, 2007), the 'urban question' is one of the most difficult for the government and people of Ghana. But the urban question threatens to be even more difficult because this year, Ghana is expected to experience an 'urban explosion'[2]. As such, we should examine the 'vision' of the new mayors who have been appointed by the Atta-Mills government, as the team to bring change to Ghanaian cities. To do this, it is important to assess the policies of the mayors and examine their distributional effects.

---

[1] A more popular version of this paper is available on *City Mayors* as 'Ghana mayors beautify their cities, while urban poor suffer' (see this popular version at http://www.citymayors.com/government/ghana-mayors.html)

[2] When more people will live in cities in Ghana than in rural areas

What Is Happening?

It is not possible, in this short piece, to assess all the mayors in Ghana. So I will focus on the mayors of Accra and Kumasi, the two most populous cities in Ghana. The mayor of Accra, Alfred Vanderpuije, is an educationist; that of Kumasi, Samuel Sarpong, is a businessman/educationist/criminologist[3]. But both have similar priorities: beautifying their cities and attracting businesses to them.

Beautifying cities has become the stock-in-trade of the mayors in Ghana. The mayor of Kumasi has promised to strictly enforce the sanitation by-laws and decongest the city. Already, the *Route Operational Permit* (ROP) is being implemented. ROP is a one year renewable 'contract' between the city authorities and urban commercial drivers in which the drivers are compelled to ply only particular routes they have paid for.

On his part, the mayor of Accra has taken, perhaps, a more militant pose in his attempt to beautify the city. At a press conference in Accra, he gave notice to hawkers and people dwelling in 'illegal' structures to leave before 15th June, 2009 (Koomson, 2009). Also, he threatened to regard as 'criminals' people in the city who sleep in lorry parks because of the lack of accommodation if they do not leave their only 'homes' by 15th June, 2009. A clean city, in the wisdom of these two mayors, would attract many businesses.

As it turned out, the demolition took place, but on a slightly later date, Monday, 29 June (note that the demolition delayed, not because of consultation with hawkers, but because the mayor reportedly travelled to the USA to attend a graduation ceremony of one of his children). The 29 June demolition exercise has been described variously as 'schocking' and a day of 'grief and mourning'. But what is more shocking is the insensitivity of the mayor of Accra, who justified the exercise as 'merely enforcing the by-laws of Accra' (Gadugah, 2009). Though, the informal sector, pejoratively called 'illegal sector', contributes substantially to the city economy of both Accra and Kumasi, these mayors are disinterested in supporting the sector to function more effectively

Instead, they have embarked on a campaign to attract more formal businesses with big capital outlays. In turn, the mayors have tried to talk up 'investor confidence' in their cities. Here, they have publicly declared their plans to make Accra and Kumasi business friendly and secure. But because private capital abhors highly regulated markets, privatisation in Accra and Kumasi has been high on the mayors' agenda. Indeed the *Route Operational Permit* (ROP) in Kumasi may be seen as the attempt to fence off the 'commons' and bring into market transactions public goods. However, the privatisation of municipal services in Accra and Kumasi did not start with these new mayors. They are building on what past mayors and governments have done (see Grant, 2009 for a more detailed account of the 'globalisation' of Accra).

So the main policy thrust of the mayors of Accra and Kumasi is to continue the beautification of these cities and attract businesses to them. It is useful to analyse how these policies have affected urbanites. A thorough distributional analysis of these policies is not possible in this short piece, but we could attempt a preliminary analysis of these policies with a view of preparing the grounds for more thorough studies in the future.

---

[3] I thank Justin Kwame Wiredu of Asenta Properties, Kumasi, for this information.

Who Gains; Who Loses? Does This Distribution Matter?

Urban 'clean up' and 'decongestion' in Accra and Kumasi is a direct attack on the urban poor. This is because those areas and structures called 'filthy' and 'illegal' are the very places from which the majority of the poor and underemployed make a living. In Ghana, 7 out of every 9 people work in the informal sector, dominated by migrants (usually from northern Ghana) and children (Obeng-Odoom, 2007; Obeng-Odoom, 2009a). The failure of the city authorities to provide alternative sites for making a living, such as well positioned markets of adequate size and the disinterest to directly tackle urban poverty raise questions about the motive of these clean up exercises. Who gains when poverty abounds in cities? Who gains when the poor live in fear of ejection?

I suggest that the urban clean up exercises and the deliberate disinterest in tackling urban poverty are designed to benefit private businesses and politicians. With the former, private businesses, it is important to have a large pool of desperate underemployed and unemployed people in cities so that wage levels can be kept low and labour can be disciplined with the threat of being sacked. Such conditions are necessary for capital accumulation. For the city authorities, it is beneficial to deliberately provide inadequate markets so that there would be room for patronage. In his stimulating paper, 'Corruption in Public Institutions in Ghana', Akwasi Prempeh found that in 2005, while it was possible for city authorities in Kumasi to obtain about 86.2 billion cedis in revenue from rates, the city authorities may have deliberately reported that they only had the potential to collect 9.5 billion cedis (Prempeh, 2008). Because of such corrupt tendencies, we cannot rule out the possibility that even the *Route Operational Permit* (ROP) would be another 'window of opportunity' for the city authorities in Kumasi.

But the city authorities are not the only beneficiaries of the attack on the poor. Central government also gains. Central government typically allows the city authorities to unleash hardship on the poor and the marginalised prior to elections. But, just when the next elections are in sight, it dramatically announces its opposition to the activities of the city authorities' and orders the mayors to stop the 'beautification of the cities'. By such a messianic gesture, central government is able to gain some votes from some unsuspecting voters. But does this distribution of gains and loses matter?

The distribution of gains and losses resulting from mayoral activities in Ghanaian cities is problematic because it creates divided cities. Empirical studies by Frank Stilwell (Stilwell and Jordan, 2007) Augustine Fosu (Fosu, 2009) and Nancy Birdsall (Birdsall *et al.,* 1994) show that inequality may not be conducive for economic growth, poverty reduction and happiness. Social conflicts like peaceful workers' uprising and violent demonstrations may be prevalent in unequal cities because as opulence is displayed in the full glare of poor people, there could be a revolt. Such protests could take time off 'productive work' and so reduce the demand and supply for goods and services. Elsewhere, such discontent may make mayors unpopular and therefore may be voted out of office. But mayors in Ghana owe allegiance to every interest group except the common people they claim to serve. It may be useful to analyse why this is the case.

Mayors in Ghana: Accountable to Whom?

Mayors in Ghana are appointed directly by central government; not elected popularly by the people. According to the Local Government Act, Act 462, 1993, 70% of local councillors must be elected and only 30 per cent appointed by central government. The spirit of the law is that central government would use its 30% mandate to provide 'technocrats' in the local governments. In and of itself, this law has a noble objective because there are some professionals who could contribute to urban economic development but may not be able to the demands of political campaigns. What makes this law undesirable is that the appointees of the central government are more powerful than the elected councillors. For example, the most powerful person in the local government set up, the mayor, is appointed by the central government[4]. So mayors in Ghana are accountable to the central government, its agents, and private businesses (which may fund the activities of central government). It is therefore hard to escape the conclusion that mayors in Ghana are not 'public servants'; only 'agents' of private and state capital.

CONCLUSION

There is a case for introducing direct elections into local governance in Ghana, as a way of holding mayors in check. But, as I argued in a recent paper, *Transforming Third World Cities Through Good Urban Governance* (Obeng-Odoom, 2009b), merely introducing direct elections in local governance in Ghana would not guarantee improvement in what Manuel Castells (1982) calls 'the means of collective consumption' (like urban housing, schools and roads). Resolving the urban question is clearly difficult. It goes beyond mere elections to ensuring deep accountability to ordinary people. Perhaps, it is time to change track from state and private capitalism to democratic socialism.

REFERENCES

Birdsall, N., David, R. & Richard, S. (1994). 'Inequality and Growth Reconsidered' Paper for the Annual Meetings of the *American Economics Association*, Boston, Janaury 3-4, 1994.

Castells, M. (1982). *City, class, and power*, St. Martin's Press, New York.

Fosu, A. (2009). 'Inequality and the impact of growth on poverty: Comparative evidene for sub-saharan Africa' *Journal of Development Studies, vol. 45, no.5*, 726-745.

Gadugah, N. (2009). 'Law lecturer shocked at AMA decongestion exercise' *modernghana.com*, http://www.modernghana.com/print/224549/1/law-lecturer-shocked-with-ama-demolition-exercise.html (accessed July 1, 2009)

Grant, R. (2009). *Globalizing city. The urban and economic transformation of Accra, Ghana*, Syracuse University Press, New York.

---

[4] According to Article 243 (1) of the 1992 Constitution of Ghana, a person nominated by the president as DCE shall only be appointed after the approval of at least two-thirds majority of members of the assembly present and voting.

Koomson, F. (2009). 'AMA boss issues ultimatum to hawkers' *myjoyonline.com* http://news.myjoyonline.com/news/200906/30871.asp (accessed July 3, 2009)

Obeng-Odoom, F. (2007). 'Enhancing urban productivity in Africa'. *Opticon 1826 Journal, vol.2, no. 1.* 1-9.

Obeng-Odoom, F. (2009a). 'The Future of Our Cities' *Cities, vol 26, no.1*, 49-53.

Obeng-Odoom, F. (2009b). 'Transforming Third World cities through good urban governance: fresh evidence' *Theoretical and Empirical Researches in Urban Management*, February, *vol. 10, no. 1.*, 46-62.

Prempeh, A. (2008). 'Corruption Public Institutions in Ghana: The Case of the Property Tax Unit of the Kumasi Municipal Assembly' *modernghana.com* http://www.modernghana.com/news/193064/1/corruption-public-institutions-in-ghana-the-case-o.html (accessed 3-07-09).

Stilwell, F. & Jordan, k. (2007). Economic Inequality and (Un) happiness, *Social Alternatives, Volume 26, Number 4.*

In: Built Environment: Design, Management and Applications   ISBN: 978-1-60876-915-5
Editor: Paul S. Geller, pp. 341-344   © 2010 Nova Science Publishers, Inc.

Chapter 16

GENTRIFICATION IN GLOBALIZATION: FORCES TO BE EXPLORED IN CHINESE CITIES

Jun Wang

National University of Singapore, Singapore

GENTRIFICATION IS NOW GLOBAL

Gentrification in Chinese Cities Emerged after Its Door Was Open to the Global

In Shanghai, it is said that "residents living within the Inner Loop speak English …those residing outside the Outer Loop speak Shanghai dialect". Although a bit exaggerated, the saying reflects an emerging spatial social segregation with foreign enclaves in the central city contrasting local neighborhoods along its fringe. The first foreign enclave is the Gubei New District which was planned to house foreign expatriates. Before 2001, foreigners are only allowed to buy housing stocks called *Waixiaofang* (housing for sale to foreigners). Since developers would select privileged lands for Waixiaofang projects to ensure higher profit return, this policy might have confined foreigners to certain areas. As a matter of fact, Gubei New District became the main cluster of foreign enclaves for years. After 2001 when *Waixiaofang* was lifted, it was expected that foreigner residents would disperse and mix with the indigenous. However, things do not develop that simply.

To date, there are mainly two large clusters of foreign enclaves: one is the Jinqiao Area at Pudong District; the other refers to the core of central Puxi, which is constituted by Gubei area in Changning District, Jing'an Temple Area in Jing'an District, the area along Huaihai Road, and Xujiahui at Xuhui District. However, it must be kept in mind that, de-concentration of foreigners does not appear; instead, it is better described as a sprawling of foreign enclaves into multiple clusters, with each occupying a privileged place. Here comes a concern raised for many developing cities, the international new middle class and the "new or rehabilitated residential enclaves which they choose to colonize" (Atkinson and Bridge 2005).

It is argued that, formation of foreign enclave in Shanghai, at its core, is gentrification, which is generally described as the process by which higher status residents displace those of lower status. Superficially, many features associated with gentrification are also observed in foreign enclaves in Chinese cities: beside displacement of the poor defined by traditional gentrification, multiple rounds of displacements have been observed here as many foreign households have moved from one cluster to another, replacing the previous replacers – a change called super-gentrification in the West.

Gentrification has been one heated area since its emergence in the 60s in London. Since 1960s when gentrification was first coined by Glass (1964), the evolving process is classified into three waves (Smith 2002). What Glass portrayed has been classified to be the first wave of gentrification, when middle-class households immigrated to inner-city neighborhoods sporadically and spontaneously. The 2nd wave of "anchoring phase" of gentrification started from the 1970s, closely interwoven with the economic restructuring in Post-Fordist societies. After the 1990s, the third wave is labeled as the generalization of gentrification, "smell of gentrification" sprawls at a scale unseen before, expanding to second tier cities (Smith 2002). During the past decades, gentrification has gone through mutations in many dimensions. For instance, rehabilitation of old buildings in the inner city used to be a key feature whilst later studies find that gentrification may also be processed through newly built buildings, and may not necessarily occur in the inner city but extend to the suburbs (Slater 2006; Smith 1996); artists who used to be the pioneer gentrifiers are displaced by affluent middle class followers (Ley 2003); and in some areas, middle class families who ever displaced working class now face another round of displacement themselves by super-rich financiers (Lees 2003).

Gentrification attracts a vast body of literature, arguments on the cause of gentrification mainly approached from two perspectives – the consumption side and the production side. Studies from the consumption side put a lot of efforts on the gentrifiers and their demands, focusing on the new middle class and their alternative lifestyle in the post-modern society (Ley 1996; Zukin 1998). The exploration goes back to the youth culture in the 1960s in Western societies – the dominant event in the social history that has stimulated rebellion to traditional concept of family life in suburbs whilst blossom of counter-culture, street life and urbanity identity (Ley 1996). At the same time, the better access to profit return possessed by the professional and managerial labors in service society allow them to have conspicuous consumption style and urbane lifestyle centered around amenities tailored for distinctive tastes and styles (Karsten 2003; Zukin 1998). Housing stocks turn out to be products tailored for alternative identities.

Studies that approach from the production side argue that gentrification, in its nature, is driven by the logic of capital flow, since inner city neighborhoods which have suffered from long term disinvestment ask for lower transaction price but may have potentially high value in the market, and then offer the opportunity of profit-making through rent gap (Smith 1986, 1987, 1996, 2002; Smith and deFilippis 1999). Given the emergence of super-gentrification, it is argued that investment opportunities are driven now by super-profits on highly valued locations, rather than by comparisons with devalorized land. The change of the state's role from managerialism to entrepreneurialism (Harvey 1989, 2005) is taken as one crucial component in the 3rd wave of gentrification. Adoption of neo-liberal policy is blamed since rent gap is purposely manipulated through state's promotion of large-scale planned urban renaissance together with gutting of social welfare (Atkinson 2000; Newman and Wyly 2006; Slater 2006; Smith 2002). Smith (2002) sees neo-liberal governance as one basic dimension distinguishing the 3rd wave gentrification from its previous versions, arguing that gentrification now is taken as a global urban strategy.

Entering its 3rd wave, gentrification is now global. However, major attention is put on internal structural changes of advanced capitalist cities that influence the production and consumption theses; less effort is deployed to project the neighborhoods changes to a broader platform where global flux of capital, population and ideology interact (Rofe 2003). The latter is, however, crucial for developing cities like Chinese cities. The seemingly similar outputs challenges the literature based on the West for the distinctive urban context of Chinese cities. In this light, an investigation into gentrification in the context of global flux of capital, international elites, and values will supplement the debates through another reflection on production and consumption. Attention is also called for unexpected consequences arising from the instant city-making process in China, like displaced poor, widened social spatial segregation and dominating of imported value system.

REFERENCES

Atkinson, Rowland. (2000). "The hidden costs of gentrification: displacement in central London." *Journal of Housing and the Built Environment, 15(4)*, 307-326.

Atkinson, Rowland & Gary, Bridge. (2005). "Introduction." In Gentrification in a global context : the new urban colonialism, (Eds). *Rowland Atkinson and Gary Bridge*. London; New York: Routledge.

Glass, Ruth. (1964). *London: aspects of change*. London: MacGibbon & Kee.

Harvey, David. (1989). "From manageralism to entrepreneurialism: The transformation in urban governance in late Capitalism." *Geografiska Annaler, 71(1)*, 3-17.

Harvey, David. (2005). *A brief history of neoliberalism*. Oxford; New York: Oxford University Press.

Karsten, Lia. (2003). "Family gentrifiers: challenging the city as a place simultaneously to build a career and to raise children." *Urban Studies, 40(12)*, 2573-2584.

Lees, Loretta. (2003). "Super-gentrification: the case of Brooklyn heights, New York City." *Urban Studies, 40(12)*, 2487-2509.

Ley, David. (1996). *The New Middle Class and the Remaking of the Central City*. Oxford: Oxford University Press.

Ley, David. (2003). "Artists, aestheticisation and the field of gentrification." *Urban Studies 40(12)*, 2527-2544.

Newman, Kathe. & Elvin, K. Wyly. (2006). "Gentrification and Displacement Revisited: A fresh look at the New York City experience." *Urban Studies, 43(1)*, 23-57.

Rofe, Matthew W. (2003). "I want to be global', theorising the gentrifying class as an emergent elite global community." *Urban Studies, 40(12)*, 2511-2526.

Slater, Tom. (2006). "The eviction of critical perspectives from gentrification research." *International Journal of Urban and Regional Research, 30(4)*, 737-757.

Smith, Neil. (1986). "Gentrification, the frontier, and the structuring of urban space." In *Gentrification of the City*, (Eds.) Neil Smith and Peter Williams. Boston, London & Sydney: Allen & Unwin.

Smith, Neil. (1987). "Gentrification and the Rent Gap." *Annals of the Association of American Geographers, 77(3)*, 462-465.

Smith, Neil. (1996). *The New Urban Frontier: Gentrification and the Revanchivist City*. London: Routledge.

Smith, Neil. (2002). "New Globalism, New urbanism: Gentrification as Global Urban Strategy." *Antipode, 34(3)*, 427-450.

Smith, Neil. & James, deFilippis. (1999). "The Reassertion of Economics: 1990s Gentrification in the Lower East Side." *International Journal of Urban and Regional Research, 23(4)*, 638-653.

Zukin, Sharon. (1998). "Urban Lifestyles: Diversity and Standardization in Spaces of Consumption." *Urban Studies, 35(5-6)*, 825-839.

INDEX

horizon, 310

hormone, 285, 306

household, xiii, 38, 40, 43, 155, 199, 201, 204, 221, 222, 223, 224, 228, 307, 308, 309, 312, 313, 314, 316

Housing and Urban Development, 165

human, ix, 2, 51, 53, 56, 59, 67, 90, 111, 121, 123, 124, 125, 128, 131, 133, 134, 137, 138, 141, 142, 143, 145, 169, 188, 191, 206, 207, 282, 285, 286, 293, 306

human capital, 207

Human Kinetics, 275

humanity, 60

humans, x, 122, 123, 124, 128, 133, 134, 167, 168, 169, 188, 189, 255, 283, 285

humidity, 301

Hungarian, 173

hybrid, 2, 12, 48, 58

hydro, 284, 287

hydrocarbon, 284, 285

hydrogen, 284

hypothesis, 56, 124, 126, 134, 273, 310

I

ICE, 288, 305, 306

identification, 265

identity, 2, 25, 26, 163, 256, 342

ideology, 5, 22, 60, 343

Illinois, 74

ILO, 164

image analysis, 124, 130, 135, 140, 144

images, ix, 29, 49, 71, 88, 93, 102, 105, 121, 123, 125, 127, 128, 129, 131, 132, 133, 134, 135, 257

imagination, 2

implementation, xii, 26, 86, 90, 111, 114, 118, 124, 141, 258, 262, 324, 325, 327

in situ, 288, 312

incentive, 16

incidence, 264

incineration, xii, 281

income, 34, 36, 150, 151, 152, 154, 155, 159, 161, 162, 163, 164, 200, 201, 203, 204, 205, 208, 210, 222, 223, 268, 310, 311, 314, 315

incomes, 37, 199, 268

indices, 227

indigenous, 206, 341

Indigenous, 169, 192, 194, 195

individualism, xiii, 307

industrial, vii, xii, 1, 2, 3, 5, 7, 9, 10, 11, 12, 13, 14, 15, 16, 18, 19, 20, 22, 23, 27, 28, 30, 31, 35, 36, 40, 48, 57, 90, 102, 111, 183, 281, 282, 284, 285

industrialization, 2, 5, 19, 22, 26

industry, vii, 1, 2, 3, 5, 15, 16, 18, 20, 21, 22, 26, 28, 30, 76, 109, 112, 114, 148, 156, 164, 207, 282, 306, 321, 322, 323, 325, 332, 333

inert, 287

inertia, 168, 190

infection, 285

Information System, 264, 305

information systems, 277

Information Technology, 15, 23, 29, 332

infrastructure, 23, 24, 34, 107, 148, 151, 152, 155, 209, 222, 263

inhalation, 283

inherited, 149

initiation, 54

injustice, 208

innovation, ix, 17, 89, 90, 93, 110, 119, 333

Innovation, 85, 86, 87, 119, 140

inorganic, 283, 284, 290

inspection, 43, 191, 248, 257

intangible, 173

integration, ix, xiii, 90, 99, 264, 314, 316, 319, 320, 322, 323, 324, 328, 329, 330, 331, 332

integrity, 320, 326, 330

interactions, xii, 173, 261, 316, 322, 329, 331

interdisciplinary, 86, 322, 330

intergenerational, xiii, 307, 308, 311, 312, 313, 314, 315, 316, 317

Internet, 305

intervention, 24, 203, 270, 274, 317

interviews, 20, 221, 246, 321, 325

intrinsic, 127, 158, 219, 301

intrinsic motivation, 219

Investigations, 142

investigative, 264

investment, 22, 35, 40, 52, 54, 151, 155, 156, 203, 314, 342

iron, 15, 68, 284

island, 77, 150

ISO, 301, 304, 305

Israel, 194

Ivory Coast, 172

J

Japanese, 7, 131, 141, 170, 176, 191, 193

joining, 20, 296, 315

journalists, 10

judge, 16, 310

junior high, 200

justice, 57

wildlife, 285
wine, 26
word of mouth, 14
workforce, 5
working class, 34, 49, 342
World Bank, 151, 199, 204, 207, 214

X

xylene, 284, 292

Z

zinc, 284